A Primer of
Human Genetics

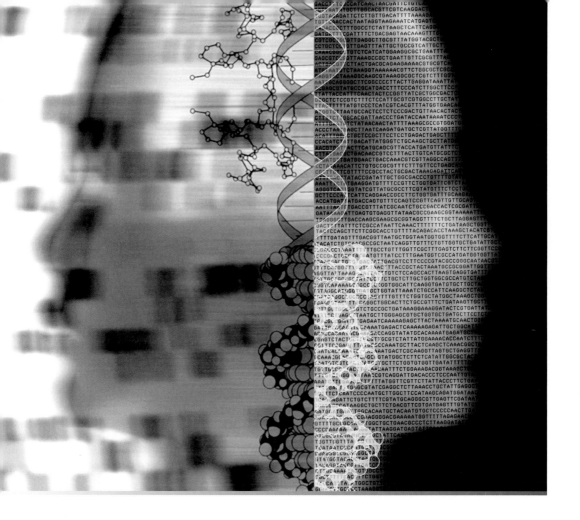

A Primer of
Human Genetics

Greg Gibson
Georgia Institute of Technology

Sinauer Associates, Inc. • Publishers
Sunderland, Massachusetts U.S.A.

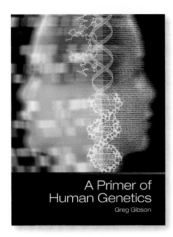

A Primer of
Human Genetics
Greg Gibson

Cover image
© Mehau Kulyk/Science Source.

A Primer of Human Genetics
Copyright © 2015 by Sinauer Associates, Inc.
All rights reserved. This book may not be reproduced in
whole or in part without permission of the publisher.
For information address:

Sinauer Associates, Inc., P.O. Box 407,
Sunderland, MA 01375-0407 U.S.A.

Phone: 413-549-4300

Fax: 413-549-1118

Email: publish@sinauer.com; orders@sinauer.com

Website: www.sinauer.com

Photo Credits
Figure 5.3 Genetic carrier screening: © michaeljung/istock.
5.3 Non-invasive prenatal testing: © Astroid/istock. 5.3
Prenatal paternity testing: © Kais Tolmats/istock. 5.3 Preimplantation genetic screening: © PaulFleet/istock. 5.3 Miscarriage testing: © IuriiSokolov/istock. 5.3 Fetal sex testing ©
katrinaelena/istock. 16.1 Infant: © sam74100/istock. 16.1
Reproductive-age adult: © sborisov/istock. 16.1 Older adult:
© Ammentorp Photography/istock.

Library of Congress Cataloging-in-Publication Data
Gibson, Greg, 1963- , author.
  A primer of human genetics / Greg Gibson.
     p. ; cm.
  Includes bibliographical references and index.
  ISBN 978-1-60535-313-5
  I. Title.
  [DNLM: 1. Genetics. 2. Genetic Phenomena. 3. Genetic
     Techniques. QU 450]
  RB155
  616′.042--dc23   2014028786

Printed in China
5   4   3   2   1

In memory of Walter Gehring,
who taught me genetics, through little humans with wings.

Contents

PART III | Diseases

Preface

I t is little exaggeration to say that, over the past decade, the field of human genetics has seen some revolutionary changes in the way it is practiced. It is not just the switch from gene-centric to genome-enabled research—it is the combination of the way that studies are organized, the technologies that have emerged, and the integration with bioinformatics that has more than incrementally changed our perception of human genetics. Students these days are more comfortable with Perl and Python than they are with plasmids and pedigrees. Yet most textbooks were written before the genome era.

Contemplating this situation while wandering the streets of Barcelona fresh off an overnight flight from Atlanta a couple of years ago, the concept of this book took shape. For some inexplicable reason, upon returning to my hotel, I actually put it down in a Word document and emailed it to Andy Sinauer, publisher of my previous textbook (written with Spencer Muse, *A Primer of Genome Science*), almost 20 years earlier at a similar revolutionary stage of biological inquiry. Barely before the delicious aromas were rising from a mid-afternoon paella by the sea, I had my positive answer, and the project was born.

The timeline has been a condensed one, wrapped around day jobs as a researcher, professor, and journal editor, so this is by no means a comprehensive textbook of human genetics. Nor it is meant to be—rather, the objective is to provide an overview. For students, who in any case are used to getting their information from the internet and Ted Talks, the book should be treated precisely as the title suggests, a primer, and hence a launching pad for further investigation. Perhaps as well, it might be used to help teachers flip their classrooms. The opinions in this book are that of one geneticist, and not even a clinical geneticist at that. The process of having each of the chapters reviewed hit home to me just how much diversity of opinion there remains in a field that, to an outsider, probably looks rather uniform: pick your disease, gather lots of cases, genotype or sequence them, and plug in the latest algorithms from the Broad and Sanger Institutes, or so it seems. No, there are so many undercurrents of thought and practice, from those who believe that only protein-coding variants are relevant in the clinic to those who see genetics as a collection of hundreds of thousands of polymorphisms that, shaped by the environment, merely influence tendencies. I have tried to provide an overview of many of the current thought trends, but inevitably have underplayed the hand of genetics as practiced prior to 2010. That is not meant as a slight, but simply results from a desire to write a book with a contemporary slant to correct the imbalance in current offerings.

I assume readers are familiar with the type of genetics that is taught in introductory biology classes. It will be possible to read the book without a more profound knowledge of molecular genetics and cell biology, but I suspect classes in those would help. Statistical and quantitative genetics are more directly relevant to much of the content, but I have deliberately avoided equations and mathematical derivations since pre-med students do not need them. Nor do budding journalists, lawyers, business people, teachers, genetic counselors, or anyone else who may be taking a course where human genetics is of interest. Those who wish to go on to practice in the field will need more advanced treatments, but it is my hope that the overview provided here will serve as a firm foundation.

The book is written in three parts. It begins with Foundations. Chapter 1 lays out the major models of inheritance, which ought to be taught in high school, but actually get displaced by Mendelian genetics and the central dogma, and consequently the chapter may take a little more effort than is usually demanded of an introduction. Chapters 2 and 3 survey the content of the human genome and its evolution, leading to a discussion of the genetics of normal variation for how we look and behave. The first section concludes with a forward-looking chapter on personalized medicine based on genomics. Five chapters in Part 2, Tools, lay out the technologies that have driven the revolution: genome-wide association studies, whole genome sequencing, gene expression profiling, epigenomics, and integrative or systems biology (yes, including the microbiome). I make no apologies for the technical tone to these chapters, since it is useful to describe them from a methodological perspective. My hope is that teachers will enhance the content with their own favorite case studies, using the text simply as a platform upon which to build their class. The final section, Diseases, consists of six chapters surveying what we know of the genetics of disease in six domains: immunology, metabolic health, cardiovascular disease, oncology, psychiatric disorders, and aging (with cognitive decline as a prominent example). Here experts will surely disagree with the emphases, and progress will quickly make some of the findings outdated; it was tempting to ask my editor to keep inserting references to fabulous new papers appearing in September 2014, a month before publication date. Alas, a line had to be drawn pretty much at the end of April, and revisions will need to await another edition.

Acknowledgments

There are numerous people I would like to thank. First and foremost my wife, Diana, for her constant support and the willingness to put up with "occasional" bouts of grumpiness. Ditto for everyone at Sinauer Associates, particularly my editors, Danna Lockwood, who nurtured the project through quickly and professionally, and Azelie Fortier. Jane Murfett provided tremendous feedback on, and suggestions for, the figures. Precision Graphics deserves many thanks as well, for rendering crisp, beautiful figures, and for paging each chapter. Special thanks to Joan Gemme for creating the book design, and Chris Small for his leadership in building the book. Of course, many thanks go to Andy Sinauer, publisher extraordinaire. Norma Sims Roche managed to find good cause to revise every one of something like 2000 paragraphs, thankfully (students, contemplate that next time you get a paper back from a professor). None of what remains in error is in any way their fault, and too many of their suggestions, as well as those of two dozen expert reviewers, have not been accommodated. I plead a desire to keep the book from becoming twice as long, a wish to stick to a common format, and the impossibility of integrating so many great and diverse ideas. Two undergraduates in the class of 2013 BIOL 4545 at the Georgia Institute of Technology, Anna Morocco and Taylor Fischer, were kind enough to provide insightful feedback on the first draft. This book owes so much to all of my students, past and present, who have helped shape my views and constantly enrich each day. It is dedicated to the memory of my PhD thesis advisor, Professor Walter Gehring, who taught me fly genetics as well as the joy to be had from building a diverse and eclectic group of colleagues. If the book manages to entice one more student into a life dedicated to understanding human genetics, thereby helping each and every person maximize their humanity, I shall be honored.

GREG GIBSON
ATLANTA, GEORGIA
OCTOBER 2014

Media and Supplements

TO ACCOMPANY A PRIMER OF HUMAN GENETICS

eBook

A Primer of Human Genetics is available as an eBook, in several different formats, including VitalSource CourseSmart, Yuzu, and BryteWave. The eBook can be purchased as either a 180-day rental or as a permanent (non-expiring) subscription. All major mobile devices are supported. For details on the eBook platforms offered, please visit www.sinauer.com/ebooks.

Instructor's Resource Library
(Available to qualified adopters)

The Instructor's Resource Library includes a collection of visual resources from the textbook for use in preparing lectures and other course materials. All textbook figures and tables are included in both JPEG (high- and low-resolution) and PowerPoint formats. Figures have all been sized and formatted for optimal legibility when projected.

PART I

Foundations

1

Conceptual Foundations

Ten years after the completion of the first draft of the human genome sequence, the landscape of human genetics research has taken a bold new form. Five thousand dollars and a few weeks well spent will now pinpoint the genetic cause of a Mendelian condition that previously would have taken many years of painstaking and expensive effort to identify. Genome-wide association studies are identifying hundreds of loci that explain some of the susceptibility of adults to essentially all the common diseases that afflict them and which are such a pervasive source of human morbidity worldwide. In the realms of cancer and pediatric congenital disorders, whole-genome and transcriptome sequencing are rapidly becoming a normal part of patient care, ushering in the new era of personalized genomic medicine. It is by any measure an exciting time to be a geneticist.

What combination of events has enabled this scientific revolution? Perhaps most important has been the development, led by companies such as Affymetrix and Illumina, of powerful technologies for interrogating genetic variation and profiling gene function. These technologies would be useless, however, without advances in computing power that have made a mockery of Moore's Law and have caused the cost of sequencing to drop more than exponentially, facilitating whole-genome analysis even by willing undergraduates. In parallel, the rediscovery of mathematics and statistics by biologists has given rise to the new discipline of bioinformatics, which has quickly become as prominent on campuses as biochemistry and cell biology. The large scale of contemporary genetics research is also a key aspect of the revolution, and management skills are at a premium as mega-consortia, in most cases international, pool resources to analyze six-figure-sized population samples. All of these advances build on the ever-crucial foundation of clinical genetics and require the rigorous validation provided by contemporary cell and molecular biology. Human genetics is a team sport.

We will consider all these aspects of genetics in this primer, but must start by reviewing some basic terms and establishing the conceptual framework. This book assumes that readers have a fundamental understanding of the basic principles of genetics. These principles include the central dogma (that information flows from DNA to RNA to proteins) as well as the notions that the three-letter genetic code matches codons to amino acids; that every person carries two versions of each gene, which are known as alleles; and that during meiosis, the processes of recombination and chromosomal segregation lead to independent assortment of genes. **Box 1.1** reintroduces some other fundamental ideas of quantitative genetics that may not be covered in introductory or molecular genetics classes. A glossary at the end of the book provides more complete definitions of hundreds of genetic terms.

This chapter will present four models that pertain to both normal genetic variation and the genetics of disease (**Figure 1.1**; Manolio et al. 2009). In order of increasing genetic complexity, these four models are Mendelian, rare alleles of major effect, common disease–common variant, and infinitesimal. The environment, of course, plays at least as important a role as genetic variation in establishing individual identities. The tools for measuring how genes and environment interact are just being developed, but the basic theory has been in place for almost a century. It starts with the concept of heritability so we will begin with a review of that concept.

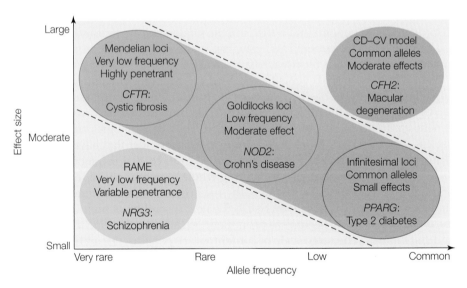

Figure 1.1 The Spectrum of Genetic Contributions to Disease.   Most disease-associated genetic variants fall along a spectrum from very rare with large effects (Mendelian loci) to common with very small effects (the infinitesimal model). The common disease–common variant (CD–CV) model predicts many common variants with moderate to large effects, while the rare allele of major effect (RAME) model predicts the opposite. In the middle of the spectrum are so-called Goldilocks alleles, with modest effects and low but not rare frequencies. One example of each type is shown. (After Manolio et al. 2009.)

BOX 1.1
Some Fundamental Principles of Quantitative Genetics

- *Genotype and Phenotype.* **Genotype** refers to the DNA sequence that an individual has at a gene. At each of the millions of polymorphic sites in the genome, every one of us is either homozygous for the **minor** (less common) **allele**, homozygous for the **major** (more common) **allele**, or heterozygous. **Phenotype** refers to the observable traits of an individual, most often to an individual's appearance (blonde hair, brown eyes, body mass index [BMI] 25.6, for example) or disease status (diabetic, subject to migraines), but sometimes to biochemical or physiological measurements. Quantitative genetics is the study of how genotypic differences influence phenotypic differences. The vast majority of phenotypes are complex: many genes interact with the environment to influence the trait.

- *Mendelian Traits.* A **Mendelian trait** is one that segregates in simple ratios of phenotype classes due to the action of the alleles at a single locus. If the trait is **additive**, then heterozygotes have an intermediate phenotype, and the segregation ratio in the progeny of two heterozygous parents is 1:2:1. If it is **dominant**, then heterozygotes and one homozygous class all have the same phenotype, and the ratio is 3:1 (the recessive allele is simply the inverse of the dominant one). In incomplete dominance, the phenotypes of heterozygotes are closer to that of one parent, and in overdominance or underdominance, the phenotypes are more or less extreme, respectively, than either homozygote class. As more genes influence the trait, it shifts from **monogenic** to **oligogenic** (few genes) to **polygenic** (many genes).

- *Variance.* **Variance** is a statistical concept that refers to the spread of values. For a continuous trait, such as height or BMI, a sample of individuals can be characterized by the **mean** (the average value) and the **standard deviation**, which specifies how far each percentile of individuals is from the mean. If the distribution is asymmetrical,

the **median** value, which separates the top and bottom halves of the distribution, will be different from the mean. Variance is critical in genetic analysis for at least two reasons. First, it allows us to assess whether different genotypes have different effects on a trait, because any statistical test depends on the size of the sample groups, the difference between the means of the groups, and the variance within the groups (actually, the ratio of the within-group to the total variance). Second, it provides a framework for evaluating the relative contributions of genotypes, environment, family history, population structure, and so forth to the variation that is observed. In particular, we can evaluate the effect of removing one of these sources of variation. A simple relationship is that the amount of variation in the population attributed to a single polymorphism is the product of the heterozygosity of the site and a measure of the **substitution effect** ($2pq\beta^2$).

- *Types of Mutation.* Mutations and polymorphisms affect either the coding region of a gene or the regulatory regions. If they are in the coding region, they may be **synonymous** (that is, they change the nucleotide, but not the amino acid, that is encoded) or **nonsynonymous** mutations (also called replacements), or they may affect splicing. Nonsynonymous single nucleotide polymorphisms (SNPs) may be deleterious or benign, depending on the effect that the substitution has on the structure and function of the gene product. **Nonsense** mutations introduce a premature stop codon, and **missense** mutations alter the amino acid that is encoded.

- *Quantitative Trait Locus.* A **quantitative trait locus (QTL)** is a region of the genome that is associated with a complex trait; it is basically the oligogenic or polygenic analog of a Mendelian locus. In general, gene mapping studies alone do not pinpoint the precise gene (let alone the nucleotide) that is responsible for an effect, and

(Continued)

> **BOX 1.1** (continued)
>
> often they simply mark an interval that can be anywhere from several kilobases to a megabase or more. Hence, we refer to a **locus** instead of a gene, and since the effect is quantitative rather than discrete, the term QTL is adopted.
>
> - *Genetic risk classification.* One of the objectives of genetic analysis is to be able to give an individual an estimate of their risk of contracting a particular disease or condition. This can be done most simply by adding up the number of alleles that increase the trait value or risk and reporting this **genetic risk score (GRS)** relative to the known risks for people with similar scores in the remainder of the population. More sophisticated measures weight each allele by its effect size, combine odds ratios, or employ machine learning algorithms to identify the best polygenic predictors. The idea of a GRS, though,
>
> is that individuals in the upper and lower deciles, for example, have different quantifiable risks of disease. These risks are proportional to the amount of variance explained by the GRS. If the contribution of the risk alleles explains only 10% of the variance, there is certainly a discernable difference in liability, but it is of questionable relevance to an individual. If it explains 50% of the variance, the GRS approaches what might be called "prediction," though this is a loaded term that must also be evaluated for accuracy. Rarely can genotypes predict disease, but it is possible to make statements such as "your risk of diabetes is 35%, compared with the population prevalence of 20%." In this book, I use the term "classification" to refer to the idea that GRS can classify individuals into different categories of risk without necessarily assigning a numerical value associated with prediction.

Heritability

Heritability is the proportion of the variation in a population that is attributable to genotypic differences. It is an estimate of the relative contributions of genetics and the environment to variation among individuals in their phenotypes (observed traits).

Mathematically, heritability (abbreviated H^2) is the ratio of the genetic to the total phenotypic variance, namely, $H^2 = V_G/V_P$, where $V_P = V_G + V_E$. This partitioning assumes that all sources of variance can be reduced to genetic (V_G) and environmental (V_E) effects, as described further below. The phenotypes may be discrete, such as disease status; categorical, such as number of digits; or continuous, such as height or a biochemical measure.

There are several key aspects of the concept of heritability that need to be emphasized, since much confusion has been introduced by the popular usage of the term (Visscher et al. 2008):

1. *Heritability is not a statement about individuals.* A heritability of 50% for diabetes does not imply that half the reason why someone has diabetes is genetic, with the other half being environmental. Rather, it suggests that there would be half as much diabetes in the population if everyone were genetically identical.

2. *Heritability is only a statement about a single population.* A heritability of 80% for height does not imply that most of the average difference

in height between Germans and French is due to genetic differences. Differences between populations could be completely environmental even though variation within each population is largely genetic. Heritability estimates alone should not be used to draw inferences about genetic divergence between groups.

3. *Heritability is not the same as inheritance.* **Inheritance** is the correspondence between children and their biological parents. It can be due to environmental, including cultural, factors that are shared by family members, and **epigenetic inheritance** is now also recognized. The only way to confidently interpret heritability is to measure the genotypic contribution, which has been possible only since the mid-2000s.

4. *Very low heritability does not imply very little genetic contribution.* Low heritability may be due to either relatively high environmental variance (hence, a large denominator, V_P), or an absence of variance in the genes that contribute to the phenotype. For example, the number of cervical vertebrae is the same in everyone; that number is certainly established by genes, but there is no variability attributable to those genes. A corollary is that the genotypic differences discovered in population screens reveal only variable genes. Many of the most important genes, including drug targets, are not polymorphic and will be discovered only through other approaches, including model organism research.

A distinction is often drawn between narrow sense (h^2) and broad sense (H^2) heritability. The former refers only to additive genetic effects (V_A). It is a mathematical convenience to assume that heterozygotes have phenotypes intermediate between those of the two homozygote classes (**Figure 1.2A**), and there is now strong empirical support for this claim as a generalization. Theory also suggests that most of the genetic variation in populations is additive, and that it is the additive component that is acted on by natural selection. However, we can also recognize **dominance variance** (V_D; **Figure 1.2B**) and **interaction variance** (V_I, or **epistasis**, in which the effect of one genotype is influenced by one or more other genotypes; **Figure 1.2C**), not to mention genotype-by-environment interactions ($G \times E$). In broad sense heritability estimation, $V_G = V_A + V_D + V_I + V_{G \times E}$. There continues to be much debate concerning how much of the genetic variance in a population is nonadditive and how important the interaction terms are for personal genetic risk assessment.

The environmental term can also be partitioned, though there is less consensus on how to do so. We can recognize fixed environmental factors such as cultural, geographic, or climate variables that tend to be shared by members of a small population; more individualized environmental factors such as diet, exercise, or education levels that vary within and among families; stochastic effects that may be identifiable (triggers and stressors such as an infection, car accident, death in the family, or losing a job at a critical time); and the types of random events that give rise to asymmetry

(A) Additive: Heterozygotes have phenotypes that are intermediate between those of homozygotes.

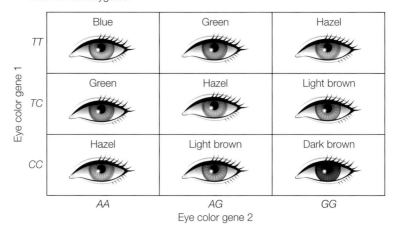

(B) Dominance: In gene 1, allele C is dominant. In gene 2, allele G is dominant.

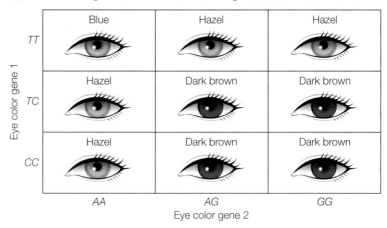

(C) Epistasis: Expression of each gene is affected by the alleles of the other gene.

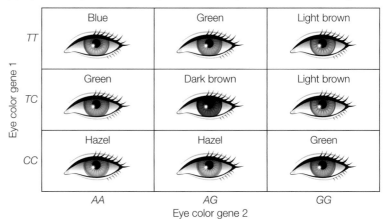

◀ **Figure 1.2 Additivity, Dominance, and Epistasis.**   Suppose there are two large-effect genes influencing eye color, which grades from blue to green to hazel to light and dark brown. According to a purely additive model (A), heterozygotes for either gene have an intermediate pigmentation between the two homozygote classes. With dominance (B), the heterozygotes resemble either of the homozygote classes (here T is recessive to C for gene 1 for lighter pigmentation, and G is dominant over A for gene 2). In complex epistasis (C), the effect at either gene is a function of the allele at the other gene.

and observable differences even between identical twins. Only some of these effects can be incorporated efficiently into V_E and G × E estimation, but they all continue to be the subject of epidemiological research.

Classically, heritability has been estimated by quantitative geneticists in three ways. **Parent-offspring regression** is the most straightforward, as the slope of the regression is equivalent to the heritability (**Figure 1.3A**). It is most appropriate in agricultural settings, where environmental contributions are controlled and cultural transmission is ignored. This approach assumes that all inheritance is genetic, so adjustments must be made in human studies. Second, a statistical method known as **analysis of variance** (**ANOVA**) is used to estimate how much of the variance is within and between groups of genetically similar individuals (**Figure 1.3B**). Derivatives of analysis of variance are used in twin studies, which continue to be a very important aspect of human genetic research. Identical twins (maternal twins, in which one fertilized egg splits to generate two siblings) share twice as much genetic material as nonidentical twins (fraternal twins, in which two eggs are independently fertilized), and so are more likely to share traits if those traits have a genetic component. Fraternal twins resemble one another more than non-twin siblings because they share the same womb during gestation and presumably have more similar upbringings. Some twin studies consider twins reared apart in order to minimize shared environmental factors. The third method is realized heritability, in which the ratio of the response to selection to the selection differential is measured (**Figure 1.3C**), but it is much more useful in evolutionary genetics and will not be considered further here.

To some extent, heritability estimation is declining in importance now that geneticists have the tools to actually study genotypes. The concept was historically useful because it helped frame debates about nature versus nurture, though this was always a false dichotomy, as genes and environment always work together. Before embarking on expensive genetic analyses, it is important to know how large the genetic contribution is expected to be, but as we will see, very powerful genome-wide studies uncover only a small proportion of the genetic variance in any case. Furthermore, there is no guarantee that highly heritable diseases will yield genes that have a measurable influence: rare alleles can have a large effect in families, but do not affect enough individuals to have a large effect on the population. Nevertheless, the concept of heritability remains a critical one, as it reminds us that genetics is only one part, sometimes small and sometimes large, of the variation in the trait we are interested in.

(A) Parental-offspring regression

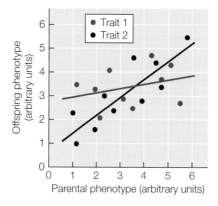

(B) Twin studies

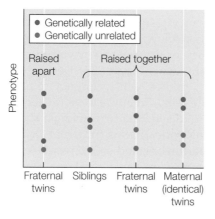

(C) Realized heritability

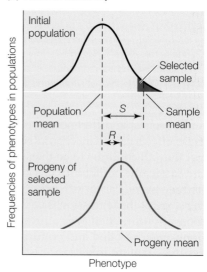

Figure 1.3 Estimation of Heritability. (A) Classically, heritability is measured from the correlation between the phenotype of children and that of their parents (either one, or the average). Here, the trait represented by the black points has higher heritability. (B) In human genetics, high heritability is inferred if two maternal (identical) twins tend to be more alike than two fraternal twins. Generally, non-twin siblings will be slightly more dissimilar than nonidentical twins, even though they share the same proportion of genes, because their environments are more different. Similarly, twins or siblings reared apart will be more different than those who grow up together. (C) In evolutionary biology and ecology, heritability is estimated as the ratio of the response to selection (R) to the selection differential (S).

Mendelian Genetics

For most of the twentieth century, practical achievement in human genetics was dominated by research on Mendelian diseases (Botstein and Risch 2003). These conditions segregate in families, typically with ratios of 3:1 (that is, one in four individuals have the recessive phenotype, such as a disease), so that it could be inferred that a mutation at a single locus is responsible for the trait. The teaching of genetics in high school, and in many introductory undergraduate biology courses, continues to lean heavily on Mendelian examples, despite the fact that few of us actually experience them in real life. Even blue eye color, detached earlobes, and hitchhiker's thumb have turned out to be more complex. Yet single-gene abnormalities could be mapped and the responsible genes cloned soon after molecular biology became established, and they illustrate many of the fundamental principles of human genetics.

The name "Mendelian" comes from the discoverer of the laws of genetic transmission, Gregor Mendel, whose observations of variation in garden peas led to the concepts of dominance and recessivity. Suppose that there are two versions of a genetic locus (what we now call the **alleles** of a gene). Those two versions can be denoted as + and –, or wild type and mutant type, respectively. If two individuals who are both heterozygous at this locus (+/–) have children, then there are four possible combinations of genotypes among those children: +/+, +/–, –/+, and –/–. For the most part, it does not matter whether the + or – allele is transmitted from the mother or the father, so the two heterozygotes, +/– and –/+, are regarded as genotypically equivalent, and individuals with these genotypes are referred to as **carriers**. (We will consider exceptions to this generalization in Chapter 9 when discussing epigenetics). Children with the +/+ geno-type are homozygous wild-type individuals. Those with –/– genotypes are homozygous mutant individuals and are expected to have the disease. If the carrier phenotype resembles the wild-type phenotype, then the children of these parents will have a ratio of 3:1 normal to disease and the condition is recessive, which is the most common case.

There are also circumstances in which the +/– combination of alleles leads to disease, in which case the condition is dominant, and the ratio of normal to disease phenotypes in the children of a heterozygous parent and a wild-type parent is 1:1 (half the progeny are affected). In one such circumstance, the mutation results in gain of function of the gene product, perhaps a hyperactive receptor or channel, such that the single copy is sufficient to promote abnormality. In another, the mutation causes loss of function to a point where there is not enough active gene product to support normal function, resulting in **dominant haploinsufficiency**. Dominant gain of function is demonstrated by the familial cardiac arrhythmia called long-QT syndrome, which in some cases is due to aberrant hyperactive potassium channels (although interestingly, ineffective channels can cause the same syndrome in a recessive manner). A good example of dominant haploinsufficiency is Marfan syndrome, once thought to have afflicted Abraham Lincoln, which is a connective tissue disorder in very tall people due to loss of function of the *FBN1* gene that encodes fibrillin.

Mendelian ratios are altered in the case of sex-linked traits. Fathers who have a recessive mutation on the X chromosome and who have a trait such as color blindness will not transmit the condition to their sons, who receive their Y chromosome from Dad and a normal X from Mom (if she has two normal Xs). Affected homozygous mothers will transmit

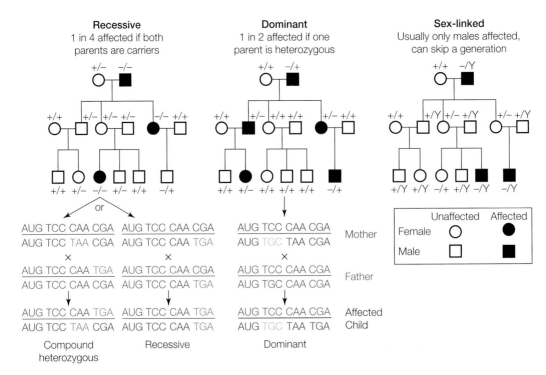

Figure 1.4 Mendelian Transmission within a Three-Generation Pedigree.
Instances of recessive, dominant, and sex-linked Mendelian transmission are shown
superimposed on the same hypothetical pedigree. Mendelian segregation ratios for autosomal mutations are either 3:1 or 1:1. The two mechanisms shown below the affected
grand-daughter in the Recessive model for 3:1 transmission, in which both of her parents
are carriers and both transmit the same mutation to one-fourth of their children, who
will have the disease, are (1) compound heterozygous transmission, in which the two
parents are carriers for different mutations that fail to complement one another if the child
receives both (in this case, two different stop codons), and (2) true recessive transmission
of the same mutation. Dominant 1:1 transmission is from one affected parent to half the
children, irrespective of the genotype of the other parent, who will not usually be a carrier.
For example, substitution of cysteine for serine may result in a hyperactive protein that
leads to a more severe phenotype than loss of function. Sex-linked inheritance usually
involves recessive mutations on the X chromosome that are hemizygous in males, who
show the disease.

the trait to all their sons and none of their daughters (assuming the father
is unaffected), but half the sons of those heterozygous daughters will be
affected. Similarly, a carrier daughter who receives the mutation from her
affected father will transmit the allele to half her sons. Traits such as color
blindness are thus said to skip a generation and to be transmitted from the
maternal grandfather, though it should be recognized that this condition
is actually more complex, involving several genes. **Figure 1.4** illustrates
how careful assembly of family pedigrees can elucidate the Mendelian

nature of single gene disorders, and it also contrasts recessive Mendelian transmission with **compound heterozygous** modes of inheritance, which I will describe shortly.

One of the best-known Mendelian diseases is cystic fibrosis (CF), which afflicts approximately 1 in 10,000 newborn children. This condition results from a defect in the protein product of the *CFTR* (cystic fibrosis trans-membrane conductance regulator) gene, which alters chloride ion transport, among other functions, and primarily affects the pancreas and lungs. Cystic fibrosis illustrates several key features of many Mendelian diseases (Zielenski and Tsui 1995), starting with the notions of penetrance and expressivity. **Penetrance** is the proportion of individuals with the genotype who have the trait; in the case of CF, it is 100%. That is, all children with mutations in both copies of *CFTR* will get the disease. **Expressivity** is the severity of disease in those people who have it, and in the case of CF, it is highly variable. Happily, some individuals with CF can now expect almost normal life spans, particularly with the aid of midlife lung transplantation, although others will die in the first few months of life due to more severe symptoms. Variable expressivity can be due to environmental exposures and quality of therapy, to genetic background that modifies the effect of the mutant protein (compensating in some cases, exacerbating in others), or to genetic heterogeneity.

Genetic heterogeneity refers to the observation that some—probably most—genetic diseases can be due to any of several different mutations (McClellan and King 2010). For example, one relatively common mutation, ΔF508 (resulting in a deletion of the phenylalanine residue at position 508 in the CFTR protein), accounts for 70% of the mutant alleles in Caucasians with CF, but over 1000 other mutations in the CFTR protein have also been documented. Homozygotes for ΔF508 constitute approximately one-half of all individuals with CF, whereas most of the remainder are "compound heterozygotes." These people are heterozygous for two different disease-causing alleles, one transmitted from each parent, but since neither allele results in normally functional CFTR protein, the compound genotype is equivalent to being homozygous for one mutation (see Figure 1.4). There are actually six recognized classes of CF attributable to these different combinations of mutations, each with characteristic expressivity of specific CF-related phenotypes. Genetic heterogeneity can also be due to mutations in different genes: scoliosis, or curvature of the spine, for example, can be due to homozygosity for mutations in at least three different genes involved in development of the somites that give rise to vertebrae. Intergenic compound heterozygosity—namely, double heterozygosity at different genes that together are insufficient to provide normal function—could be a much underappreciated source of morbidity, since it is very difficult to study.

Cystic fibrosis, like most Mendelian diseases, is a rare condition, since natural selection prevents the mutant allele from attaining a frequency greater than 1%. According to the well-established Hardy-Weinberg rules, genotype frequencies can be estimated from allele frequencies simply by using the binomial expansion of the formula $(p + q)^2 = 1$, where $q = 1 - p$ is the frequency of the mutant allele and p is the frequency of the wild-type allele. In the case of CF, q^2 individuals are homozygotes—namely, $0.01^2 = 0.0001$, or 1 in 10,000—and $2pq = 2 \times 0.01 \times 0.99 = {\sim}0.02$, so 1 in 50 people are carriers. Thus one in 2500 marriages will be between carriers, and one-fourth of their children will be affected. In a class of 50 students, at least 1 is likely to be a carrier.

Similar reasoning is applicable to the vast majority of Mendelian diseases. It is worth noting that the ratio of carriers to affected people is often about 200 to 1. This observation has two public policy implications. One is that even though all children of a marriage between, say, two individuals with CF will have the disease, they will still be a tiny fraction of the "mutant" gene pool. The second is that any genetic engineering to remove the mutation from the gene pool would have to include all carriers, but since everyone carries mutations for multiple Mendelian conditions, that engineering would need to include everyone's genomes, which is unlikely to be practical or beneficial.

A second well-known recessive Mendelian disease is sickle-cell anemia, which results from homozygosity for a mutation in the hemoglobin gene that causes red blood cells to assume a rigid, sickle shape (Bunn et al. 1982). Heterozygous carriers are offered some protection against malaria, and they are said to have sickle-cell trait, which is unfortunately confusing, since it is usually the disease that is called the trait. The loss of fitness in homozygotes offsets the "overdominant" advantage to carriers and keeps the allele at a frequency of less than 5% in most countries, but there is substantial genetic heterogeneity, with different mutations observed in different regions of sub-Saharan Africa. Various other modes of resistance to malaria have also arisen, including alteration of hemoglobin production resulting in the thalassemias, notably in the Mediterranean region.

Investigations into human Mendelian disorders actually began with Sir Archibald Garrod, an English physician who in 1908 first called attention to the possibility that inborn errors of metabolism could be due to segregation of recessive mutations (Lanpher et al. 2006). He studied a tetrad of enzyme defects that are responsible for abnormal urine color and smell as well as albinism, but realized that the same explanation could account for thousands of other conditions. A well-known example is phenylketonuria (PKU), which results from the loss of the enzyme that metabolizes phenylalanine into tyrosine. Left untreated, PKU causes severe intellectual deficiency and seizures, but a strict diet low in phenylalanine can prevent these symptoms, providing a poster-child example of how personalized genetic analysis can positively influence medical care.

Today, the **Online Mendelian Inheritance in Man** (**OMIM**) website (**Figure 1.5**), curated at Johns Hopkins University and maintained by the U.S. National Library of Medicine, lists over 7000 Mendelian conditions. This essential resource is an outgrowth of the database of diseases pioneered by Victor McKusick, which was printed in 12 book editions between 1966 and 1998 (Hamosh et al. 2005; McKusick 2007). At time of this writing,

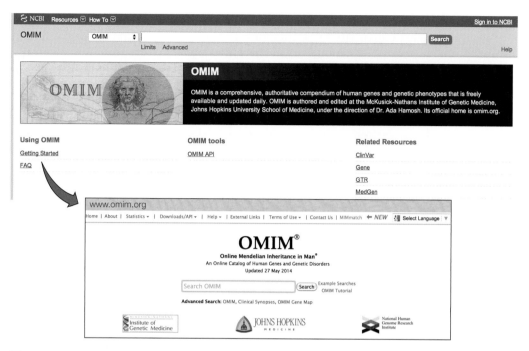

Figure 1.5 Online Mendelian Inheritance in Man (OMIM). This NCBI-sponsored website is the central database for all Mendelian diseases. Users can search either by gene (Gene Map) or by condition (Morbidity Map and Clinical Synopsis). A tutorial is available at www.openhelix.com/OMIM

in January 2014, the molecular lesion has already been defined for more than half of the 7000 suspected Mendelian diseases, and the locus has been mapped to a candidate gene for another 1600. Eleven of the diseases are Y-linked, over 500 are X-linked, 28 are mitochondrial, and the remainder are autosomal. This means that of the order of one-third of all known human genes are associated with simple genetic conditions, an observation that guarantees the ongoing importance of Mendelian genetics in research and medical practice.

Rare Alleles of Major Effect

There are two other circumstances under which rare variants can have a major effect on human health without being contributors to recognized Mendelian conditions. One is when a variant is so rare that it is present in just one individual, namely, as a **de novo mutation**. The other is when an effect is large, but by no means completely penetrant. Collectively, such **rare alleles of major effect** (**RAME**) could conceivably account for a substantial amount of genetic variance for disease susceptibility as well as normal trait variation.

Mutation has long been appreciated as a powerful engine producing genetic diversity in populations. Estimates from mutation accumulation experiments in model organisms indicate that in each generation, typically of the order of 0.1% of the standing variation attributable to environmental factors is generated by new mutations. This means that, starting with a single pair of individuals, within a few hundred generations there would be as much genetic variability in the derived population as is found in nature for many traits. Selection against the deleterious consequences of most new mutations keeps mutant allele frequencies low, and this balance between mutation and selection helps to establish the steady state of variability.

Correspondingly, it is now clear that every person carries a burden of rare mutations, most of which have appeared in recent human history as the size of the human population has exploded (Nelson et al. 2012; Tennessen et al. 2012; Fu et al. 2013). **Figure 1.6** shows the number of variants (relative to reference human genome "HuRef19") per individual that are synonymous, missense, predicted to be deleterious, to be deleterious and unique to the individual, to affect a splice site, or to have a nonsense (stop) codon. Africans tend to have 20% more variants in each class due to their longer population history, which has allowed more mutation accumulation, but the total number of predicted deleterious variants per genome is very similar among all human populations. On average, approximately 1.5% of an individual's genes (313 genes) carry a potentially deleterious rare allele found in fewer than 1% of people. These alleles thus represent an appreciable source of variants that are likely to contribute to disease.

The visible effect of mutation for the typical person is not obvious. However, new direct estimates from whole-genome sequencing confirm that

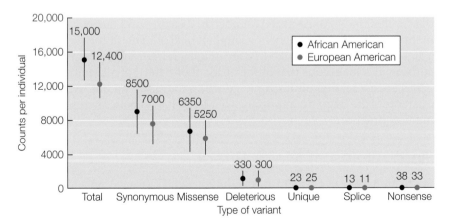

Figure 1.6 Distribution of Deleterious Mutations per Genome.   The plot shows the mean number and range (minimum to maximum indicated by bars) of coding variants in each indicated category for African Americans and European Americans. The data are estimated from whole-exome sequencing of 2440 individuals as part of the NHLBI Exome Sequencing Project. (After Tennessen et al. 2012.)

each person has dozens of novel variants that are not present in either parent, including an average of slightly more than one new amino acid–altering mutation (Campbell and Eichler 2013; de Ligt et al. 2013). Most of these mutations will be benign, but assuming that approximately 5% are sufficiently damaging to alter protein function, and that 20% of all proteins are essential for good health, it is expected that as many as 1% of all individuals will have a de novo mutation that meaningfully affects their well-being. In addition, regulatory mutations and structural changes will add to the burden. This estimate is consistent with the proportion of children born with congenital abnormalities, including immunological, metabolic, and craniofacial defects and failure to thrive. Some defects will emerge in the first few years of life as psychological syndromes, including autism and schizophrenia.

A critical parameter relating to the influence of de novo mutation on disease risk is the mutational target size. Muscular dystrophy, for example, has a relatively high prevalence in large part because the key gene that is disrupted in people with the disease, dystrophin, encodes one of the longest proteins in the human genome. There is thus almost ten times more opportunity for new mutations to disrupt the coding sequence of this gene than that of other genes. If we extend this concept to include the full set of genes that might be disrupted to generate a particular phenotype, it is easy to see that different diseases will have different mutation target sizes and hence different chances of being caused by new mutations. This is referred to as variation in the **mutational burden**. In addition, different genes have different deleterious mutation rates because some proteins are more tolerant of amino acid changes than others. As the most complex organ in the body, the brain is thought to be most susceptible to the effects of de novo mutations, and emerging research suggests that of the order of 10% of autism, schizophrenia, and intellectual deficiency is attributable to them. Of particular concern is the observation that mutation rates increase with the father's age as spermatogenesis occurs throughout the life of a man. This increase is likely to account for the so-called paternal age effect on psychiatric disease.

In order for new autosomal mutations to affect health, they must usually exert their effect in a dominant fashion, either by gain of function or by haploinsufficiency. (If a mutant copy on the other chromosome is inherited from a carrier, de novo mutations can also cause compound heterozygosity). Effect sizes will be large, for the most part increasing risk at least fivefold over the genetic background, if not considerably more. However, again borrowing from research on flies and mice, we know that the genetic background heavily modifies the effects of even large-effect mutations. A mutation that produces severe blistering of the eye in one strain of flies can be completely suppressed if crossed into another strain. Consequently, effect sizes are likely to be heterogeneous, and the consequences of the same mutation will range from undetectable to debilitating in different people. Note that for a disease with a prevalence of 1%, a new mutation that increases risk 20-fold would result in only one-fifth of all carriers having the disease. Highly penetrant Mendelian effects must be much greater than this, but it is thought that most protein-altering mutations have much smaller effects in the heterozygous state.

De novo mutations generally cannot contribute to heritability because they are observed in only one person and are not inherited from someone who also had the variant. The only way that they can contribute to resemblance between siblings is if they are found in identical twins, or if the mutation occurs early enough in germ line development to be present in two or more siblings. Consequently, most of the genetic component captured by heritability estimates must be due to segregating variants, namely, polymorphisms that are also present in the parents. **Polymorphisms** are places in a sample of genome sequences where two or more different nucleotides (or small deletions) are found at the same nucleotide location. They arise as de novo mutations, but by the time they are observed in multiple individuals they are no longer called mutations, unless they unambiguously cause a phenotype.

The vast majority of polymorphisms, particularly in humans, are rare. As we will see in Chapter 3, the minor allele, which is the less frequent one at a locus, often represents no more than 5% of all alleles in the population, and in fact it usually represents no more than 1%. Consider any of several hundred diseases that have a lifetime prevalence of 1%, and suppose that a number of genes, each with polymorphisms at a frequency of 1%, contribute to this disease. Further imagine that each of these polymorphisms has a **genotype relative risk** (**GRR**) of 5, meaning that an individual carrying the allele has a fivefold greater risk of disease than people without the allele, somewhere in the vicinity of a 5% risk. From Hardy-Weinberg theory, we can calculate that 2% of all people will carry each polymorphism, and that these polymorphisms will each "cause" the disease in $0.02 \times 0.05 = 0.1\%$ of all people (that is, 2% of babies will have 5 times more than the baseline 1% risk). Ten such polymorphisms could, by this simple calculation, account for essentially all of the disease, which is observed in 1% of people. Alternatively, there might be 10 genes, each harboring a series of even rarer mutations that have a GRR of 5, that cumulatively account for 1% of the alleles. Similarly, 20 genes with alleles at an average frequency of 0.5% could explain most of the genetic component of risk.

It is for this reason that **whole-exome** and **whole-genome sequencing** (**WES** and **WGS**, respectively) are being pursued vigorously as the most direct route to discovery of rare variants that may have major effects (Ramu et al. 2013). As sequencing costs have dropped and computational tools have become more accessible, there are few technical obstacles to this effort, but two considerable conceptual challenges remain. The first is prediction of whether a particular variant is likely to be disruptive, and the second is establishing that it actually is disruptive. Tools such as Phen-Gen (phen-gen.org), eXtasy (homes.esat.kuleuven.be/~bioiuser/eXtasy) and VAAST (www.yandell-lab.org/software/vaast.html) take lists of rare variants in a genome sequence and deliver a prioritized set of those that are likely to be disruptive, but there is currently no way to be sure that they really are responsible for disease in a particular person. It is estimated that between one-fourth and one-half of all rare pediatric congenital diseases might be diagnosed simply by DNA sequencing. The issues surrounding accurate diagnostics are discussed in later chapters, as they require a combination of evolutionary and structural insight, statistical genetic know-how, and cell and molecular biological validation. Suffice it to say that since the effects of most variants are not expected to be fully penetrant, the search is for variants that are enriched in affected individuals relative to healthy controls, but given their low frequencies, this is a daunting challenge.

Nor should it be assumed that rare alleles act alone. In fact, in order to account for the elevated concordance observed among family members that generates high heritability estimates for many rare diseases, it is convenient to assume multiplicative models, as shown in **Figure 1.7**. A single dominant rare variant with a GRR of 5 should be found in only half of all siblings, and hence fewer than 5% of these children should have a disease whose rate is only 1% in the general population. Thus, there must be other genetic (and environmental) factors increasing the risk in affected families. If the effects of two variants are multiplied together (or added on the logarithmic scale of risk), then they might have a joint GRR of 25, but since they assort randomly, they still would not explain the high concordance observed. Depending on the frequencies and number of segregating polymorphisms, more likely models require higher-level interactions between five or more variants. In families in which multiple risk polymorphisms are brought together by chance (or perhaps by **consanguinity**—marriage among relatives), different siblings

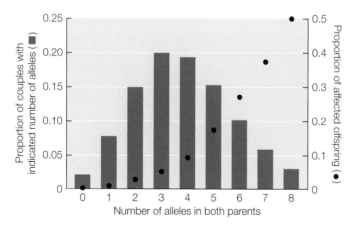

Figure 1.7 Multiplicative Effects of Rare Mutations in Families.   Assuming 100 genes, each of which has an allele with a frequency of 1% that affects a disease, we can compute the expected number of alleles in both parents shown as a histogram, which will typically be 3 or 4. However, high sibling relative risks (indicated by a high proportion of affected offspring, black dots) arise under a multiplicative model only once 7 or more such alleles are combined, and only a small number of families will have such combinations under random mating.

might have different combinations of variants, but the combined effect could be significant heritability due to a combination of rare variants of major effect. Unfortunately, we do not yet have any empirically verified instances of this scenario, as the RAME discovered to date have been isolated polymorphisms.

Common Disease–Common Variant

When **genome-wide association studies** (**GWAS**) were first introduced in the mid-2000s as a powerful new approach to disease variant discovery, they were motivated largely by the notion that common diseases would be caused by common variants. Whereas rare variants were considered likely to cause rare diseases, the thought was that more complex diseases that are found in 5% or more of the population—diabetes, asthma, coronary disease, depression, cognitive aging—must be due to more common polymorphisms. These polymorphisms could not have large effect sizes; otherwise, natural selection would remove them from the gene pool. Yet decades of quantitative trait locus (QTL) mapping in agriculture and in model organisms has found that within any cross, more than half the genetic variance can usually be ascribed to between 10 and 20 loci. Linkage mapping in human pedigrees was less powerful, so it rarely found individual genes, but it was at least consistent with this view. If heritability was due to, say, 50 genes, each of which accounted for 2% of the genetic variation in the population, then GWAS would find them. That, in a nutshell, is the **common disease–common variant** (**CD–CV**) hypothesis (Pritchard and Cox 2002).

As it turns out, such common variants are few and far between. Good examples include complement factor H (*CFH*) for age-related macular degeneration, which was the very first variant discovered by GWAS, and the *FTO* locus for body mass index (BMI). By contrast, most common variants discovered have average effects that are much more modest, only rarely increasing risk by more than 50% (GRR = 1.5) and generally by less than 20%, hence accounting for much less than 1% of the phenotypic variance in the population. Even in cases in which one or a few CD–CV alleles have been identified, the remaining variance must be due to rare alleles or common variants with small effects. There are no cases of a disease in which the majority of the heritability is explained by a few dozen common variants with modest effect sizes. This finding has given rise to what is popularly called the "missing heritability problem" (Eichler et al. 2010), which is outlined in **Figure 1.8**.

There are many possible explanations for this failure to discover common variants of relatively large effect. One is that genetic variants discovered in the offspring of a cross between two parents, or inferred in a pedigree, are just a small proportion of the variants in the total population. Their contributions are magnified in that family, but diluted across multiple families, where they do not segregate. Another is that GWAS necessarily captures average effects, and variants that are critical in one family or individual may have minuscule effect sizes in others, so these variants cancel out. Proponents of the importance of broad sense heritability, who consider

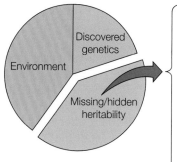

Technical
- Heritability overestimated in families
- Causal SNPs not adequately tagged
- Trait/Disease measurement error

Broad sense
- Epistasis and genotype-by-environment interactions obscure additive effects
- Individuals are not averages

Non-SNP effects
- Epigenetic modification (e.g., methylation)
- Copy number variation
- Somatic mutation

Hidden heritability
- Common variant effect sizes too small
- Rare variants don't impact enough people
- Goldilocks alleles are not genotyped

Figure 1.8 The Missing Heritability Problem.   The pie chart represents the proportion of the average phenotype explained by each of the factors labeled. There is a very large gap between the proportion of the phenotype explained by discovered common variants (SNPs), typically 5%–15%, and the amount expected from classical methods, typically 25%–75%. Four classes of potential explanations for the missing heritability are shown: technical issues, claims that broad sense heritability is more important than additive effects, genetic effects not represented by SNPs on genotyping chips, and the idea that variants are more hidden than missing due to low statistical power.

epistasis and genotype-by-environment interactions to be pervasive, favor this argument. A third possibility is a technical explanation relating to loss of resolution due to incomplete linkage disequilibrium, discussion of which is deferred to Part II of this book.

The simpler explanation is that human geneticists have overestimated the likely effect sizes of causal variants. Either causal variants, on average, rarely have effect sizes promoting risks greater than a GRR of 1.2, or those that do are too rare in the population to account for a sufficient amount of the variance to allow them to be discovered. These considerations have given rise to the notion of "Goldilocks alleles." The idea is that polymorphisms whose frequency is in the range of 0.5%–2% could have a GRR in the range of 1.2–1.5 and contribute substantially to heritability, but would not have been discovered in the first wave of GWAS that focused on more common variants. Their effects would be large enough to contribute to disease in the CD–CV framework, but small enough to survive natural selection sufficiently to reach the "just right" Goldilocks frequencies. Their discovery will require targeted sequencing or fine-scale genotyping with technologies being introduced in the mid-2010s. Simulation studies of whole-genome evolution under various scenarios of population growth, migration, and coupling of selection to disease demonstrate that the empirical data are consistent with a very wide range of architectures of effect sizes and allele distributions (Agarwala et al. 2013).

The Infinitesimal Model

The **infinitesimal model** was first proposed by the great biostatistician R. A. Fisher in 1908 and expanded on in his book *The Genetical Theory of Natural Selection*, which was first published in 1930 and serves as the foundation for the field of statistical genetics. This model is essentially the notion that hundreds, if not thousands, of loci contribute to complex traits. The term "infinitesimal" does not imply that the effects of these loci are vanishingly small, but rather that they are so small that they cannot be measured in reasonable samples, up to and including every individual human! Their existence can only be inferred. In the first half of the twentieth century, before the structure of DNA was known and when the concept of nucleotide variation was yet to be developed, this model reconciled Mendelian principles with continuous traits, which many biometricians denied could be influenced by discrete genetic factors. Today, the core idea is well validated by the observation that GWAS of hundreds of thousands of individuals (a sample size ten times as large as the effective population size of humans throughout most of our early history as a species) have discovered well over 100 variants influencing height and body mass, yet they explain no more than 20% of the variance of each trait (Ku et al. 2010; Visscher et al. 2012).

We can consider two versions of the infinitesimal model, the hard and soft versions. The soft version is the all-encompassing model, which assumes that allele effect sizes are expected to cover the full range from undetectably small to fully penetrant. Larger-effect alleles will tend to be less common and might be said to "cause" disease in some individuals, whereas small-effect alleles can have any frequency and only contribute to risk, never alone being sufficient to promote disease. It is difficult to refute the soft infinitesimal model. The hard version is what is left after eliminating RAME and Goldilocks CD–CV variants: it assumes common variants with a GRR typically in the range of 1.05–1.2, each of which explains a fraction of a percent of heritability and most of which may never be discovered.

There is much debate over what fraction of variability is attributable to common polymorphisms proposed by the hard infinitesimal model and whether this fraction varies across classes of disease. A probably false dichotomy arose for a time because geneticists studying immune and metabolic disorders or normal variation tended to adhere to the infinitesimal model, whereas psychiatric geneticists emphasized the primacy of rare variants. Some observers find the infinitesimal model depressing, as it withdraws hope that we can ever fully describe the genetic basis of a disease or trait and also implies that genetic analysis may never be sufficiently predictive to have clinical utility. Rare variants, by contrast, may often be clinically actionable. The counter-perspective is that our obligation as scientists is to understand the nature of reality, and if we are patient and diligent, we will eventually capture sufficient variance to make risk classification useful and to identify novel biological pathways.

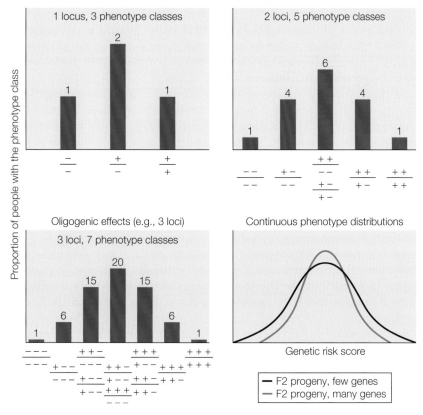

Figure 1.9 The Genetic Basis of Complex Phenotypes. Complex traits that are influenced by hundreds of genes and the environment can nevertheless be modeled as an extension of simple Mendelian genetics. For a single gene with two alleles at equal frequency, the ratio of genotypes is 1:2:1. If two such genes are combined, the ratio is 1:4:6:4:1. As more loci are added, fewer people are at the extremes, and most people have an intermediate number of alleles that increase or decrease the trait. The discrete categories blend into one another, and the effects of the environment add variability, resulting in an approximate normal distribution of phenotypes. F2, second filial generation.

Fisher's central insight is illustrated in **Figure 1.9**. This insight assumes additive effects, though they are not required. For a single gene with two alleles, there are three genotype classes that have three levels of gene activity, with the heterozygotes intermediate. If both alleles are common, half of all people will be heterozygotes and have the intermediate phenotype. An example might be eye color, where hazel eyes are intermediate between blue and brown. Since such discrete classes are rarely observed, the environment can be invoked to explain the blending of the classes into one another. Now suppose that there is a second locus located on a different chromosome, and thus transmitted independently, that has a similar effect. Now there are five classes of genotypes, which we can write as --/--, +-/--, +-/-+ (this is the same as ++/-- for most purposes), ++/+–, and ++/++. If both alleles are common at both loci, then the ratios of these classes are 1:4:6:4:1, and again the environment can help blend them toward one another. Continuing in the same vein, adding more and more loci, the discrete genotype classes merge into a normal frequency distribution.

As the number of loci increases, the number of individuals at either of the extremes drops, and an increasing number of people have intermediate normal phenotypes. Even if there is a bias toward the less common variants having activity in the same direction, homozygotes for –/– alleles will tend to be compensated by +/+ genotypes at other loci. The frequency distribution of the sum of – alleles will be somewhat normal, even if skewed. Under a **threshold liability model**, individuals with a count of alleles greater than some set number will be at higher risk of disease than individuals with fewer alleles. Just as some smokers never get lung cancer and some never-smokers die of emphysema, some individuals in the upper percentile of genetic risk scores will never get the disease, while some in the lower percentile will. Under the infinitesimal model, there may be thousands of variants generating the risk distribution. The task for geneticists is to identify a sufficient fraction of them that they can usefully classify risk and define biological mechanisms.

Since common variants of small effect are difficult to detect, what is the evidence for their existence? One line of evidence is extrapolation from discoveries made in the first phase of GWAS experiments (Park et al. 2010; Chatterjee et al. 2013). Given the observed distribution of effect sizes, and making reasonable assumptions, we can make fairly accurate predictions of the number and effect sizes of new discoveries that will be made as larger samples are considered. Different predictions of how much variation will be explained are obtained for cancers than for immune disorders, and again for psychiatric diseases. Extrapolation to samples of hundreds of thousands predicts hundreds of variants, most with a GRR less than 1.1 (that is, risks per allele of less than 10%), for essentially all diseases.

A more comprehensive approach, known as **SNP-based heritability estimation** (Yang et al. 2010), is implemented in the GCTA (Genome-wide Complex Trait Analysis) software (www.complextraitgenomics.com/software/gcta; Yang et al. 2011a). The idea is that individuals who share more of their genotypes should resemble one another more phenotypically, as illustrated in **Figure 1.10**. Thus, comparison of genotypes at hundreds of thousands of SNPs genome-wide allows an estimation of relatedness. Regression of this SNP similarity measure on how similar each pair of individuals is for a quantitative trait provides an estimate of the genotypic contribution to the trait. So long as 5000 or more unrelated individuals are sampled, GCTA generates a robust estimate of heritability. Applied to height, genotypes explained just over 50% of the variance, and after some statistical adjustments, this estimate was revised upward to close to the 80% expected from classical estimates. An interpretation of this result is that almost all the variation in height can be explained by common variants, even though only about 10% of it is explained by 95 well-validated loci.

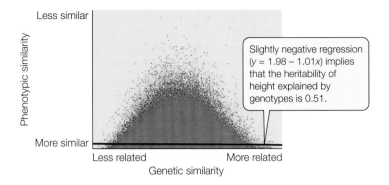

Figure 1.10 SNP-based Heritability Estimation. With 25 million pairwise comparisons of 5000 unrelated people's genetic similarity (estimated genome-wide from their SNPs) and phenotypic similarity (estimated as the squared deviation in z-score), the regression of phenotypic on genetic similarity appears modest, but provides a powerful estimate of heritability. (After Yang et al. 2010.)

SNP-based heritability estimation has been extended to dozens of other traits and diseases, including schizophrenia, which is not influenced by any common variants of even moderate effect. It suggests that common variants contribute in an infinitesimal manner to biochemical traits, such as the level of von Willebrand blood clotting factor, and to generalized intelligence. Chromosome partitioning, in which the analysis is performed for each chromosome separately, typically indicates that the amount of variance explained is proportional to the length of the chromosome and, presumably, the number of genes on that chromosome (Yang et al. 2011b). Similar methods motivated by Bayesian reasoning and adjustments for genotype correlations (linkage disequilibrium) are progressively increasing the accuracy of SNP-based heritability estimation. Critics note that analyses of BMI and other traits have not explained as much variance as was found for height, and they suggest that there is a true gap between the expected heritability and that detected, constituting a true missing heritability problem. Whether or not the gap will be filled by technical advances and improved measurement, there is little doubt that common variants of small effect make a substantial contribution to the heritability of a very wide range of human traits.

Conclusions and Synthesis

The four classes of models described above have been on geneticists' minds for the better part of half a century, but it is only since about 2010 that the tools and resources have been available to evaluate their fit to the actual architecture of diseases. Yet with few exceptions, only a minority of the genetic variability for disease has been ascribed to specific genes, so the jury is still out on the relative contributions of rare and common variants. Some authors even argue that certain complex diseases, such as schizophrenia, are caused by complexes of hundreds of almost Mendelian genes that have similar enough consequences to be classified as a single condition. A safer argument is to adopt the soft version of the infinitesimal model, recognizing that complex disease almost certainly has a complex genetic basis.

Several authors have compiled arguments for and against the prevalence of rare alleles of major effect and common alleles of small effect, ranging from evolutionary arguments through insight from classical genetics to evaluations of the new empirical genomic data. Resolution of this debate is not just an academic exercise, but has critical importance as translational genetic strategies are developed that seek to advise individuals on their genetic predispositions or target novel biochemical pathways for therapeutic intervention. Different assumptions also contribute to different decisions about which research strategies to use and which expensive technologies to adopt in ongoing searches for more causal genes, as well as in fine mapping of the contributing polymorphisms.

My own view, shaped (some would say clouded) by 20 years of studying complex traits in fruit flies, is that the genetic background is always a potent modifier of the effects of even large-effect mutations. The straw-man arguments represented by the RAME and hard infinitesimal models are readily reconciled by supposing that thousands of genetic variants of small effect account for the majority of the heritability for most diseases, but that rare variants of large effect are often superimposed on this variation, providing one source of perturbation that pushes an individual into disease (Gibson 2012). Environmental triggers, acting in conjunction with all classes of genetic factors, are also likely to be important. Whether or not this perspective is correct, statistical geneticists face ongoing challenges in reconciling the discoveries of contemporary genomics with both heritability estimates and demographic data on the incidence of disease within and among families. Ultimately, systems geneticists must also confront the challenge of discovering how polymorphisms work together to modulate the activity of biochemical and physiological pathways. Only then will it be possible to understand how genotypic variation quantitatively maps onto phenotypic variation.

Summary

1. Heritability is the proportion of the phenotypic variance in a population that is attributable to genotypic differences.

2. Heritability estimates do not tell us how much of the risk of disease to an individual is genetic and how much is environmental, and they do not inform us about genetic divergence between populations.

3. Twin studies, particularly comparisons of the resemblance between identical and nonidentical twins, continue to be an important way for geneticists to estimate the extent of genetic contributions to traits.

4. Mendelian conditions segregate in families with typical ratios of unaffected to affected siblings of 3:1 (when the gene is autosomal recessive). There are thousands of such conditions in humans, half of which have already been traced to a single gene.

5. Penetrance is the proportion of individuals with a genotype who have a disease, and expressivity is the severity of disease in affected individuals. Both are modified by genetic and environmental factors.

6. De novo mutations are a major source of human morbidity and probably account for a large proportion of the congenital syndromes that afflict a small percentage of all children.

7. Genotype relative risk (GRR) refers to the risk that a person with a particular genotype has of developing a disease, relative to the risk in people without that genotype. Large-effect mutations or polymorphisms have GRR in excess of 5 but are almost all rare. Small-effect mutations have GRR of less than 2 and may sometimes be more common.

8. The rare allele of major effect (RAME), Goldilocks allele, and infinitesimal models ascribe different contributions of alleles across the range of frequencies and effect sizes to different diseases. There is increasing evidence for contributions of all classes of variants.

9. Genome-wide association studies (GWAS) are used to identify genetic loci that contribute to complex diseases. They utilize measurements of hundreds of thousands of variants in thousands of affected people and controls.

10. The missing heritability problem refers to the fact that variants discovered by GWAS explain much less of the observed genetic variance than initially expected. A simple explanation is that effect sizes are small or allele frequencies low, and that the heritability is not so much missing as hidden beneath stringent statistical thresholds.

11. Broad sense heritability includes such effects as genotype-by-genotype interactions (also called epistasis), genotype-by-environment interactions, and epigenetic inheritance. The importance of these factors as determinants of individual risk of disease remains to be determined.

12. SNP-based heritability estimation provides direct evidence that common polymorphisms can explain at least half of the genetic variance for a wide variety of traits and diseases.

References

Agarwala, V., Flannick, J., Sunyaev, S., GoT2D Consortium, and Altshuler, D. 2013. Evaluating empirical bounds on complex disease genetic architecture. *Nat. Genet.* 45: 1418–1427.

Botstein, D. and Risch, N. 2003. Discovering genotypes underlying human phenotypes: Past successes for Mendelian disease, future approaches for complex disease. *Nat. Genet.* 33 Suppl.: 228–237.

Bunn, H. F., Noguchi, C. T., Hofrichter, J., Schechter, G. P., Schechter, A. N., and Eaton, W. A. 1982. Molecular and cellular pathogenesis of hemoglobin SC disease. *Proc. Natl. Acad. Sci. U.S.A.* 79: 7527–7531.

Campbell, C. D. and Eichler, E. E. 2013. Properties and rates of germline mutations in humans. *Trends Genet.* 29: 575–584.

Chatterjee, N., Wheeler, B., Sampson, J., Hartge, P., Chanock, S. J., and Park, J. H. 2013. Projecting the performance of risk prediction based on polygenic analyses of genome-wide association studies. *Nat. Genet.* 45: 400–405.

de Ligt, J., Veltman, J. A., and Vissers, L. E. 2013. Point mutations as a source of *de novo* genetic disease. *Curr. Opin. Genet. Dev.* 23: 257–263.

Eichler, E. E., Flint, J., Gibson, G., Kong, A., Leal, S. M., et al. 2010. Missing heritability and strategies for finding the underlying causes of complex disease. *Nat. Rev. Genet.* 11: 446–450.

Fisher, R. A. 1930. *The Genetical Theory of Natural Selection.* Clarendon Press, Oxford UK.

Fu, W., O'Connor, T. D., Jun, G., et al., NHLBI Exome Sequencing Project, and Akey JM. 2013. Analysis of 6,515 exomes reveals the recent origin of most human protein-coding variants. *Nature* 493: 216–220.

Gibson, G. 2012. Rare and common variants: twenty arguments. *Nat. Rev. Genet.* 13: 135–145.

Hamosh, A., Scott, A. F., Amberger, J. S., Bocchini, C. A., and McKusick, V. A. 2005. Online Mendelian Inheritance in Man (OMIM), a knowledgebase of human genes and genetic disorders. *Nucleic Acids Res.* 33(Database issue): D514–D517.

Ku, C. S., Loy, E. Y., Pawitan, Y., and Chia, K. S. 2010. The pursuit of genome-wide association studies: Where are we now? *J. Hum. Genet.* 55: 195–206.

Lanpher, B., Brunetti-Pieri, N., and Lee, B. 2006. Inborn errors of metabolism: The flux from Mendelian to complex diseases. *Nat. Rev. Genet.* 7: 449–459.

Manolio, T. A., Collins, F. S., Cox, N. J., Goldstein, D. B., Hindorff, L. A., et al. 2009. Finding the missing heritability of complex diseases. *Nature* 461: 747–753.

McClellan, J. and King, M-C. 2010. Genetic heterogeneity in human disease. *Cell* 141: 210–217.

McKusick, V. A. 2007. Mendelian Inheritance in Man and its online version, OMIM. *Am. J. Hum. Genet.* 80: 588–604.

Nelson, M. R., Wegmann, D., Ehm, M. G., Kessner, D., St Jean, P., et al. 2012. An abundance of rare functional variants in 202 drug target genes sequenced in 14,002 people. *Science* 337: 100.

Park, J. H., Wacholder, S., Gail, M. H., Peters, U., Jacobs, K. B., et al. 2010. Estimation of effect size distribution from genome-wide association studies and implications for future discoveries. *Nat. Genet.* 42: 570–575.

Pritchard, J. K. and Cox, N. J. 2002. The allelic architecture of human disease genes: Common disease-common variant…or not? *Hum. Mol. Genet.* 11: 2417–2423.

Ramu, A., Noordam, M. J., Schwartz, R. S., Wuster, A., Murles, M. E., et al. 2013. DeNovoGear: *de novo* indel and point mutation discovery and phasing. *Nat. Methods* 10: 985–987.

Tennessen, J. A., Bigham, A. W., O'Connor, T. D., Fu, W., Kenny, E. E., et al. 2012. Evolution and functional impact of rare coding variation from deep sequencing of human exomes. *Science* 337: 64–69.

Visscher, P. M., Brown, M. A., McCarthy, M. I., and Yang, J. 2012. Five years of GWAS discovery. *Am. J. Hum. Genet.* 90: 7–24.

Visscher, P. M., Hill, W. G., and Wray, N. R. 2008. Heritability in the genomics era—Concepts and misconceptions. *Nat. Genet.* 9: 255–266.

Yang, J., Benyamin, B., McEvoy, B. P., Gordon, S., Henders, A. K., et al. 2010. Common SNPs explain a large proportion of the heritability for human height. *Nat. Genet.* 42: 565–569.

Yang, J., Lee, S. H., Goddard, M. E., and Visscher, P. M. 2011a. GCTA: A tool for genome-wide complex trait analysis. *Am. J. Hum. Genet.* 88: 76–82.

Yang, J., Manolio, T. A., Pasquale, L. R., Boerwinkle, E., Caporaso, N., et al. 2011b. Genome partitioning of genetic variation for complex traits using common SNPs. *Nat. Genet.* 43: 519–525.

Zielenski, J. and Tsui, L. C. 1995. Cystic fibrosis: Genotypic and phenotypic variations. *Annu. Rev. Genet.* 29: 777–807.

CHAPTER

2

The Human Genome

An international project to sequence the entire human genome and involving scientists in Europe, Asia, and America was initiated in 1990. It was billed as "one of the great feats of exploration in history—an inward voyage of discovery rather than an outward exploration of the planet or cosmos" (www.genome.gov/10001772), and captured public attention with an announcement of completion of the first draft from the White House on June 26, 2000. The total cost was approximately $3 billion, remarkable given that a single genome can be sequenced for $1000 just over a decade later, but many would argue that the economic benefits have been vastly greater. Actually, two different draft genomes were generated in parallel, one by the international consortium (IHGSC 2001) and one by a maverick company, Celera (Venter et al. 2001). The former was stitched together from nine anonymous genomes, the latter mainly from Craig Venter the Celera CEO, himself. This chapter summarizes what we now know about the structure and content of the human genome, including the number and nature of genes and the types of variation that it harbors.

Chromosome Content

The human genome consists of 23 pairs of chromosomes. The 22 autosomes range in length from 250 megabases (Mb) to just under 50 Mb, the smallest being chromosome 21. The sex chromosomes are an X chromosome, usually present in two copies in females, that is 155 Mb long, and a much smaller (60 Mb) Y chromosome that is present in males, though the two do share extensive pseudo-autosomal regions. The mitochondrial DNA is 16.5 Mb in length. The total genome length is just over 3 gigabases (Gb), of which 47% encodes repetitive elements, including transposable elements; 40% is unique sequence that includes genes and intergenic regions (although only 2% actually encodes proteins); and 8% is a complex mixture of mainly hypervariable sequence elements known as **heterochromatin**. The remaining 5% consists of segmental duplications that vary in copy number among individuals.

Each chromosome has a centromere, which organizes chromosomal segregation during mitosis and meiosis, and two arms. The long arm, by convention, is called the q arm. The short one, called the p arm, is just a stub in five of the human autosomes. In interphase cells, one of the X chromosomes is inactivated and tightly packaged into a Barr body, which can be used to determine the sex of a fetus following amniocentesis.

Cytogenetics is the study of the structure and function of chromosomes. Chromosomes can be visualized by **karyotype analysis**, which typically uses white blood cells arrested in metaphase by treatment with colchicine and stained with Giemsa dye. A common application of karyotype analysis is the detection of gross chromosomal rearrangements and **aneuploidy**, which is an abnormal number of chromosomes. There are only a handful of known common benign human chromosomal inversions or translocations. Most visible chromosomal aberrations are associated with cancer or with syndromes caused by the gain or loss of large chromosome segments. **Figure 2.1** shows a typical stylized image of the normal human chromosomes derived from karyotype analysis.

There are three common human conditions attributed to changes in chromosome number. Klinefelter syndrome is where males have two X chromosomes in addition to their Y (XXY). This condition occurs in fewer than

Figure 2.1 Schematic Human Karyotype, Showing the Relative Sizes of the 22 Chromosomes and XY Sex Chromosomes. By convention, the shorter p arms are shown above the longer q arms. The bottom row illustrates four chromosome abnormalities. The Philadelphia reciprocal translocation, which is formally designated t(9;22) (q34;q11), is observed in 95% of people with chronic myelogenous leukemia. It results from fusion between the q arms of chromosomes 9 and 22 at the *ABL1* gene and *BCR* regions.

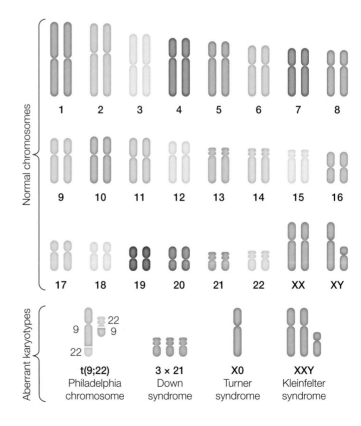

1 in 500 men, and it is associated with a series of mild symptoms that first appear in puberty but may go unnoticed. These symptoms may include partial feminization, including larger than usual breasts, less facial and body hair, and in some cases smaller testicles or increased likelihood of infertility. Women with Turner syndrome have just one copy of the X chromosome (X0). This condition is less common than Klinefelter syndrome, occurring in fewer than 1 in 2000 women, but it has a slightly stronger phenotype, resulting in short stature with a broad chest and neck, low-set ears, and infertility due to reduced ovarian function, as well as morbidity from heart disease, diabetes, and cognitive deficits. The only common autosomal polyploidy is trisomy 21, or Down syndrome, which occurs in approximately 1 in 700 births in the United States. The symptoms are intellectual disability with an average IQ of 50, small body stature, and distinct facial characteristics, as well as dozens of other features that have about 50% penetrance. A half-dozen other trisomy syndromes involving different chromosomes have been described, but they are very rare, and it is thought that nondisjunction of the larger autosomes generally results in spontaneous abortion.

Gene Content

Despite the fact that an essentially complete sequence of the human genome has been drafted for over a decade, there is constant revision of the estimated number of genes. The Wikipedia Human Genome page, drawing on information from the European Bioinformatics Institute, lists 20,051 confirmed or suspected protein-coding genes, almost 13,000 **pseudogenes** (which are expressed but likely nonfunctional genes or gene fragments), 1756 microRNAs, and 5206 other mostly small nuclear noncoding RNAs. Some authors suspect that the true number of functional noncoding RNAs is much greater than 10,000. The average gene density genome-wide is one gene every 150 kb, but the density on chromosome 13 is half of this, while that on chromosomes 17 and 19 is three times greater. The Y chromosome encodes no more than 53 proteins and thus has a remarkably low density of genes.

Part of the difficulty in precisely annotating the gene content of the genome is that the definition of a gene needs to be updated as new functional elements are discovered. MicroRNAs were unknown when the draft genome sequence was completed in 2001, and the discovery that there are hundreds of long noncoding (lnc) RNAs that have important regulatory functions is even more recent. Deep sequencing of RNA now suggests that transcripts are generated from the majority of the chromatin. Most of these transcripts are present at low abundance, but have discrete tissue distributions that are often conserved across mammals, suggesting that they are functional. Most genes have very complex patterns of short RNAs read from the opposite strand near their termini, which may themselves also be considered novel genes. Pseudogenes are defined on the basis of the accumulation of mutations that either disrupt the protein (premature stop codons, microdeletions) or prevent its expression, and they generally have a pattern of missense substitutions that is not consistent with evolutionary constraint. In classical genetics, genes are defined functionally, and while it is unlikely that many of the newly recognized genes are associated with strong phenotypes, even pseudogenes or rare noncoding transcripts may have subtle effects. The term "junk DNA," which was once used to describe parts of the genome devoid of obvious functional genes, must now be used more cautiously.

It is clear that, by comparison with other primates and mammals, there is nothing particularly unusual about the content of the human genome (IHGSC 2001; Venter et al. 2001). *Homo sapiens* does not have a raft of novel genes beyond what would be expected of any close species comparison. The diversity and number of copies of olfactory and immune receptors probably contribute to specific human attributes, but for all intents and purposes it is the way that our existing genes are used, and the variation among them, that makes us human. Our upright posture, unique mental capacities, facial features, and behavioral repertoire trace their origins to changes in the sequences of genes that are also present in our close relatives.

The classification of gene function is known as **gene ontology** (**GO**). The GO Consortium (Gene Ontology Consortium 2000; www.geneontology.org) has constructed a controlled vocabulary that classifies gene functions in a hierarchical manner in each of three domains. Biological GO terms refer to the processes that a gene influences—for example, anatomical structure development at the highest level of the hierarchy, then heart development, and perhaps development of the ascending aorta as the categories grow more specific. Molecular GO terms refer to the biochemical function of the gene product; for example, transcription factors that refine to homeodomain proteins and specific Hox family members, or enzymes that can be oxido-reductases and dehydrogenases, each of which have specific substrates. Cellular component GO terms define where inside the cell (or outside, for secreted proteins) the gene product is found, likewise having hierarchical structure from intracellular to nuclear to nucleolar, or from cytoplasmic organelle to lysozome. The GO Consortium's AmiGO tool is just one of the user-friendly bioinformatic resources that exist to help investigators explore the likely function of any gene of interest (Cabon et al. 2009). The treelike structure of its GO terms is shown in **Figure 2.2**. Other ontologies also exist, including those that classify genes according to diseases that they influence or the biochemical and physiological pathways that they occur in. Some of these can be explored with open-source tools, such as the popular KEGG database and the growing WikiPathways, whereas others require fee-for-access tools, such as Ingenuity Pathway Analysis.

In the molecular domain, just over three-fourths of all human proteins have been assigned a likely function, most on the basis of comparison with structurally similar proteins that have been characterized in bacteria, yeast, or other model organisms. The most common classes of proteins, according to the PANTHER classification scheme (www.pantherdb.org), are transcription factors (2067); transferases (1512); nucleic acid binding (1466); transporters (1096); receptors (1076); signaling molecules (961); enzyme modulators (857); oxidoreductases (550); proteases (476); hydrolases (454);

Biological process

All [716478]
    GO:0008150 : biological_process [563081]
        GO:0032502 : developmental process [51920]
           GO:0048856 : anatomical structure development [52222]
               GO:0001568 : blood vessel development [4180]
                  GO:0060840 : artery development [423]
                     GO:0035904 : aorta development [160]
                        GO:0035905 : ascending aorta development [20]

Molecular function

All [716478]
    GO:0003674 : molecular_function [577197]
        GO:0001071 : nucleic acid binding transcription factor activity [18771]
           GO:0003700 : sequence-specific DNA binding transcription factor activity [18743]
               GO:0000981 : sequence-specific DNA binding RNA polymerase transcription factor activity [4244]
                  GO:0000982 : RNA polymerase II core promoter proximal region sequence-specific DNA binding transcription factor activity [1208]
                     GO:0001078 : RNA polymerase II core promoter proximal region sequence-specific DNA binding transcription factor activity involved in negative regulation of transcription [337]

Cellular component

All [716478]
    GO:0005575 : cellular_component [505276]
        GO:0005623 : cell [298439]
           GO:0005622 : intracellular [262296]
               GO:0043299 : intracellular organelle [209322]
                  GO:0043231 : intracellular membrane-bounded organelle [175805]
                  GO:0005634 : nucleus [83664]

Figure 2.2 AmiGO Gene Ontology Trees for the HES-1 Human Transcription Factor that is Involved in Development of the Ascending Aorta.   Each tree shows the successive refinement of categories, each of which is given a seven-number GO term. The number of genes associated with each term, shown in square brackets, includes all annotated genes across many different species. These trees can be obtained by searching at amigo. geneontology.org, where you will see that each gene has multiple alternative ontologies in each of the domains. The search can be refined by species or by the database that annotates the genes.

and cytoskeletal proteins (441). Genes can also be classified according to whether their products are for housekeeping, meaning that they are required in all cells for basic functions such as metabolism or cell structure; have specialized functions related to the relevant cell type (muscle myosin, pancreatic β cell insulin, eye lens proteins); or are involved in regulation of development, cell division, and cell differentiation. Many genes are **pleiotropic**, which means that they have dual or multiple functions in diverse tissues. Generally, pleiotropy reflects different applications of the same molecular function in different cellular contexts.

There are three major online resources that investigators use to study the structure of individual genes: the University of California at Santa Cruz (UCSC) genome browser (genome.ucsc.edu), the U.S. National Center for Biotechnology Information (NCBI) browser (www.ncbi.nlm.nih.gov), and the European Bioinformatics Institute (EBI) browser (www.ebi.ac.uk). There is considerable overlap in their content, and the databases also cross-link to one another, so the decision as to which one to use is driven by personal preferences. Another useful open-access resource is the Israeli Weizmann Institute's Genecards database (www.genecards.org), which compiles links to all of the above resources as well as summaries of genetic, protein, polymorphism, drug, expression, evolutionary, and antibody data.

It is important to recognize that there are at least four commonly used nomenclature systems for gene names. To make matters worse, many genes have pseudonyms, reflecting the fact that they have been discovered or annotated independently by different groups. For example, the Human Genome Organization Gene Nomenclature Committee (HGNC) has given the gene encoding Homeobox B2 the official gene symbol *HOXB2* and the HGNC ID number 5513. The same gene can also be found in the literature as *HOX2H* or *Hox-2.8*. The EBI Ensembl gene identity is ENSG00000173917, and the NCBI Entrez identity is GeneID 3212. To complicate matters further, the Online Mendelian Inheritance in Man entry gives it yet another number (142967), as does the Human Protein Reference Database (00854) and the UniProt Knowledgebase (P14652). The so-called Reference Sequences (RefSeq) corresponding to the best-characterized gene structure have names beginning with NP_ for proteins (NP_002136), and NM_ (NM_20145.1) for mRNA transcripts, the latter having multiple entries for different splice forms.

All of the major browsers show the local structure of each gene. An increasingly large number of options allow users to zoom in or out and to display tracks of the information that they are most interested in. A screenshot from the UCSC browser for *HOXB2* is shown in **Figure 2.3** as an example. These views tend to change every few months as new information is added, but some essential elements can be highlighted. The top few rows show the inferred structure of the gene with respect to exons and the direction of transcription relative to the nucleotide sequence, which is numbered from the telomere of the p arm. The *HOXB2* gene is located at cytological

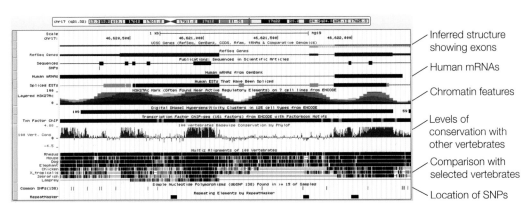

Figure 2.3 Screenshot of the UCSC Browser Page for *HOXB2* at 2 kb Resolution.
The full page, including hundreds of tracks for comparative genomics and visualization of functional elements, can be found by entering *HOXB2* into the search tool at the genome. ucsc.edu gateway and following the top UCSC gene link. (From genome.ucsc.org)

band 17q21, at kb 46,621 on chromosome 17 and is transcribed toward the centromere. It has only one major transcript isoform, so only one mRNA species is shown. The next few tracks show aspects of the chromatin, such as where histones bind and whether they are modified, as well as the locations of DNAse hypersensitive sites (DHS) and transcription factor binding sites. The black and gray bars toward the bottom of the screenshot show conserved sequence elements in a variety of vertebrate species (primates, rodents, dogs, birds, frogs, and fish) as well as a summary score. Finally, the bottom two rows show the locations of known sequence polymorphisms, which are themselves annotated in the dbSNP database (www.ncbi.nlm.nih. gov/snp), as well as simple sequence repeats. It is easy to follow links to more detailed information, including the DNA sequence in something called FASTA format, and if you scroll down the page, you can access hundreds of other features that may be of interest.

Gene Structure

The structure of a typical higher eukaryotic gene is shown in **Figure 2.4**. Isoforms are alternative messages, and subsequently proteins, derived from the same locus. They arise from alternate usage of **exons**, which are the sequence fragments that are found in mature messenger RNAs (mRNAs). Exons are separated on the genomic DNA by **introns**, which are transcribed and then spliced out of the primary transcript and do not contribute to the coding of gene products. It is difficult to say how many isoforms there are per gene, but an estimate of an average of 4 is more likely to be low than high, and many genes have 20 or more alternative transcripts. There is more than ten times as much intronic as exonic DNA in the human genome, which implies that a typical mRNA of up to 5 kb in length is transcribed over a stretch of 50 kb or more. Some genes have over a hundred exons and may give rise to thousands of protein isoforms; the DSCAM neuronal guidance protein in *Drosophila* is a good example (though there is no evidence for such complex splicing in the human homolog, which is associated with Down syndrome; Yamakawa et al. 1998).

Alternative splicing can involve 5′ exons, internal exons, or 3′ exons. Many, if not most, genes have alternative start sites for transcription at the 5′ end, which may be regulated in tissue-specific or temporal manners but are spliced to common coding exons. Similarly, the 3′ untranslated regions (UTRs) at the end of transcripts can include alternative 3′ exons. Internal alternative exon splicing involves the skipping of specific exons and can be highly complex. A gene with four exons—say, A, B, C, and D—can give rise to four different proteins based on alternative uses of just the internal two exons: ABCD, ABD, ACD, and AD. Each additional alternatively spliced exon potentially doubles the number of isoforms, though in reality only a fraction of all possible transcripts are generated. Some exons also have alternative splice junctions that can give rise to the inclusion or exclusion of a small number of amino acids at a critical location in the protein.

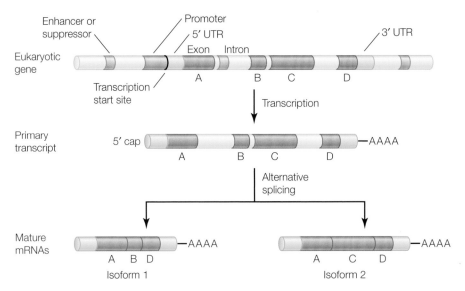

Figure 2.4 Basic Eukaryotic Gene Structure.   The coding region, which can be over 100 kb in length, includes multiple exons (blue). The exons are separated by introns (yellow), which are transcribed as part of the primary message, but spliced out during production of mature RNAs. Alternative splicing generates different mRNA isoforms, two of which are shown corresponding to removal of exon C and exon B, respectively. Enhancer and suppressor elements (green) can be 5′ or 3′ to the coding region or in introns. They are distinguished from promoter elements (red) that are proximal to the transcription start site (black) within 1 or 2 kb. The 5′ and 3′ ends of the primary transcript include untranslated regions (UTRs; orange), and many genes have alternate versions of these elements as well.

The structure of transcripts encoded at each locus is generally annotated using a combination of bioinformatic prediction and experimental verification (Brent 2005). One class of powerful bioinformatic tool is called a Hidden Markov model (HMM). HMMs employ a statistical algorithm to identify probable genes. They search DNA sequences for features that are commonly associated with genes, such as promoters, open reading frames, and splice junctions, and assign probabilities that the features are included in a transcript given the constraint that they must be arranged in a specific order along the chromosome. Different HMM implementations agree well on their predictions of the presence of a gene, but these models are less precise in their annotation of exon boundaries and alternative isoforms. Genome sequence alone cannot indicate which tissues different isoforms occur in, so deep sequencing of RNA is used to generate empirical evidence. Complete annotation, however, requires the sampling of a diversity of tissues under a range of conditions in multiple individuals, and consequently new transcripts at a locus are often found with the addition of each new study.

The strongest experimental data are obtained by complete sequencing of full-length **cDNAs** (complementary DNA copies of messenger RNAs), but this method is laborious and expensive, so most efforts have focused on sequencing short RNA fragments. Expressed Sequence Tags (ESTs) are fragments of cDNA, but they indicate only that a particular sequence is part of an exon. With the advent of high-throughput automated sequencers, it is now trivial to generate over 50 million **paired-end reads**—that is, sequences of 100 base pairs (bp) from either end of a 500 bp cDNA fragment—from a single sample, providing an average of 5000 reads per gene. Each fragment may provide evidence that two or more exons are present in the same transcript, and graph theory can be used to computationally assemble maps of the likely frequency distribution of transcripts. A popular piece of software for doing this is Cufflinks (Trapnell et al. 2012; see Chapter 8), which, in addition to estimating the abundance of each exon, and hence each gene, in the sample, also estimates isoform abundance. It should be noted that different algorithms can generate quite different estimates, which in part reflects the complexity of transcription. A study of several hundred lymphocyte cell lines (Lappalainen et al. 2013) documented 146,498 exons in 16,084 expressed genes (an average of 9 per gene) and inferred that 129,805 splice junctions are used in the generation of 67,603 different transcripts present in at least half the samples. Interestingly, most of the variation involved relatively rare isoforms, in which alternative 5′ and 3′ exon use was pervasive. There was also considerable interindividual diversity and even among-population diversity in splicing, much of which is under genetic control.

Variation at the ends of transcripts can be studied explicitly using the techniques of 5′ and 3′ RACE (reverse amplification of cDNA ends) as well as CAGE sequencing (cap analysis of gene expression). The latter technique has led to a reconception of the nature of **promoters** (Carninci et al. 2006; **Figure 2.5A**). Whereas textbooks usually state that there is an AT-rich "TATA box" motif 30 bp upstream of the transcription start site of most genes, it is now understood that this is mainly true of genes that are transcribed under tight regulation of where and when they are expressed in tissues. As many as 80% of all transcripts are actually derived from TATA-less promoters, usually associated with **CpG islands**, and instead of there being a unitary start nucleotide, transcription commences over at least a 30 bp region with multiple alternative first nucleotides.

The regulation of splicing is achieved in part by sequence motifs in the vicinity of splice junctions (**Figure 2.5B**). The consensus elements are a GU at the 5′ end of the intron, a "branch site" centered on an A nucleotide near the 3′ end of the intron that is followed by several pyrimidines (C or T), and then an AG at the 3′ junction. Splicing occurs in a molecular complex known as the **spliceosome**, which consists of a series of short RNAs and accessory proteins (Wang and Burge 2008). The cellular machinery must use some sort of code to regulate when and where splicing occurs, but investigators are not yet able to predict isoform generation accurately.

Initiation of translation, too, is more complex than first thought. Ribosomes are usually assembled in the vicinity of the 5′ methyl cap on mRNAs and then use the first available AUG to initiate protein synthesis. It turns out, however, that approximately half of human transcripts have short

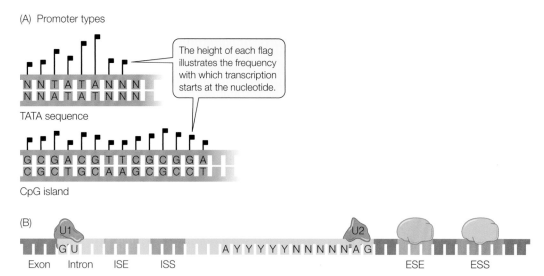

(A) Promoter types

NNTATANNN
NNATATNNN

The height of each flag illustrates the frequency with which transcription starts at the nucleotide.

TATA sequence

GCGACGTTCGCGGA
CGCTGCAAGCGCCT

CpG island

(B)

U1   G U   AYYYYYNNNNNN AG   U2

Exon   Intron   ISE   ISS   ESE   ESS

Figure 2.5 Regulation of Transcription. (A) Two types of promoters are now recognized: those in which transcription is initiated in a focal 10 bp region centered on a TATA-like sequence, and those with a broad transcription start site spread over 30 bp enriched for CpG dinucleotides (a CpG island). (B) Splicing requires a GU motif at the 5' end of an intron and an AG at the 3' end following a branch-point AYYYYY sequence, where Y is a pyrimidine (C or U). Intronic and exonic splice enhancer (ISE, ESE) and splice suppressor (ISS, ESS) elements also act by binding to regulatory proteins and RNA-protein complexes.

upstream open reading frames (**uORFs**, a series of codons starting with AUG that could encode a tiny peptide before a stop codon is encountered); these reduce translation of the true protein by between 30% and 80%. Hundreds of these uORFs are polymorphic, in some cases contributing to rare diseases (**Figure 2.6**; Calvo et al. 2009). It is believed that in normal translation, the ribosomes read through the short messages encoded by these uORFs, but internal ribosome entry sites (IRES) provide an extra mode of regulation of gene activity, notably in the context of the unfolded protein response, an important stress survival mechanism (Wek and Cavener 2007).

One final feature of gene structure is that genes can be nested within one another. Some short genes are located within the introns of other genes and are usually transcribed in the opposite direction from the genes that encompass them. The classical conception of genes as "beads on a string" is also challenged by the realization that regulatory regions can jump over intervening genes, acting over tens or even hundreds of kilobases. For example, persistent expression of the lactase enzyme, which is responsible for tolerance of milk, into adulthood is facilitated by a handful of mutations in an enhancer element that is actually found in an intron of the *MCM6* gene, 14 kb upstream of the lactase-phlorizin hydrolase (*LPH*) gene's transcription start site (Ingram et al. 2009). Very recently, techniques for cross-linking physically adjacent chromosome fragments, then sequencing the

(A)

| | | | | | Reduction in protein expression |

5′ CAP   uORF   UTR   CDS   Reduction in protein expression

*cuauugauguggacuccuggauaggcagcuggaccaacgAUGAAC...   In-frame uORF

*cuauugauguggacuccuggauaggcagcuggaccaacggAUGAAC...   Out-of-frame uORF   } (41%)

*cuauugauguggacuccuggauuggcagcuggaccaacgAUGAAC...   Read-through uORF   (18%)

*cuauugauguggacuccuggauuggcagauggacuaacgAUGAAC...   Multiple uORFs   (26%)

(B)

Gene name	Function	Disease associated with SNP
F12	Coagulation factor	N/A
IRF6	Interferon regulation	N/A
CFTR	Cystic fibrosis transmembrane conductance	N/A
THPO	Thrombopoietin	Thrombocythemia
CDKN2A	Cell cycle kinase inhibitor	Melanoma
HR	Transcription regulator	Hypotrichosis (abnormal hair pattern)

Figure 2.6 Upstream Open Reading Frames (uORFs) in Human Genes. (A) Almost 50% of all human 5′ untranslated regions (UTRs) contain an uORF, of which the majority would encode peptides that would terminate before the normal translation start site (AUG). These uORFs can either be in or out of frame (as in the top and second rows, respectively), and read-through from the former may lead to functional proteins, as seen in 18% of transcripts. One-fourth of transcripts represent more than one uORF. (B) Examples of the 500 genes with polymorphisms that generate or abolish an uORF are listed, including three that have been associated with rare diseases. Most uORFs are expected to lead to a more than 30% reduction in protein levels. (See Calvo et al. 2009 for details.)

juxtaposed fragments, have revealed that there are thousands of places in the genome where sequences megabases apart on the same chromosome consistently touch (Dekker et al. 2013). There is evidence that these interactions are important for gene regulation, and hence for a broader conception of gene structure that includes elements dispersed over large stretches of a chromosome, and possibly even on other chromosomes.

Gene Regulation

Regulation of gene activity is essential for ensuring that only the appropriate genes are expressed in the correct tissue or cell type at the correct time and at optimal levels of abundance. Over a third of the genome is dedicated to the regulation of gene activity, which occurs at multiple levels, affecting transcription, translation, and protein function. Transcriptional regulation, also called gene expression regulation, is mediated by elements proximal to the promoter and by distal enhancers and suppressors, as well as by chromatin structure. Translational regulation involves ribosome initiation, mRNA translocation, and mRNA decay, and it is mediated by both proteins and microRNAs. Protein function is regulated by posttranslational modifications, including phosphorylation, glycosylation, ubiquitination, and protein complex formation. It will not be considered further here; suffice it to say that measuring transcript abundance does not equate to measuring protein abundance, and measuring protein abundance does not equate to measuring protein activity.

The first mechanism of transcriptional regulation, initially elucidated in bacteria and yeast, is the binding of the transcription initiation complex to promoter-proximal elements. Positive regulation involves the binding of transcription factors to the DNA. These help to recruit or release the transcription initiation complex, allowing transcription to proceed once the appropriate cue is obtained (Lee and Young 2000). Studies have also shown regulation by pausing of transcript elongation shortly downstream of the transcription start site, providing a rapid mechanism for turning on gene expression (Kwak and Lis 2013). Promoter activity can be studied experimentally by using genetic engineering to manipulate promoter sequences and then observing the effect of nucleotide sequence changes on reporter genes, either in vitro or after reintroducing the recombinant DNA constructs into cells. The role of polymorphisms in regulating gene expression has also been inferred statistically from the observation that many SNPs that are associated with transcript abundance—so-called eSNPs or eQTL—map within a kilobase on either side of the promoter (see Chapter 8).

Tissue-specific and quantitative regulation is also provided by more distal regulatory elements that are known as **enhancers** or **suppressors**, depending on whether they activate or repress transcription. These elements may be located upstream, downstream, or within the gene body (often in the first intron) and can act over many tens of kilobases in an orientation-independent manner. Binding of transcription factors to an enhancer, accompanied by looping of the chromatin that brings the enhancer physically adjacent to the promoter, facilitates interactions with RNA polymerase and the transcription initiation complex. Suppressors also bind transcription factors, but either prevent enhancer activity or actively inhibit transcription. Other classes of regulatory elements include **insulators**, which isolate enhancer complexes and ensure that they act only on the appropriate target promoter; and Polycomb-group or Trithorax-group binding sites, which regulate chromatin structure (Gerasimova and Corces 2001).

Chromatin modification provides higher-order regulation of gene expression, mediating access by transcription factors to the regulatory sequences. It includes regulation of the assembly of histones on DNA, packaging of chromatin into coils and supercoils (ultimately turning off an entire chromosome; for example, one of the X chromosomes in women), and organization of three-dimensional nuclear structure. Two common mechanisms of chromatin modification are methylation and histone acetylation, which are discussed further in Chapter 9. Chromatin modification is probably responsible for epigenetic transmission and imprinting.

Posttranscriptional regulation is increasingly seen to be mediated by **microRNAs (miRNAs**; Pasquinelli 2012). There are 1756 well-documented miRNAs, a few dozen of which are expressed at much higher abundances than the remainder. In humans, miRNAs exert their effects mostly by suppressing translation, though they can also influence mRNA stability. Most miRNAs bind to complementary sequences in the 3' UTR of their target genes. The complementarity does not have to be perfect, and given that this binding occurs over fewer than 15 bp, computational prediction of targets is not trivial, and different software yields varying results. Nevertheless, several online resources, such as miRBase (www.mirbase.org), compile lists of likely targets and support genomic explorations of mRNA-miRNA regulatory interactions. Only a minority of predicted interactions show the negative correlation between miRNA abundance and mRNA abundance expected if the former represses the latter. This observation is attributed to feedback loops and regulatory cross talk as well as to the fact that some regulation is not at the level of mRNA abundance. Cancer studies in particular have established the core role of miRNA-mediated regulation of gene activity (Calin and Croce 2006).

RNA Genes

There are almost 7000 non-protein-coding genes in the human genome, and this number is only likely to increase as new RNA functions are discovered. For the purposes of this discussion, we will consider three broad classes of RNA genes: those involved in translation, posttranscriptional regulators, and **long noncoding RNAs (lncRNAs)**, which have diverse functions.

The ribosome has long been known to be a protein-RNA molecular machine. There are 532 **ribosomal RNA (rRNA)** genes, and there are almost 500 nuclear **transfer RNA (tRNA)** genes, which encode the adapters that guide the correct amino acid into the growing protein chain. In addition, over 1500 **small nucleolar RNA (snoRNA)** molecules are used to modify ribosome assembly by methylating and pseudouridylating rRNAs. Some have very specific functions, such as a brain-specific snoRNA that is required for the processing of serotonin 2C receptor mRNA, whose deletion leads to Prader-Willi syndrome, with symptoms of food craving and motor deficits (Doe et al. 2009). Another, even larger set of **small nuclear RNAs (snRNAs)** have heterogeneous roles in splicing and other steps in the processing of primary transcripts, maintenance of telomeres, and regulation of transcription. They also function as components of ribonucleoproteins.

Only in the twenty-first century has the critical role of miRNAs been recognized. They were discovered in nematodes soon after experiments showed that double-stranded RNA can specifically knock down the function of complementary genes. Estimates of the number of true miRNAs vary widely, depending on whether the detection of miRNA tags in a library of small-RNA sequences is regarded as sufficient evidence. Most miRNAs are present at very low abundances, so there is considerable sampling variance in which ones are observed to be expressed in different people. The Human MicroRNA Disease Database (HMDD; Lu et al. 2008) listed 591 miRNAs associated with 396 diseases as of January 2013, but genome annotations suggest three times this number of likely miRNAs.

Mature miRNAs are 21–24 nucleotides (nt) long, having been processed from longer transcripts by the Drosha and Dicer complexes (Winter et al. 2009; **Figure 2.7**). They then associate with the Argonaute protein to assemble an RNA-induced silencing complex (RISC) with target mRNAs, which leads to translational repression or to cleavage or deadenylation of the mRNA. Another important class of small regulatory RNAs are the **Piwi-interacting RNAs (piRNAs)**, which are 26–31 nt single-stranded molecules that suppress retrotransposons, particularly in the germ line.

Current estimates suggest that there may be over 1200 lncRNAs that have diverse functions. These RNAs are transcribed from intergenic regions despite their lack of an open reading frame. In addition, as many as half of all genes contain long RNAs near the 5′ and 3′ exons that are transcribed, often in the antisense direction, as shown in **Figure 2.8A**. Operationally, lncRNAs are defined as noncoding RNAs greater than 200 nt in length, since this is the size cutoff used in the preparation of small RNAs for miRNA sequencing. Evidence that lncRNAs are functional comes from

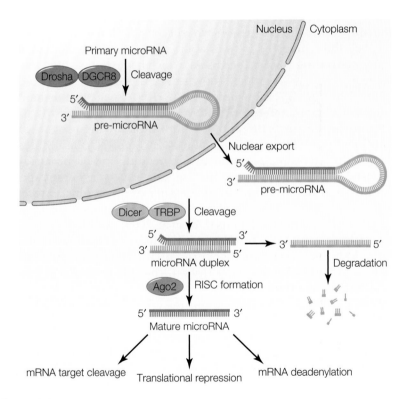

Figure 2.7 The Processing of microRNAs. MicroRNA genes are transcribed in the nucleus. Their products are cleaved into pre-miRNAs by the Drosha enzyme complex, then exported into the cytoplasm. There, the Dicer enzyme complex cleaves the internal loop, allowing the nonfunctional strand to be degraded, and the mature miRNA complexes with Argonaute (Ago2) to form the RISC. (After Winter et al. 2009.)

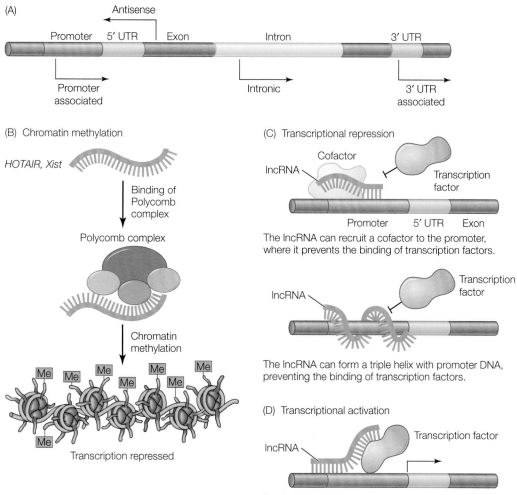

Figure 2.8 Long Noncoding RNAs (lncRNAs).   (A) Deep sequencing reveals that in addition to the primary transcript, most genes produce many unprocessed RNAs, which may be derived from sequences near the promoter, within the introns, or at the 3′ end of the gene, and may be transcribed in either the sense or the antisense orientation. (B–D) Four of many mechanisms of lncRNA function in gene regulation. *HOTAIR* and *Xist* complex with Polycomb-group proteins to modulate heterochromatin methylation (B). Transcriptional repression can be achieved by the tethering of a lncRNA to the promoter, where it recruits cofactors, or by the hindrance of binding of transcription factors through the generation of a triple helix (C). Similarly, transcriptional activation can be achieved by recruitment of transcription factors to the promoter (D).

conservation of expression patterns from humans to rodents, conservation of at least small sequence stretches (since these RNAs are noncoding, the RNA secondary structure is more important than the primary sequence), and increasing evidence of phenotypes due to lncRNA mutations. One of the first lncRNA genes to be characterized was the *Xist* gene, whose product inactivates one of the X chromosomes in females by recruiting a protein complex to methylate histones in the heterochromatin. Other lncRNAs act by a variety of mechanisms, including tethering transcription and chromatin factors to the local DNA, forming tertiary complexes with the double helix, and even serving as scaffolds for assembly of cytoplasmic protein complexes (Mercer et al. 2009; **Figure 2.8B–D**). A database of lncRNAs is available at www.lncrnadb.org (Amaral et al. 2011).

Repetitive Elements

Given that **repetitive elements** account for approximately one half of the sequenced portion of the human genome, a brief discussion of these repeated sequences is in order. Undoubtedly they contribute to some disease conditions, and there is accumulating evidence that they collectively influence stress responses and genome stability, but because individual repetitive elements are only occasionally associated with disease or other phenotypes, they appear only briefly in the remainder of this primer.

The five major classes of repetitive elements are shown in **Figure 2.9** (Treangen and Salzberg 2012). The most common are long interspersed elements (LINEs) and short interspersed elements (SINEs), each with over 1.5 million copies spread throughout the genome. The LINEs are up to 8 kb in length, the SINEs less than 300 bp, and collectively they account for 36% of the DNA in each human cell. A small number of LINEs retain the capacity for self-replication via RNA intermediates, so these elements are a class of retrotransposons capable of contributing to expansion of the genome. SINEs do not transpose autonomously, but they are transcribed, and the most common type, Alu elements, are often spliced into active transcripts whose activity they can modulate.

Two less abundant classes of human transposons are long terminal repeat–containing (LTR) retrotransposons and DNA transposons, but they still account for 9% and 3%, respectively, of the reference human genome HuRef19, with around half a million copies each. To put this number in perspective, there are more than 20 copies of each of these elements for every protein-coding gene. LTR retrotransposons are characterized by repeats at least 100 bases long oriented in the same direction at their two termini (hence the name long terminal repeats). The LTRs facilitate transcription of the elements into RNA, which is then reverse-transcribed into DNA and inserted at a new location in the genome. DNA transposons move by a different mechanism that involves excision of the element and its reinsertion at a new location. New insertions can be mutagenic, but most human DNA transposons are immobile, so they are not thought to be a major source of new variation.

The fifth class of common repeats is the microsatellites, which are very small hypervariable elements, typically di- or trinucleotides, but sometimes longer. There are over 400,000 of these elements, one for every 6 kb of DNA,

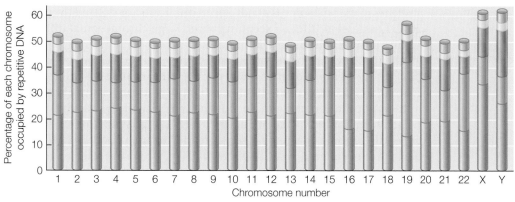

	Percentage of human genome	Number in human genome
■ LINEs	21	1.5 million
■ SINEs	15	1.8 million
■ LTR rtps	9	720,000
▢ DNA tps	3	465,000
▨ Microsatellites	3	430,000
■ Duplicates	5	2270

Figure 2.9 Repetitive DNA Content by Chromosome.   Each color represents one of the five major classes of repetitive elements. The graph shows their relative proportions on each chromosome, and the key gives their proportions in the human genome and their total numbers. Approximately 5% of the Y chromosome is tandem repeats. (After Treangen and Salzberg 2011.)

interspersed throughout genes. Most have considerable allelic diversity: people differ with respect to the number of copies. In some cases microsatellites are included in coding regions; for example, repeats of the CAG codon generate long polyglutamine repeats in some transcription factors. Longer simple repeats up to 100 bp in length are called minisatellites, short tandem repeats (STRs), or variable number tandem repeats (VNTR). A set of 13 tetranucleotide STRs has a distinguished history in forensic genetics because the combined copy numbers of these so-called CODIS variants are sufficient to identify most people with extraordinarily high confidence (Butler 2006).

Sequence Variation

Just as important as the gene content of the human genome is the content of variation among individuals, as it is ultimately a major source of phenotypic variation. The vast majority of variation is transmitted through the germ line and hence is present in all the cells of an individual. There is recent evidence, however, that in 1% of people, the blood is mosaic for somatic segmental aneuploidy—namely, some of the cells have large stretches of chromosome that have been duplicated or deleted (Jacobs et al. 2012; Laurie et al. 2012). Except in cancer, large chromosomal differences among individuals are rare in humans.

The most frequent class of variant is single nucleotide polymorphisms (SNPs). The genome of an average person harbors almost 4 million SNPs. As explained in the next chapter, most polymorphisms are rare in the human population as a whole, but the majority of the variants in each person have a minor allele frequency over 5%. It is typical for the sequencing of the genome of a new person unrelated to any previously sequenced person to reveal more than 30,000 new SNPs, which are referred to as **private variants**. If every living person were to be sequenced, it is likely that a polymorphism would be found at every site in the genome. Humans are actually not particularly polymorphic compared with other species (fruit flies, for example, have a SNP every 30 bp), but our species has vastly more variation than would be needed to generate the phenotypic diversity observed among humans.

The NCBI maintains a SNP database, dbSNP (www.ncbi.nlm.nih.gov/SNP), that documents the location, frequency, type, and predicted function of each newly annotated variant. In humans, each variant is referenced with an identifier beginning with rs, for reference SNP, and a pseudo-random number. The Ensembl browser (www.ensembl.org/info/genome/variation/index.html) provides a user-friendly way to access this database. The sample screenshot in **Figure 2.10** shows that rs1042822 is a T/G polymorphism located in the 3′ UTR of the HOXB2 transcript. The frequency of the minor allele, T, is 0.19 in Europeans (EUR), 0.06 in East Asians (ASN), and 0.07 in Yoruban Africans (AFR). Just in case you are interested, the dbSNP entry also lets you know that this SNP is heterozygous in Jim Watson, while Craig Venter is homozygous for the minor allele (but note that it refers to the A/C polymorphism on the opposite strand). Various links tell you that the variant is not associated with a disease or phenotype. The minor allele is also inferred to be the derived allele, since the G is found in nonhuman primates and thus is said to be ancestral. Across the genome, the majority of derived alleles are minor alleles, but some derived alleles reach a high frequency at least in some populations. The naive expectation that disease-associated variants will also be minor alleles is also often false, suggesting that a substantial fraction of new mutations in the human lineage are protective against disease. Results capturing association of each SNP with disease or other phenotypes are also reported where available, or users can follow links to resources such as SNPedia.com or GeneWiki that summarize data from the literature.

Functionally, SNPs can be classified according to the type of nucleotide change or its predicted consequence. A/G and T/C polymorphisms are **transitions**, as they change a purine to a purine or a pyrimidine to a pyrimidine. All other changes (A or G to T or C and vice versa) are **transversions**. Most SNPs have just two alternative alleles, but there are so many in the genome that inevitably some sites have been mutated on

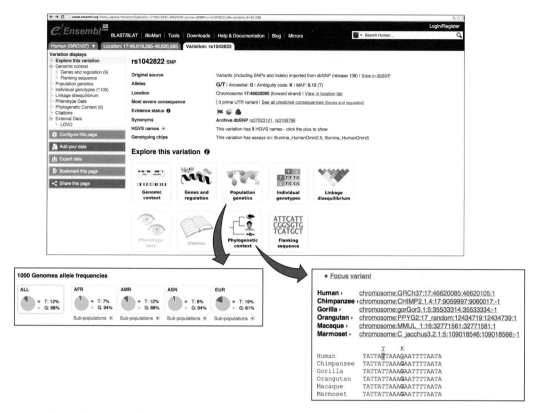

Figure 2.10 The Ensembl SNP Viewer.   The main page for each reference SNP provides some basic information on the source of the data, alternative names, and the location of the variant, then links to nine resources. Results from two of these resources are shown in the remainder of the figure: Population Genetics, showing the allele frequency across HapMap populations and Phylogenetic Context, showing the sequence in other primates. (From www.ensembl.org)

multiple occasions, and three, or even all four, nucleotides are observed at those sites. SNPs in the coding region of an exon can be **nonsynonymous** if they lead to a replacement of one amino acid with another, or synonymous if the alternative codon specifies the same amino acid due to the degeneracy of the genetic code. **Synonymous** substitutions are usually expected to be functionally silent, but they may influence gene expression, translation, or mRNA stability and thus may still affect traits. Other SNPs within the gene body may be located in the 5′ UTR, in an intron, or in the 3′ UTR, where they may influence translation initiation, splicing, or miRNA-mediated regulation, but it is very difficult to predict their function from the location alone. Intergenic SNPs may be located in the promoter, in distal regulatory elements, in spacer DNA 5′ or 3′ to genes, or in "gene deserts" where there are no annotated genes for hundreds of kilobases. For example, several SNPs in the 8q24 interval are associated with a variety of cancers, probably exerting their effects by long-range regulation of transcription of the *MYC* oncogene 300 kb away (Grisanzio and Freedman 2010). Computational annotation of regulatory SNPs is a major challenge for bioinformaticians.

Some polymorphisms involve the loss or gain of a single nucleotide, which is usually highly deleterious if it occurs in an exon, as it disrupts the reading frame. In contrast, 3 bp insertions or deletions are often tolerated because they add or remove a single codon. Short insertion/deletion polymorphisms, or **indels**, are common; each individual has up to 100 indels that affect coding regions. The term **simple nucleotide polymorphism** is sometimes used to include SNPs and indels.

Individual genotypes are determined (or "called") using one of two approaches, genotyping chips or sequencing. The chips produced by Illumina or Affymetrix contain between 500,000 and 2.5 million SNP genotyping assays that are individually over 99% accurate. They provide a comprehensive overview of the common genetic variants in a genome and can be used to impute (infer) several million other common SNPs. For a thousand dollars, it is now possible to sequence an entire genome, which provides comprehensive coverage of common, rare, and private alleles. The genotypes are called by comparison with a haploid reference genome, currently HuRef19, which is a composite genome derived from several anonymous donors. HuRef19 is not a consensus genome: it does not represent the major allele at all sites. Rather, it has a typical distribution of allele frequencies—and in fact, the hypothetical individual it represents would be at elevated genetic risk for type 1 diabetes (Chen and Butte 2011). Consequently, some of the alleles in any person's genome that are different from HuRef19 are actually the major alleles in the human population.

The other major class of variation in the human genome is copy number variation. It is estimated that approximately 12% of the genome is covered by copy number variations (CNVs). The size of these CNVs range from 1 kb to more than a megabase; those at the higher end of this range affect multiple genes and cause many congenital syndromes. Deletion of a long stretch of DNA results in hemizygosity for that region—namely, a single copy—and consequently the genome sequence will appear to homozygous, but with fewer reads than most of remainder of the genome. Duplication of a long stretch usually occurs in tandem (that is, the extra copy is located adjacent to the normal one), and individuals are likely to have three (or occasionally more) copies rather than two. Since the dosage of most genes is a contributing determinant of gene activity, both losses and gains of copies can contribute to phenotypes. Copy number variations up to 1 Mb in length cumulatively affect as much as 3% of the euchromatin of each individual. CNVs that are not extremely rare are also called CNPs, for copy number polymorphisms.

The first global survey of CNVs (Redon et al. 2006), which compared 270 individuals from Europe, Nigeria, and East Asia in the HapMap collection, documented almost 1500 CNVs affecting 2500 genes. Estimates of the number of CNVs per individual continue to be modified as new technologies increase the resolution to detect deletions and duplications involving just tens of kilobases of DNA, but it is clear that everyone has several dozen genes affected by CNVs. A comprehensive survey of eight common diseases by the Wellcome Trust Case Control Consortium found little evidence for a major contribution of common CNVs to genetic variance (WTCCC 2010). On the other hand, many studies of rare diseases, notably schizophrenia and autism, as well as of extreme phenotypes have established a clear role for rare CNVs in the etiology of these conditions (Grozeva et al. 2010).

The number of SNP or indel variants per genome is summarized in **Table 2.1**. The data are derived from phase 1 of the 1000 Genomes Project (2010, 2012), which has made low coverage draft sequences available for 1092 people from 14 populations. They document 38 million SNPs in all, in addition to 1.4 million small indels, and 14,000 CNVs. There are up to 3000 derived nonsynonymous SNPs per genome, as well as 30 stop codon gains that would truncate the proteins involved. A similar number of frameshift and splice site mutations means that every person is functionally heterozygous for at least 50 genes, but most of us are homozygous for a small number as well. It is more difficult to estimate the extent of functional regulatory polymorphisms, but at least 1000 genes per genome have a polymorphism in a suspected transcription factor binding site, and about one-fourth of these result in the gain of a binding site. Each person also carries between 15 and 40 variants that are documented in the human gene mutation database (www.hgmd.org) as likely to be disease promoting (Stenson et al. 2012). The implications for personalized medicine remain to be understood.

TABLE 2.1   Load of sequence variants in individuals from the 1000 Genomes Project

Variant type	Number of derived sites per individual in frequency bin		
	<0.05% (Very rare)	0.5%–5% (Rare)	>5% (Common)
All sites	30,000–150,000	120,000–680,000	3.6–3.9 million
Synonymous	29–120	82–420	1300–1400
Nonsynonymous	130–400	240–910	2300–2700
Nonsense (stop codon gain)	4–10	5–19	24–28
Indel	2–4	20–48	127–138
Splice site	3–6	4–11	5–10
Constrained UTR[a]	120–430	300–1400	3500–4000
Constrained ncRNA[a]	4–17	14–70	180–200
Affecting known transcription factor binding site	23–83	94–359	750–830
Other conserved[a]	2–10	8,000–42,000	130,000–150,000

Source: WTCCC 2012.

[a]Conserved elements are defined as having a GERP score >2 (this is a measure of evolutionary constraint).

Summary

1. The normal human genome consists of 22 pairs of autosomes and 1 pair of sex chromosomes.

2. The total length of the genome is over 3 Gb, about half of which consists of repetitive DNA. Only 2% encodes proteins.

3. The roughly 20,000 human genes have diverse functions in regulation, cellular differentiation, and metabolism. The most common functions are transcription factors, transporters, receptors, intracellular signaling, and enzymes

4. Gene ontology (GO) is used to classify genes according to biological role, molecular function, and cellular component. Other classification systems document disease associations and physiological pathways.

5. There are three commonly used genome browsers, maintained by UCSC, the NIH/NCBI, and the European EBI. Hundreds of other open-access resources provide additional information on the human genome.

6. Most genes have a complex structure that includes multiple exons that can be alternatively spliced to generate more than half a dozen protein isoforms. There is also considerable complexity at the 5′ and 3′ termini.

7. Gene regulation is mediated at multiple levels, from chromatin accessibility to transcriptional regulation by transcription factors that bind to enhancers and promoters, as well as post-transcriptional regulation.

8. About 7000 non-protein-coding genes encode known RNA products. The RNA component of the genome includes many classes of short nuclear RNAs that are essential for translation, long noncoding RNAs that have diverse roles, and thousands of microRNAs that help regulate transcription.

9. Repetitive DNA consists mostly of short and long interspersed elements (SINEs and LINEs) but also includes LTR retrotransposons as well as DNA transposons, most of which are no longer mobile.

10. Microsatellites are hypervariable repeats of very short nucleotide sequences. They are an important tool in forensic analysis.

11. The most numerous type of variation in the human genome is single nucleotide polymorphism (SNP). However, copy number variation (CNV) affects a larger fraction of nucleotides, since each deletion or duplication includes thousands of bases.

12. Each person has several dozen potentially deleterious mutations, including stop codons and indels, which if homozygous would likely cause morbidity or lethality, and may also have dominant effects on health.

References

Amaral, P. P., Clark, M. B., Gascoigne, D. K., Dinger, M. E., and Mattick, J. S. 2011. lncrnadb: A reference database for long noncoding RNAs. *Nucl. Acids Res.* 39 (suppl 1): D146–D151.

Brent, M. R. 2005. Genome annotation past, present, and future: How to define an ORF at each locus. *Genome Res.* 15: 1777–1786.

Butler, J. M. 2006. Genetics and genomics of core short tandem repeat loci used in human identity testing. *J. Forensic Sci.* 51: 253–265.

Cabon, S., Ireland, A., Mungall, C. J., Shu, S. Q., Marshall, B., and Lewis, S. 2009. AmiGO: Online access to ontology and annotation data. *Bioinformatics* 25: 288–289.

Calin, G. A. and Croce, C. M. 2006. MicroRNA-cancer connection: The beginning of a new tale. *Cancer Res.* 66: 7390.

Calvo, S. E., Pagliarini, D. J., and Mootha, V. K. 2009. Upstream open reading frames cause widespread reduction of protein expression and are polymorphic among humans. *Proc. Natl. Acad. Sci. U.S.A.* 106: 7507–7512.

Carninci, P., Sandelin, A., Lenhard, B., Katayama, S., Shimokawa, K., et al. 2006. Genome-wide analysis of mammalian promoter architecture and evolution. *Nat. Genet.* 38: 626–635.

Chen, R. and Butte, A. J. 2011. The reference human genome demonstrates high risk of type 1 diabetes and other disorders. *Pac. Symp. Biocomput.* 2011: 231–242.

Dekker, J., Marti-Renom, M. A., and Mirny, L. A. 2013. Exploring the three-dimensional organization of genomes: Interpreting chromatin interaction data. *Nat. Rev. Genet.* 14: 390–403.

Doe, C. M., Relkovic, D., Garfield, A. S., Delley, J. W., Theobold, D. E., et al. 2009. Loss of the imprinted snoRNA mbii-52 leads to increased 5htr2c pre-RNA editing and altered 5HT2CR-mediated behavior. *Hum. Mol. Genet.* 18: 2140–2148.

Gene Ontology Consortium. 2000. Gene Ontology: Tool for the unification of biology. *Nat. Genet.* 25: 25–29.

Gerasimova, T. I. and Corces, V. G. 2001. Chromatin insulators and boundaries: Effects on transcription and nuclear organization. *Annu. Rev. Genet.* 35: 193–208.

Grisanzio, C. and Freedman, M. L. 2010. Chromosome 8q24-associated cancers and MYC. *Genes Cancer* 1: 555–559.

Grozeva, D., Kirov, G., Ivanov, D., Jones, I. R., Jones, L., et al. 2010. Rare copy number variants: A point of rarity in genetic risk for bipolar disorder and schizophrenia. *Arch. Gen. Psychiatry* 67: 318–327.

Ingram, C. J., Mulcare, C. A., Itan, Y., Thomas, M. G., and Swallow, D. M. 2009. Lactose digestion and the evolutionary genetics of lactase persistence. *Hum. Genet.* 124: 579–591.

International Human Genome Sequencing Consortium (IHGSC). 2001. Initial sequencing and analysis of the human genome. *Nature* 409: 860–921.

Jacobs, K. B., Yeager, M., Zhou, W., Wacholder, S., Wang, Z., et al. 2012. Detectable clonal mosaicism and its relationship to aging and cancer. *Nat. Genet.* 44: 651–658.

Kwak, H. and Lis, J. T. 2013. Control of transcriptional elongation. *Annu. Rev. Genetics* 47: 483–508.

Lappalainen, T., Sammeth, M., Friedländer, M. R., t' Hoen, P. A. C., Monlong, J., et al. 2013. Transcriptome and genome sequencing uncovers functional variation in human populations. *Nature* 501: 506–511.

Laurie, C. C., Laurie, C. A., Rice, K., Doheny, K. F., Zelnick, L. R., et al. 2012. Detectable clonal mosaicism from birth to old age and its relationship to cancer. *Nat. Genet.* 44: 642–650.

Lee, T. I. and Young, R. A. 2000. Transcription of eukaryotic protein-coding genes. *Annu. Rev. Genetics* 34: 77–137

Lu, M., Zhang, Q., Deng, M., Miao, J., Guo, Y., et al. 2008. An analysis of human microRNA and disease associations. *PLoS ONE* 3: e3420.

Mercer, T. R., Dinger, M. E., and Mattick, J. S. 2009. Long non-coding RNAs: Insights into functions. *Nat. Rev. Genet.* 10: 155–159.

Pasquinelli, A. E. 2012. MicroRNAs and their targets: Recognition, regulation, and an emerging reciprocal relationship. *Nat. Rev. Genet.* 13: 271–282.

Redon, R., Ishikawa, S., Fitch, K. R., Feuk, L., Perry, G. H., et al. 2006. Global variation in copy number in the human genome. *Nature* 444: 444–454.

Stenson, P. D., Ball, E. V., Mort, M., Phillips, A. D., Shaw, K., and Cooper, D. N. 2012. The Human Gene Mutation Database (HGMD) and its exploitation in the fields of personalized genomics and molecular evolution. *Curr. Protoc. Bioinformatics* Unit 1.13.

The 1000 Genomes Project Consortium. 2010. A map of human genome variation from population-scale sequencing. *Nature* 467: 1061–1073.

The 1000 Genomes Project Consortium. 2012. An integrated map of genetic variation from 1,092 human genomes. *Nature* 491: 56–65.

Trapnell, C., Roberts, A., Goff, L., Pertea, G., Kim, D., et al. 2012. Differential gene and transcript expression analysis of RNA-seq experiments with TopHat and Cufflinks. *Nat. Protoc.* 7: 562–578.

Treangen, T. J. and Salzberg, S. L. 2012. Repetitive DNA and next-generation sequencing: Computational challenges and solutions. *Nat. Rev. Genet.* 13: 3–46.

Venter, J. C., Adams, M. D., Myers, E. W., Li, P. W., Mural, R. J., et al. 2001. The sequence of the human genome. *Science* 291: 1304–1351.

Wang, Z. and Burge, C. B. 2008. Splicing regulation: From a parts list of regulatory elements to an integrated splicing code. *RNA* 14: 802–813.

Wek, R. C. and Cavener, D. R. 2007. Translational control and the unfolded protein response. *Antiox. Redox Signal.* 9: 2357–2371.

Wellcome Trust Case Control Consortium (WTCCC), Craddock. N., Hurles, M. E., Cardin, N., Pearson, R. D., Plagnol, V., et al. 2010. Genome-wide association study of CNVs in 16,000 cases of eight common diseases and 3,000 shared controls. *Nature* 464: 713–720.

Winter, J., Jung, S., Keller, S., Gregory, R. I., and Diederichs, S. 2009. Many roads to maturity: microRNA biogenesis pathways and their regulation. *Nat. Cell Biol.* 11: 228–234.

Yamakawa, K., Huot, Y. K., Haendelt, M. A., Hubert, R., Chen, X. N., et al. 1998. DSCAM: A novel member of the immunoglobulin superfamily maps in a Down syndrome region and is involved in the development of the nervous system. *Hum. Mol. Genet.* 7: 227–237.

3

Human Evolution

This chapter provides a brief overview of human evolution. Its intent is simple: to present the perspective that it is much easier, and intellectually richer, to understand human variability in light of the history of our species. We will start with three sections describing that history from our primate origins through human dispersal across the globe and the imprint it has left on human population structure. The next two sections will consider the major evolutionary forces that shape human variation genome-wide, starting with recombination and how it generates linkage disequilibrium, then moving to mutation, migration, and population expansion. Only then will we consider natural selection acting on individual polymorphisms, starting with purifying selection, but then seeing how adaptation has affected the human genome. The concluding section asks, in general terms, why it is that our genes eventually make us sick.

Homo sapiens

Humans are primates. By any criterion—behavioral, morphological, or genetic—we are closer to the great apes than we are to any other group of species on the planet. We are the only surviving member of the genus *Homo*, and we have given ourselves the species name *sapiens*, suggesting wisdom that sets us apart. In terms of **nucleotide divergence**, the fraction of sites that differ between typical members of two species, there is actually little to justify our having a genus to ourselves. Two humble fruit flies of the species *Drosophila melanogaster* are more different from one another than a human and a chimpanzee. Yet our behavioral uniqueness does seem to justify the hubris of the assignation *Homo sapiens*.

Our closest relatives are the chimpanzees, *Pan troglodytes* and *Pan paniscus* (bonobo). Along with humans, the gorilla, and orangutan, they make up the family Hominidae. That family, in turn, falls within the parvorder Catarrhini, which also includes gibbons and Old World monkeys. The other members of the order Primates, which arose between 65 and 85 million years ago, are the New World monkeys, tarsiers, and lemurs.

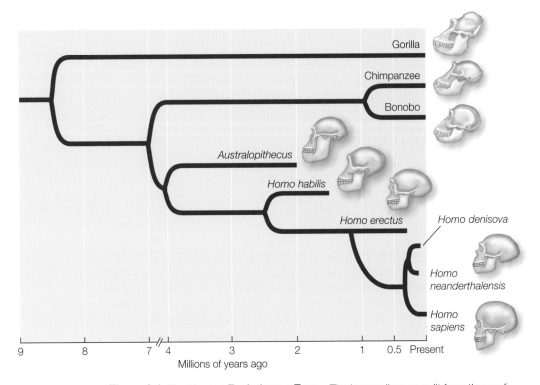

9   8   7   4   3   2   1   0.5   Present

Millions of years ago

Figure 3.1 The Human Evolutionary Tree. The human lineage split from those of gorillas and chimpanzees between 5 and 10 million years ago. No earlier than 1 million years ago, modern humans split from *Homo erectus*, the first hominoid species to explore Eurasia, and perhaps half a million years ago, from a second lineage that subsequently became Neanderthals and Denisovans. Some interbreeding with those subspecies occurred after *Homo sapiens* migrated out of Africa approximately 50,000 years ago.

The most recent common ancestor of humans and chimpanzees is most commonly dated at less than 7 million years ago (**Figure 3.1**), which is a blip in evolutionary time—only 1% of the time vertebrates have been on Earth and 2.5% of the time of mammals. Yet whether it is a short or a long time depends on your perspective. Australia, which inches northward at the same rate fingernails grow, has moved over 100 km since the genus *Homo* appeared, and even one of the oldest Hawaiian islands (Kauai) has risen above the Pacific and been substantially eroded in that time. There are approximately 35 million sequence differences between the human and chimpanzee (and another 3% of the two genomes do not align due to insertions and deletions). This number is equivalent to ten times the number of single nucleotide polymorphisms (SNPs) found in any one person and implies the fixation of five new mutations in either lineage each year.

It is interesting to ask whether it would be possible to select a new species that looks and acts just like modern humans from the current chimpanzee gene pool without allowing more than a handful of new mutations. It is a false proposition, however, considering that humans never derived from chimpanzees (we share a common ancestor), but the answer is not trivial, and the question gets even more uncomfortable if it is reversed. Some 80 of our common ancestor's genes have been lost in humans. Almost half of these genes are for olfactory receptors (though we have plenty of others!), but they also include a hair keratin and a muscle myosin used in masticatory muscles (Mikkelsen et al. 2005; Scally et al. 2012). The functions of the genes that have been added can only be speculated on. One of the most celebrated gene sequence changes between chimpanzees and humans is two amino acids in the human *FOXP2* gene that might have contributed to the evolution of the human capacity for vocalization (Enard et al. 2002). Certainly many of the differences between primate species show substantial heritability, but it would take a brave person to propose an artificial selection scheme that would lead to human intelligence, which is qualitatively distinct in many ways.

Human Dispersal: Out of Africa

There are two basal hominid genera in the archaeological record from Africa of the lineage leading toward humans: *Ardipithecus* and *Australopithecus*. Both of these hominids had small brains like those of chimpanzees, but fossils show signs of bipedal gait and adaptation to arboreal and savanna lifestyles, respectively. The first two *Homo* species, *H. habilis* and *H. erectus*, lived side by side for a million years starting 2.5 million years ago. They had brains with up to 75% of the volume of modern human brains, a clearly upright posture, and stone tool cultures. One or both of these species gave rise to *H. ergaster* in Africa and a tall and relatively naked species, known as Heidelberg Man, that migrated out of Africa into eastern and western Europe over 1.5 million years ago and into Asia, where it is known as Peking Man or Java Man, at least 700,000 years ago. Despite some regional continuity between the million-year-old fossils and tools of these hominoids and those of modern European and Asian humans, respectively, most geneticists do not consider them to be direct ancestors of *H. sapiens*.

Rather, the overwhelming evidence is that *Homo sapiens* emerged in southern Africa and dispersed across the globe in relatively recent migrations (Cann et al. 1987; Templeton 2002). Population genetic analyses imply that all contemporary humans share mitochondrial DNA that derived from a single woman, dubbed "mitochondrial Eve," who lived perhaps 120,000 years ago in a population of 10,000 people. Autosomal and Y chromosome markers are broadly consistent with this inference, which suggests that our species is only around 10,000 generations old (Jorde et al. 2000). Therefore, someone who can trace their family pedigree back to the Pilgrims actually has a record of a measurable fraction of a percent of human history.

One complication in this picture of human history is the presence of two other proto-human lineages that are distinctly more modern than *Homo erectus* but genetically well outside the range of *Homo sapiens* diversity. Bones from caves in Europe and central Asia tell us that early humans cohabited with *Homo neanderthalensis* for several thousand years until the latter mysteriously went extinct 33,000 years ago. Farther east, in the Altai Mountains of Siberia, a single specimen of a new subspecies yielded sufficient DNA to support the generation of a complete high-quality genome sequence (Meyer et al. 2012). This Denisovan individual was clearly neither Neanderthal nor modern human, but there is strong evidence that her kind interbred with the new *H. sapiens*. The genome of a second specimen from the same cave turned out to resemble that of European Neanderthals (Prüfer et al. 2014), so it is likely that these lineages also interbred. Sequence comparisons indicate that 2% of the DNA of the average Caucasian is Neanderthal and that, collectively, over one-fourth of the Neanderthal genome is present in the modern human population (Sankararaman et al. 2014; Vernot and Akey 2014). It seems that admixture between out-of-Africa *H. sapiens* and these forebears who explored Asia and Europe hundreds of thousands of years before us occurred in multiple waves over an extended period.

The major lines of evidence placing the origin of *H. sapiens* in Africa are patterns of DNA sequence diversity, which is greatest in sub-Saharan Africa. As people populate new regions in waves of migration, progressively smaller amounts of the genetic variation in the gene pool accompany them, and since each new population is younger than its source population, it has had less time to accumulate new mutations. The population with the greatest genetic variation is therefore assumed to be the oldest. The archaeological record, and its proximity to the presumed land route for exodus via northern Arabia, led many to believe that the cradle of humanity was East Africa, but southwestern Africa, in the vicinity of modern Namibia or Angola, now seems more likely. The genome sequences of several Khoi-San bushmen, which showed two people from adjacent villages to be as different from one another as any two non-Africans, and comprehensive genotyping of multiple hunter-gatherer populations confirms southern Africa as the most likely place of human origin (Schuster et al. 2010; Lachance et al. 2012). The results of computer simulations used to model this stepwise migration process seem to fit the data best when they allow for back-migration and bottlenecks in which the population is reduced to a few thousand for some time then recovers and expands (Fu and Akey 2013).

Human dispersal across the globe appears to have occurred in multiple waves, as summarized in **Figure 3.2** (Rasmussen et al. 2011; Stoneking and Krause 2011; Henn et al. 2012). The genetic history of Africa itself is unresolved, since its various population groups are under-sampled, and in any case, agrarian societies from elsewhere are thought to have resettled the continent, mostly in the past 5000 years. After at least a 50,000-year incubation period in Africa, the young species moved across southern Asia and Indonesia, interbreeding to some extent with extant Denisovans, and arrived in Australia no earlier than 40,000 years ago. Also 50,000 years ago, *H. sapiens* moved into the steppes of central Asia and across to Siberia and East Asia,

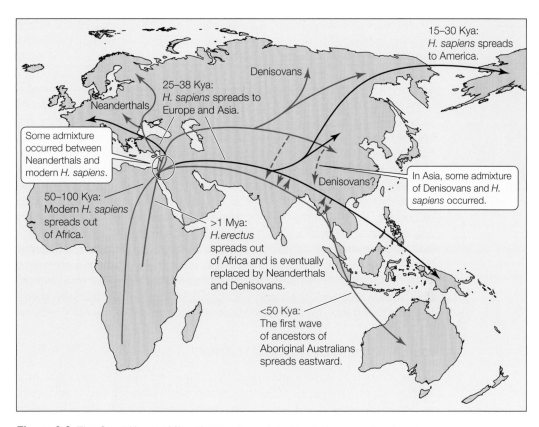

Figure 3.2 The Great Human Migrations.   Current thinking is that more than 1 million years ago (Mya) *Homo erectus* migrated out of Africa into much of Asia and Europe, eventually being replaced by Neanderthals and Denisovans (blue arrows). Between 50,000 and 100,000 years ago (50–100 Kya) *Homo sapiens* emerged as a new species in southern Africa and spread out of Africa (red lines). Some interbreeding occurred between *H. sapiens* and Neanderthals, but the populations were mostly reproductively isolated. More than 50 Kya, a major migration of *H. sapiens* established the ancestors of Australian Aborigines (dark green arrow). A second migration of *H. sapiens* 25–38 Kya colonized Europe and Asia (black arrows), and after about 30 Kya, populations of *H. sapiens* in northeastern Asia migrated across the Bering land bridge into northwestern America. Dashed blue lines indicate an admixture of Denisovanlike populations and *H. sapiens*. (After Futuyma 2013, based on Stoneking and Krause 2011 and Rasmussen et al. 2011.)

while other groups moved north and eventually west into Europe some 40,000 years ago, where they encountered the Neanderthals. Later migrations starting perhaps 30,000 years ago brought people across the Bering Strait into North America. Detailed analyses suggest that AmerIndians dispersed across the Americas via coastal routes and that two later out-of-Asia migrations account for half the genetic variance in Eskimo-Aleuts and 10% of the genomes of Canadians who speak NaDene languages. More recent migrations have seen the peopling of North Africa by Berbers in the past

10,000 years and of the Pacific islands in the past 3000 years, with the Māori arriving in New Zealand just 700 years ago, which is not too long before the European influx into North America.

A fascinating question that can be addressed with the tools of molecular anthropology is to what extent did gene flow, as opposed to purely cultural transmission of ideas, spread agriculture across Europe. Luca Cavalli-Sforza pioneered this approach in the 1990s, publishing maps documenting the spread of genes on each continent in his book *The Great Human Diasporas*. In Europe, the major component of variation spreads from the Urals in an east-to-west gradient, while a second component follows more of a south-to-north gradient. This pattern was consistent with the spread of genes accompanying the spread of agriculture, and it has been affirmed by the sequencing of ancient DNA from the 5000-year-old remains of three hunter-gatherers and one farmer from Sweden (Haak et al. 2005; Skoglund et al. 2012). The farmer's genome is much closer to those of southern Europeans, whereas those of the hunter-gatherers more closely resemble northern Europeans. Thus, farming and hunting-gathering cultures are thought to have existed side by side for over a thousand years, with little gene flow between them. It seems that it took more than ideas for agriculture to take hold.

Human Population Structure

While it is true that the vast majority of human variation is shared among all population groups, all of this migration, admixture between previously isolated populations, and rapid population growth, has left very strong patterns of **population structure** in the human genome. These patterns allow population geneticists to make confident statements about any individual's ancestry based on their genes. Anyone with even a hint of mixed ancestry can quite readily paint their chromosomes different colors according to which continent each chromosomal region derives from, and it is now even possible to tell a Dutchman with some confidence that he is from the north or south of the country.

There are two types of approaches that can be used to quantify population structure. One, implemented in the popular *Structure* software, assigns a proportion of the genome of each individual in a probabilistic manner to one or more of K populations (Pritchard et al. 2000; pritchardlab.stanford.edu/structure.html). As the user steps through models with increasing K, a graphical picture of the global structure in the data set emerges. At $K = 6$ contributing populations, each of six major continental populations are clearly demarcated (Africa, Europe, the Middle East, Central and East Asia, Oceania, and America; **Figure 3.3A**). The insets show more focused analyses of four of the populations (Rosenberg et al. 2002). A more recent study from the International Human Genome Diversity Project examined 938 people from 51 populations (Li et al. 2008). Individuals in populations that have experienced **admixture**—the interbreeding that occurs when a group migrates into an already occupied territory—are seen to share some proportion of their genomes with two or more ancestral groups.

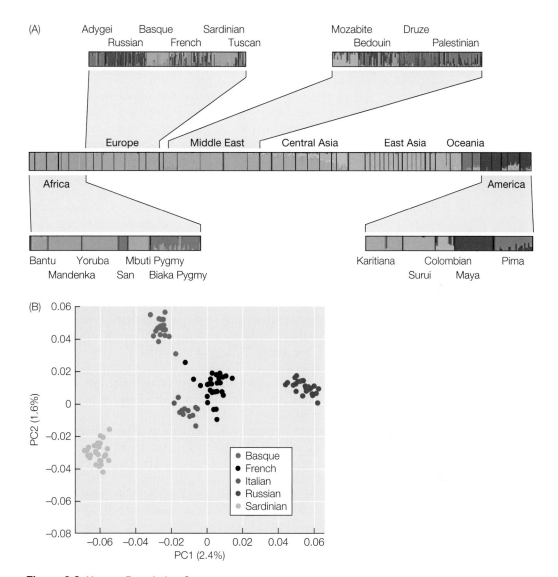

Figure 3.3 Human Population Structure.   (A) The proportional contribution of six
contributing populations to the genomes of 1056 individuals is visualized with a different
color for each source (center bar). Africans, Europeans, and East Asians are for the most
part cleanly discriminated, but Middle Easterners tend to have a mixture of DNA from
three populations. (B) Principal component (PC) analysis separates five populations within
Europe, coincidentally to some extent clustering them in a pattern that resembles their
geographic locations. (A, after Rosenberg et al. 2002; B, after Li et al. 2008.)

The second approach allows an investigator to quantify the population structure in a sample without forcing each individual to belong to
one or another group. It uses statistical principal component analysis
(PCA, or similarly, multi-dimensional scaling) to reduce the full matrix of

hundreds of thousands of SNPs to a series of "eigenvectors" that capture the major components of covariance. That is, PCA is used to quantify which polymorphisms tend to be found in the same individuals. For example, if 5000 of the SNPs share the property that they tend to have the minor allele in 30% of the sampled individuals and the major allele in the other 70%, then they may emerge from the unsupervised statistical analysis as one of the principal components of variation. Principal components (PCs) are identified until all of the variance is captured, with every SNP contributing to some degree (usually a very small one) to each component. This approach is illustrated in **Figure 3.3B**, which shows the first two PCs in Europeans from the International Human Genome Diversity Project. The percentages associated with each PC indicate how much of the variance in genotype frequencies genome-wide is captured by that component, which is generally only a few percent for the major PC and small fractions of a percent for the minor ones. This result confirms that most human variation is shared among all groups, and that human population structure involves subtle shifts in the frequencies of some alleles.

For the most part, the two analytical strategies agree. At $K = 2$, the two groups tend to have positive or negative values of PC1; at $K = 3$, the third group will often separate along PC2, and so forth, but with less commonality as more components are considered. Minor components often correspond either to close relatives who share one-half or one-fourth of their genotypes (and can dominate small data sets if care is not taken to identify them) or to small blocks of one chromosome that have for some reason attained a different frequency in a subgroup of people. A popular program for performing PCA is Eigenstrat (Price et al. 2006), available in the EIGENSOFT package at genetics.med.harvard.edu/reich/Reich_Lab/ Software.html.

A whole discipline of molecular anthropology has developed around these approaches. It is associated with a vast and fascinating literature that explores the demography of our species. Maps of genetic variability in India, the Americas, Australasia, Africa, and across Europe paint a molecular portrait of ancestry and migration and facilitate hypothesis testing regarding the contributions of genes and culture to recent human evolution. They also provide the crucial context against which genetic associations with disease must be assessed. Ignoring population structure can result in false conclusions that variants contribute to diseases that have different prevalences in different populations. Including it in analyses can help us tease apart the contributions of genetic divergence and admixture to those differences in prevalence.

The eigenvectors detected by PCA have a central role in genetic association studies because they can be used to correct for genome-wide population structure, as explained in Chapter 6 (Price et al. 2006). For some purposes, it is also useful to conduct population structure analyses one SNP at a time, or across a small region of a locus. Classical F_{ST} statistics are useful for this purpose: the idea, first enunciated by the great American geneticist Sewall Wright, is that population differentiation can be inferred by measuring the difference between the observed heterozygosity and that expected under Hardy-Weinberg equilibrium (Wright 1943; Holsinger and Weir 2009). Heterozygosity tends to be reduced as populations drift apart because random assortment of alleles is reduced and a form of inbreeding occurs, which tends to increase the number of homozygotes. By contrast, under admixture, the number of heterozygotes can actually increase, at least temporarily. Recall that under Hardy-Weinberg equilibrium, the expected frequencies of each genotype are given by p^2, $2pq$, and q^2, where p and q are the allele frequencies. Then F is defined as 1 minus the ratio of the observed to the expected heterozygote frequency ($2pq$) in the combined population. If there is no deviation, $F = 0$, and the allele frequencies are as expected in a single randomly mating population, so there is no population structure. Typically, the deviation is small in the direction of a reduction in heterozygosity, so F has a value between 0 and 0.1, implying slight divergence.

Values of F that are greater than 0.1 are regarded as a signature of strong genetic divergence. Only a small fraction of the genome shows such differentiation between any two populations, so scans along the genome can be used to identify locations where the divergence is greater than expected by chance. This is one approach to detection of selection acting on individual genes. Given the out-of-Africa migration model, we also expect F_{ST} to increase with geographic distance between populations, which it does, so genes that contradict this trend and show high deviation even between populations within a continent are particularly good candidates for local adaptations. Only a few dozen such genes have been detected, however, since selection is weak and its signatures are subtle (Pritchard et al. 2010).

Recombination and Linkage Disequilibrium

One of the most important parameters governing the distribution of diversity in the genome of any species is the recombination rate. Recombination leads to independent assortment of nucleotide variation. Two polymorphisms on different chromosomes have a recombination rate of 0.5, since any two alleles have a 50-50 chance of being transmitted together to a sperm or an egg. But variants located near one another on the same chromosome have a high probability of being transmitted together, since there are only one or two recombination events per chromosome. When variant alleles are transmitted together, they are more likely to be found together in the population at large, forming what are called **haplotype blocks**. A **haplotype** is a combination of two or more alleles in the same region of a chromosome. The haplotype structure of a population is a function of the recombination rate, population size, and number of generations of transmission (**Figure 3.4**). In humans, these factors have conspired to construct a genome that can be regarded as a mosaic of some

Polymorphic site

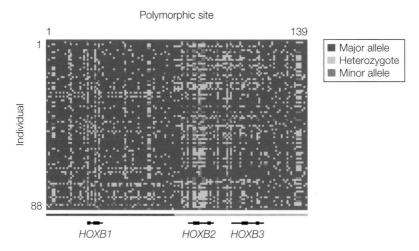

Figure 3.4 Haplotype Structure of a Typical Human Gene Region. The genotype at each of 139 SNPs from a 30 kb region of chromosome 17 encompassing the *HOXB1*, *HOXB2*, and *HOXB3* loci is shown for 88 Yoruban individuals sampled by the HapMap Project. Only variants with at least five copies of the minor allele are shown. Note the fairly sharp discontinuity in the middle of the image, demarcating a break between two haplotype blocks, shown as dark and light green bars, that is also seen in the linkage disequilibrium plot in Figure 3.5.

50,000 haplotype blocks, each typically between 50 and 150 kb in length (Reich et al. 2001; Ardlie et al. 2002).

The standard measure of haplotype structure is **linkage disequilibrium (LD)**; that is, a measure of the tendency of two alleles to be transmitted together more often than expected under independent assortment. Consider a new mutation—say, a T to C transition that arises on a chromosome that also contains an A instead of a G at a polymorphic site 50 kb away (**Figure 3.5A**). If this mutation is transmitted to two offspring in the next generation, the chances are that both of the C alleles will be found on chromosomes with an A at the second site, whereas none of the Cs will be found with a G. However, if the A/G polymorphism were located 20 Mb away, the chances are that recombination would produce some gametes with both a T and a G, as shown in the second pair of chromosomes. The probability that this will happen is related to the physical distance between the polymorphisms. As the new T increases in frequency, it will remain associated with the A for as long as it takes for recombination to allow random assortment. Such haplotype blocks form because polymorphisms increase in frequency more rapidly than recombination over short distances can allow them to assort independently.

Fast-forward a few thousand generations to a time when the T and C alleles are now each at a frequency of 0.5, and suppose that the A and G alleles at the other site are also at a frequency of 0.5. We can easily generate the expected frequencies of the two-locus genotypes by multiplying

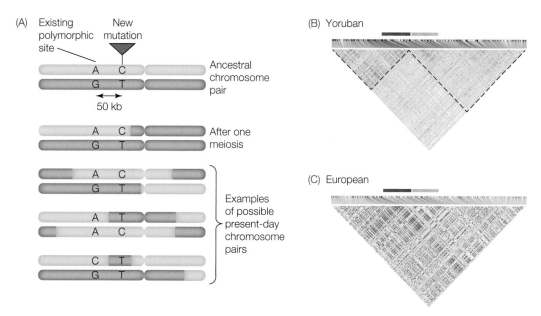

(A) LD arises due to the history of new mutations

Figure 3.5 Linkage Disequilibrium (LD). (A) LD arises due to the history of new mutations, which arise on a particular genetic background and remain with any linked polymorphisms until such time as recombination breaks up the correlations. (B, C) Haploview can be used to visualize LD as a heat map triangle of pairwise correlations (in this figure, D') between each polymorphism. In the 100 kb region of *HOXB2*, there is strong LD observed in Europeans (C), but two haplotype blocks are evident in Yorubans from Africa, although long-range LD is evident across the entire region. Also note the higher level of polymorphism (number of variants) in the Africans. The region shown in Figure 3.4 is indicated by the dark and light green bars.

together the p^2, $2pq$, and q^2 expectations and compare the results with what is observed:

<table>
<tr><td colspan="4" align="center">Expected</td><td colspan="4" align="center">Observed</td></tr>
<tr><td></td><td>TT
(0.25)</td><td>TC
(0.50)</td><td>CC
(0.25)</td><td></td><td>TT
(0.25)</td><td>TC
(0.50)</td><td>CC
(0.25)</td></tr>
<tr><td>AA (0.25)</td><td>0.0625</td><td>0.125</td><td>0.0625</td><td>AA (0.25)</td><td>0.25</td><td>0</td><td>0</td></tr>
<tr><td>AG (0.50)</td><td>0.125</td><td>0.25</td><td>0.125</td><td>AG (0.50)</td><td>0</td><td>0.50</td><td>0</td></tr>
<tr><td>GG (0.25)</td><td>0.0625</td><td>0.125</td><td>0.0625</td><td>GG (0.25)</td><td>0</td><td>0</td><td>0.25</td></tr>
</table>

The values at the left are those expected under random assortment, and are what is usually observed for two alleles at either end of a chromosome or on different chromosomes. The values at the right imply that all the A alleles are with T alleles, and that all the C alleles are with G alleles, a situation called **perfect** or **complete linkage disequilibrium**. It is more likely to be observed for two sites that are physically close together. Within a 1 Mb region of any chromosome, a situation somewhere between these two is common—namely, **partial LD**. This is illustrated by the bottom three pairs of chromosomes in Figure 3.5A. LD is usually quantified by the squared correlation between allele frequencies, r^2, or with a measure that compares the observed with the maximum possible LD between two variants, D'.

Linkage disequilibrium is easily visualized either by color-coding the genotypes of a sample across a locus (see Figure 3.4) or plotting the pairwise LD measures (or their probability) in a triangular **heat map** (Figure 3.5B,C). These figures show the application of these methods to the *HOXB2* gene in the Yoruban and European individuals from the 1000 Genomes sample, using Haploview software (Barrett et al. 2005). The actual genotypes that give rise to this haplotype structure in the Yoruban population are shown in Figure 3.4, the first striking feature of which is that the colors for any one individual (each row) are often similar to those for other individuals, because major-allele homozygotes (in red) at one SNP tend to be major allele homozygotes at all the surrounding sites (columns). This pattern is similar for the heterozygotes (in yellow) and minor-allele homozygotes (in blue). The second striking feature is a discontinuity in the middle of the figure, where there appears to be a transition from one haplotype block to another. This discontinuity is also captured in the LD plot in Figure 3.5B, where two large blocks of red (outlined with dashed lines) show much higher LD between variants within those two blocks. The same is not true in Europeans (see Figure 3.5C), who have a much less clearly defined haplotype structure and less polymorphism overall. The UCSC and Ensembl browsers also facilitate visualization of LD with similar triangle plots.

Across all human populations, LD tends to have a similar structure, with the proviso that the haplotype blocks are considerably shorter in Africa, where there is more diversity that has been present for a longer time. As we will see, this diversity makes disease association mapping more difficult in Africans and African Americans, but the overall conservation of haplotype blocks facilitates comparative studies across populations. In some cases, the differences between populations support fine structure localization of the causal variant(s). Interestingly, nonhuman primates have quite different LD structures, indicating that the recombination landscape changes with time between humans and primates. In fact, there is considerable variation among human families in the recombination rate, much of which is attributed to genetic variation at the *PRDM9* histone methyltransferase gene (Baudat et al. 2010).

Why are there such abrupt transitions between haplotype blocks? The answer is that recombination in humans does not occur at random within segments of chromosomes, but rather is concentrated in hotspots. These hotspots are locations where the recombination machinery is more likely to bind, and they are as small as a kilobase in length, occurring only once or twice every 100 kb. It is not clear what factors mark a hotspot, but they must contain features that affect chromatin structure. The restriction of recombination to hotspots has been confirmed directly by single-sperm genotyping, which shows that recombination really does tend to occur in the DNA intervals that mark discontinuities between haplotype blocks (Myers et al.

2005). Detailed maps of haplotype blocks in the human genome have been assembled by the International HapMap Consortium (2005, 2007).

Demographic Processes

There are four other evolutionary processes that have had, and continue to have, a major impact on the structure of variation in the human genome. Mutation, migration, and inbreeding are discussed in this section. Another important process that is tightly coupled to the prevalence of very rare deleterious alleles is recent population expansion, but this is not discussed further.

Mutation, as noted in Chapter 1, has a barely measurable effect on the health of most newborns, but it is nevertheless one of the key factors shaping the distribution of disease susceptibility population-wide. Population geneticists can readily estimate the product of the effective population size and the mutation rate from sequence data (giving rise to the parameter $4N\mu$, often denoted θ, where N is the number of individuals in the population and μ is the neutral mutation rate), but such estimates are biased by the filters of natural selection. More direct measurements can be obtained by comparing the genomes of offspring and their two biological parents, which confirms an overall rate of 10^{-8} mutations per base per generation (Roach et al. 2010; Conrad et al. 2011). Every new neutral mutation has a $1/2N$ chance of eventually becoming the new canonical human allele at the site, so the vast majority are lost within a few generations. Yet, collectively, they have an impact on phenotypic variability in the short time they occupy the gene pool.

Most mutation in the germ line—sperm and eggs—is attributable to errors of DNA replication during meiosis, simply because the process is imperfect. In somatic tissues, a wide variety of chemicals in tobacco, pollution, and household plastics, as well as sources of radiation from sunlight to nuclear accidents, increase the overall mutation rate. Each of these mutagens introduces a unique spectrum of changes that leave a specific footprint in sequence diversity. Some also lead to cancer, and mature tumors accumulate hundreds of mutations, in part because the machinery that normally corrects sequence errors is itself incapacitated by mutations. Somatic mutation does not, however, contribute to genome evolution.

Migration is such an important factor because its effects are genome-wide and can lead to much more rapid and larger changes in allele frequencies than natural selection. When a small group of individuals move to a new location, they necessarily take with them only a sample of the variation in the source population, and inevitably this **founder effect** will alter thousands of allele frequencies (**Figure 3.6**). If the group happen to reproduce with members of another population that had previously moved to the new territory, that admixture will affect allele frequencies at all sites that differ between the two interbreeding populations. In genetic melting pots like North America, where an increasingly large proportion of children are from mixed couples, there is also much more variance in allele frequencies than in the parent populations. Another effect of migration is that it can bring together combinations of genotypes that would never occur in

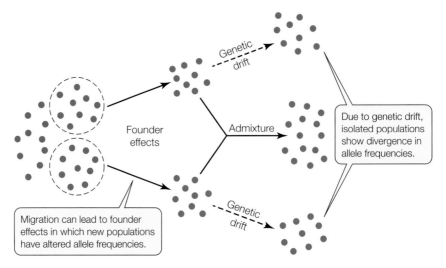

Figure 3.6 Migration and Genetic Drift.   If a small group of founders leave a source population and establish a new population, they are likely to take with them a slightly biased representation of the gene pool of the source population (represented by the proportions of the two colors). Just a small number of individuals moving between two such new populations will tend to homogenize them. If they remain isolated, however, the result will be two distinct lineages that may also drift apart genetically over time. Alternatively, admixture is what occurs when two populations come back together, which restores diversity.

more than a handful of individuals if they required sequential mutations in just one population. Hybrid individuals are less likely to be homozygous at rare variants, which may improve a population's fitness in general and reduce the incidence of rare diseases.

Perhaps less intuitively, migration is also a potent force preventing divergence between populations. In fact, it takes only a handful of individuals per generation to move back and forth between two groups of any size to prevent them from diverging. This would not be a factor if the migrants never had secondary contact with the source population, which presumably was the case for settlement of Melanesia and Polynesia, but may have been common during the initial global continental migrations. Nomads can cover hundreds of kilometers during their annual migrations, much more than the tens of kilometers per generation that would be required to prevent divergence as people spread worldwide. Individual migrants may also have an increased likelihood of contributing to future generations, as shown by a survey of church and birth registries for millions of modern residents of Quebec, which showed that the first settlers of remote valleys feeding into the St. Lawrence River contributed disproportionately to the contemporary gene pool (Moreau et al. 2011).

Inbreeding also has multiple consequences for human variation. Close inbreeding, or consanguinity, refers to marriages between relatives—most often cousins, but cross-generational or once- and twice-removed relatives are also included. Most Christian societies have taboos against incest that also prohibit consanguinity, but this is not true in Muslim countries or expatriate communities, and first-cousin marriage is quite common in Pakistan, for instance. Globally, approximately 10% of humans are related to their partner at least as second cousins (Bittles and Black 2010). Offspring of consanguineous marriages are at significantly heightened risk of autosomal recessive diseases, since it is much more likely that rare deleterious variants will be carried by both parents after inheritance from a shared grandparent. Additionally, the loss of heterozygosity at the highly polymorphic human leukocyte antigen (HLA) complex may affect competence to fight off infectious disease.

More generally, inbreeding refers to the inevitable increase in homozygosity that occurs in the absence of migration between isolated populations. It is easy to imagine how mating among close relatives leads to random fixation and loss of alleles within the group. If this occurs in multiple lineages, there may not be a change in allele frequencies overall, but there is divergence between those lineages, a process known as **genetic drift**. The same phenomenon occurs in large subpopulations on a longer time scale. It is not known to what extent nonrandom mating accelerates this tendency for genetic drift, but people tend to marry similar people with respect to such traits as height, temperament, and facial features. This phenomenon of "homophily" may also lead to a type of inbreeding, but it would tend to homogenize just those parts of the genome that contribute to those traits, whereas consanguinity and genetic drift affect the entire genome.

Purifying Selection

Purifying selection is the process that prevents deleterious mutations from becoming common polymorphisms. A new allele that causes sterility or childhood mortality in a dominant fashion will never be transmitted to the next generation. Variants that lead to severe deficiencies of the immune system, cognitive function, facial or musculoskeletal development, or any essential organ function, for that matter, have a low likelihood of being transmitted. If, however, the effect is recessive, the allele can drift to frequencies in the range of 1% before selection starts acting against homozygous individuals. Even variants that have some positive attributes in one context are likely to have deleterious consequences in others due to the pleiotropic nature of gene function. The net result is that the vast majority of mutations that are not functionally neutral are selected against. McVicker et al. (2009) estimate that purifying selection has reduced autosomal genetic diversity in hominids by between 19% and 26%.

Purifying selection is one of the reasons why the **derived allele frequency (DAF) distribution** is shifted toward a preponderance of low-frequency minor alleles. Derived alleles are variants in a contemporary population that replace or are polymorphic with the ancestral allele that was present in the preceding populations. Another reason for the low frequencies of derived alleles is that the establishment of the allele frequency distribution approximates a diffusion process. Mutations at very low frequency ($1/2N$) occasionally drift to common status by random sampling. By comparing the distributions of alleles of different types, we can infer the influence of purifying selection. Thus, polymorphisms in introns and long intergenic regions show a closer fit to expectations under neutral evolutionary theory than do those that cause amino acid substitutions, lie in the vicinity of splice sites, or are located near promoters. Nonsynonymous substitutions are much less common in the human genome than any other type of variant, as is seen for each of two HapMap populations in **Figure 3.7A**. Genome-wide, there is an average of approximately one SNP per

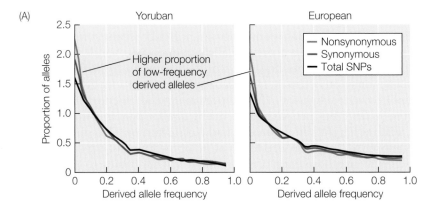

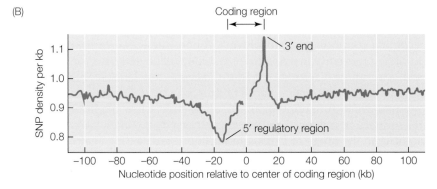

Figure 3.7 Distribution of SNPs in the Human Genome.   (A) In each of two major human population groups studied by the International HapMap Consortium, namely, Yorubans, and Europeans, the proportion of rare variants with a low derived allele frequency is greater for those that change an amino acid (nonsynonymous) than for those that are synonymous, which in turn is greater than that for all SNPs. The same is not true of variants that have reached a high frequency. (B) SNP density averages just under 1 per kilobase, but it is not evenly distributed with respect to the coding region, being reduced in upstream regulatory regions and increased in 3' regions. (After International HapMap Consortium 2007.)

kilobase per individual, though that rate varies with respect to position along the chromosome relative to the coding region (**Figure 3.7B**) and with the type of gene. Some genes are more tolerant of mutations than others (Petrovski et al. 2013), a fact that has implications for the likelihood that specific rare variants are functional or promote disease.

In fact, as more and more genomes are sequenced, it is becoming apparent that humans have a particularly high load of essentially private (found in only one person or population) nonsynonymous mutations (Nelson et al. 2012; Fu et al. 2013). This pattern is attributed to the recent explosion in the size of the human population. At the origin of the species, the effective population size seems to have been around 10,000 individuals, and even at the beginning of recorded history, 5000 BCE, the global population of humans likely numbered only in the tens of millions. In the past 2000 years, the human population has grown exponentially, reaching 7 billion in 2011. A consequence of this growth is relaxation of purifying selection and the accumulation of deleterious ultra-rare variants that are likely having a bigger impact on the incidence of morbidity in humans than they do in other species that have not experienced rapid population growth. The dire potential consequences of this mutation accumulation for human health have been noted (Lynch 2010) and debated (Lesecque et al. 2012).

There is thus much interest in developing tools for identifying deleterious variants. The strongest evidence of their presence is supplied by biochemical or cell biological assays or by testing the function of a suspected variant in a model organism such as the zebrafish or mouse (Bedell et al. 1997; Zaghloul and Katsanis 2011). It is obviously not practicable to do this for the thousands of protein-coding polymorphisms in any person's genome, however, so bioinformatic strategies must also be adopted. These strategies tend to be based on some combination of assessment of the evolutionary conservation of a suspect variant and prediction of its influence on protein structure or regulatory DNA function. There are half a dozen algorithms that are commonly used for this purpose, notably PolyPhen and SIFT (Ng and Henikoff 2003; Adzhubei et al. 2010), which show decent agreement but are by no means definitive. Variants that are predicted to be deleterious by multiple bioinformatic approaches make up a large proportion of those in the independently generated HGMD database of mutations confidently associated with disease (Cooper et al. 2006). Since only a fraction of deleterious variants are fully penetrant, and because evolutionary conservation is only a guide to function, there is a lot of noise in these assessments. Consequently, it is not possible to read a genome and assert which variants are disease-promoting, but in general, private or very rare variants that disrupt a conserved site in a critical region of a protein or regulatory element are not good news.

Adaptation

Adaptation is the opposite of purifying selection: it is the process whereby genome sequence evolution as a result of positive selection of advantageous mutations increases the fitness of the species. It has clearly been important not only in the evolution of unique human attributes, but also in local adaptation to high altitudes, frigid and dry climates, and the availability of different food sources, as well as challenges posed by different pathogens. Observed correlations between allele frequencies and geographic or environmental variables provide a compelling basis for the inference that specific variants contribute to local or global differentiation of human populations (Hancock et al. 2011; Novembre and di Rienzo 2009). Evolutionary geneticists have used sophisticated statistical methods to scan for genes that have been under selection at various phases of human history and have thereby identified thousands of sites that are likely to have experienced subtle positive selection for functions in growth, immune responses, metabolism, and cognition. They have also assembled dozens of very plausible adaptive scenarios that could be responsible for specific genetic contributions to pigmentation, dietary preferences, and some diseases.

A distinction is often made between hard and soft selection. **Hard selection** is the type that is usually presented in introductory biology classes. It is the process in which a new mutation appears that is patently good for the organism, and so has a greater than average likelihood of being transmitted in each generation. Over a period of a few hundred generations, it becomes common and eventually fixed as the new sequence that is thereafter observed in everyone. It is doubtful that this happens very often, at least not in a linear fashion. **Soft selection** is more complex, and there are two conceptions of this process. One recognizes genetic heterogeneity by noting that during the sojourn of any polymorphism from low to high frequency, other functional alleles will appear at the same gene, and all of these alleles will evolve in parallel. The other recognizes that adaptive circumstances can change with the environment and the genetic background, so that whether an allele is advantageous or deleterious can also change. Under novel circumstances, a polymorphism that was essentially neutral can become adaptive. This functionally hidden source of variability could provide as large a pool of variation as new mutations do. Soft selection leaves very different footprints on the genome than does hard selection, so it complicates the interpretation of gene sequence evolution (Hermisson and Pennings 2005; Pennings and Hermisson 2006).

Two approaches are commonly used to detect selection in humans. The simplest is to determine the ratio of nonsynonymous to synonymous nucleotide changes in a coding region. If a protein has experienced positive selection, it will have a ratio somewhat greater than 1, which is typical throughout the genome. More sophisticated "tests of neutrality," such as the HKA (Hudson-Kreitman-Aguadé) test and those developed by McDonald and Kreitman, Fu and Li, and others provide higher resolution as well as confidence and extend the analyses to comparisons of noncoding regions by including comparisons with other species or populations (Kreitman 2000). The second common approach is to search for extended runs of homozygosity. The extended haplotype homozygosity (EHH) class of tests compares the lengths of haplotype blocks surrounding the two alleles that selection is proposed to have affected (Sabeti et al. 2002). Due to the phenomenon of **genetic hitchhiking**, in which selection acting on one nucleotide also increases the frequencies of all other nucleotides that are in linkage disequilibrium with it, under hard selection there is expected to be a reduction in diversity in the vicinity of the selected allele. Consequently, homozygotes for a derived allele will share extended blocks in which most of the linked polymorphisms are on the same haplotype, whereas the ancestral allele will have accumulated variation around it. Two striking examples are illustrated in **Figure 3.8** (Voight et al. 2006). Another example is provided by the selection of new regulatory variants on multiple

(A) *SPAG4* in East Asians

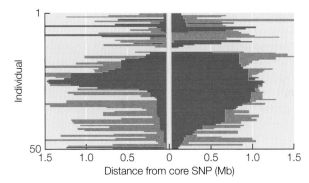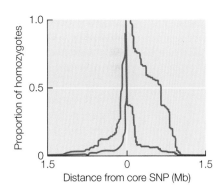

(B) *NCOA1* in Yorubans

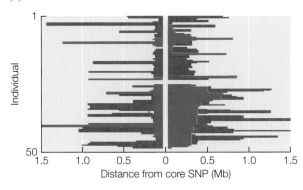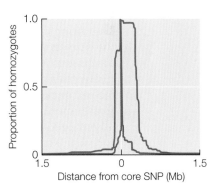

Figure 3.8 Extended Haplotype Homozgosity.   The plots at the left show the lengths of homozygous haplotypes observed in each of approximately 50 individuals at (A) the *SPAG4* locus for East Asians and (B) the *NCOA1* locus for Yorubans. The plots are centered around a core SNP that anchors the ancestral (blue) and derived (red) haplotypes. Less common haplotypes are shown collectively in gray. The plots at the right show the average distance at which each haplotype is homozygous in the indicated proportion of individuals. In both cases, it is clear that the proportion of sites that are homozygous either side of the core extends for a greater distance for the derived alleles, which is consistent with its having increased in frequency under positive selection. (After Voight et al. 2006.)

occasions in agrarian societies in East Africa and Europe over the past few thousand years that activate the lactase enzyme in adults, who are thus able to digest the milk of domesticated sheep and goats (Bersaglieri et al. 2004; Tishkoff et al. 2006).

Although dozens of such examples of strong positive selection have been observed in the human genome, most adaptation has involved more subtle combinations of processes. The major histocompatibility complex (MHC), also known as the human leukocyte antigen (HLA) complex, which is the most polymorphic region of the genome and encodes hundreds of genes essential for immune function, has a particularly complex history. This history includes balancing selection maintaining polymorphisms that protect against different viruses or bacteria. Polymorphism that is shared with other primates is also present, and introgression from Denisovan and Neanderthal genomes has occurred. There is also evidence of strong differentiation between populations on different continents, recurrent mutation, and gene conversion, among other processes. Elsewhere in the genome, soft selection is evident in parallel changes in allele frequencies in population groups that share similar environments.

Current initiatives are focused on detecting selection and understanding how it shapes variation not only in individual genes, but also in networks of genes that regulate specific processes. For example, the hundred or so SNPs that are unambiguously associated with height appear to be under selection as a group, as they vary in a coordinated manner across populations, following the north-to-south gradient in Europe. Such studies are a useful model for understanding the evolution of disease susceptibility.

Why Do We Get Sick?

This brings us to the key question: Why is there so much variation in the human genome that seems to be contributing to rare and common diseases alike? There is no one simple answer to this question, and different scientists will give very different answers according to their intellectual backgrounds. Many in the field of evolutionary medicine tend to assume that any variant that promotes disease must also have some advantageous influence that offsets the deleterious ones, and they search for such explanations (Williams and Nesse 1991). Others who believe that rare alleles of major effect are the primary source of disease would argue that mutation-selection balance is the key factor, and that some level of disease is inevitable, particularly in an expanding population with reduced purifying selection (Reich and Lander 2001). Still others see disease as a by-product of the existence of variation for the myriad physiological and developmental traits that disease-associated variants normally contribute to. So long as there is a normal distribution of susceptibility, there are bound to be individuals at the extremes, and natural selection can only limit the number of affected individuals, not prevent disease entirely (Gibson 2009).

Underlying all of these explanations is the basic expectation that natural selection should keep the frequencies of disease-associated variants low because disease is deleterious to fitness. Many diseases appear only in old age, after the affected individuals have already had children, so there should be weakened selection on variants that promote late-onset disease. Another possibility is that there are trade-offs between functions early and late in life—namely, that alleles that elevate fecundity or promote survival in infancy are the same ones that later contribute to cancer or diabetes or heart disease or cognitive decline.

Yet the emerging genome-wide association data also tell us that it is not the case that the derived variant, or the minor allele, is always the disease-associated one. There is actually a considerable amount of ongoing evolution of protection against disease, a notable example being the relatively low type 2 diabetes risk in East Asians, who have high frequencies of protective derived alleles at several of the genes most strongly associated with the disease (Chen et al. 2012). Such observations beg the question of whether there is differentiation among human populations in their disease susceptibility, and if so, what its basis is. It is very difficult to answer these questions, and there are strong public policy reasons to be careful about drawing conclusions. Two possible mechanisms for divergence in risk between populations are alteration of allele frequencies and alteration of the functional effect sizes of disease-associated alleles. The high level of replication observed in genetic association studies conducted in Europe and Asia, and to some extent in African Americans (though there are technical complications to this comparison), tells us that most variants that affect risk in one human population affect risk to a similar degree in other populations (Carlson et al. 2013; Marigorta and Navarro 2013). Much more comparative research is to be expected in the coming decades.

Just as compelling is the issue of why so many diseases seem to be rising in frequency, some in almost epidemic fashion. Whereas famine and infectious disease have dominated the landscape of human morbidity in developing countries, the World Health Organization estimates that by the year 2020, as many as 7 in every 10 deaths worldwide will be attributable to diabetes, cardiovascular disease, and other noncommunicable diseases that, along with cancer, depression, and cognitive decline, dominate morbidity in the developed world (**Figure 3.9**; Butler 2012). This shift may have a lot to do with people living longer and with better medical diagnoses. It is hard, though, to ignore the facts that autism and Alzheimer's disease were unknown in the nineteenth century, and that hypertension and metabolic syndrome have increased in prevalence more than an order of magnitude in the last two generations in countries as diverse as Fiji, China, and the United States. The most obvious explanation is that lifestyle and behavioral changes have increased the environmental component of the burden of disease. Whether this is a straightforward additive effect, where lack of exercise and high-fat diets resemble common alleles of large effect, or a more complex perturbation

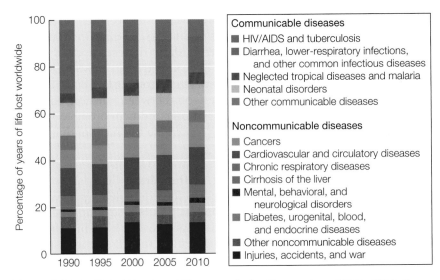

Figure 3.9 The Global Burden of Disease.   In the 20 years from 1990 to 2010, the global burden of disease shifted toward a more Western disease profile as developing countries saw a rise in the prevalence of noncommunicable diseases, particularly cardiovascular disease and cancer, and a drop in the incidence of infectious diseases, particularly enteric and respiratory bacteria and viruses. (After Butler 2012.)

of the equilibrium between genetics and environment that evolved to buffer people against disease, remains to be seen.

Summary

1. Humans are the only surviving member of the genus *Homo*, which diverged from the lineage that gave rise to the great apes no more than 9 million years ago.

2. The out-of-Africa model states that modern *Homo sapiens* evolved in southern Africa and spread globally starting 50,000 years ago, following migration routes across what is now northern Arabia.

3. Our species interbred with two other extant proto-human lineages represented by Neanderthal and Denisovan specimens in Europe, central Asia, and Siberia.

4. Although the majority of human variation is shared among all ethnicities, there is nevertheless considerable structure to the variation within and among populations, which can be detected with clustering and principal component methods.

5. *F* statistics are commonly used to detect population divergence at the level of individual genes, essentially measuring the decline of heterozygosity due to "inbreeding" in isolated populations.

6. The human genome can be regarded as a mosaic of some 50,000 haplotype blocks, which are 50–150 kb segments of DNA within which genotypes tend to be similar to one another.

7. Linkage disequilibrium is the nonrandom association of genotypes that occurs before recombination has time to generate new haplotypes. Recombination occurs in hotspots that help define the limits of haplotype blocks.

8. Mutation, migration, and inbreeding are three of the most important evolutionary processes that have shaped the structure of variation in the human genome.

9. Humans have an excess of very rare deleterious alleles as a result of our recent massive population expansion, and these alleles contribute to the burden of disease.

10. Positive natural selection facilitates adaptation and has affected allele frequencies throughout the genome, also contributing (but to an unknown degree) to the subtle differences that exist among populations.

11. Extended haplotype homozygosity due to genetic hitchhiking marks loci that have experienced strong recent directional selection, but soft selection produces different signatures.

12. The evolution of disease susceptibility differences among populations may involve changes in allele frequencies as well as changes in the effect sizes of alleles, but for the most part, risk allele identities seem to be conserved across populations.

References

Adzhubei, I. A., Schmidt, S., Peshkin, L., Ramensky, V. E., Gerasimova, A., et al. 2010. A method and server for predicting damaging missense mutations. *Nat. Methods* 7: 248–249.

Ardlie, K. G., Kruglyak, L., and Seielstad, M. 2002. Patterns of linkage disequilibrium in the human genome. *Nat. Rev. Genet*. 3: 299–309.

Barrett, J. C., Fry, B., Maller, J., and Daly, M. J. 2005. Haploview: Analysis and visualization of LD and haplotype maps. *Bioinformatics* 21: 263–265.

Baudat, F., Buard, J., Grey, C., Fledel-Alon, A., Ober, C., et al. 2010. *PRDM9* is a major determinant of meiotic recombination hotspots in humans and mice. *Science* 327: 836–840.

Bedell, M. A., Largaespada, D. A., Jenkins, N. A., and Copeland, N. G. 1997. Mouse models of human disease. Part II: recent progress and future directions. *Genes Dev.* 11: 11–43.

Bersaglieri, T., Sabeti, P. C., Patterson, N., Vanderploeg, T., Schaffner, S. F., et al. 2004. Genetic signatures of strong recent positive selection at the lactase gene. *Am. J. Hum. Genet.* 74: 1111–1120.

Bittles, A. H. and Black, M. L. 2010. Evolution in health and medicine Sackler colloquium: Consanguinity, human evolution, and complex diseases. *Proc. Natl. Acad. Sci. U.S.A.* 107 Suppl 1: 1779–1786.

Butler, D. 2012. Global survey reveals impact of disability. *Nature* 492: 322.

Cann, R. L., Stoneking, M., and Wilson, A. C. 1987. Mitochondrial DNA and human evolution. *Nature* 325: 31–36.

Carlson, C. S., Matise, T. C., North, K. E., Haiman, C. A., Fesinmeyer, M. D., et al. 2013. Generalization and dilution of association results from European GWAS in populations of non-European ancestry: The PAGE study. *PLoS Biology* 11: e1001661.

Cavalli-Sforza, L. L. and Cavalli-Sforza, F. 1995. *The Great Human Diasporas: The History of Diversity and Evolution*. Perseus Books, New York NY.

Chen, R., Corona, E., Sikora, M., Dudley, J. T., Morgan, A. A., et al. 2012. Type 2 diabetes risk alleles demonstrate extreme directional differentiation among human populations, compared to other diseases. *PLoS Genet.* 8: e1002621.

Conrad, D. F., Keebler, J. E. M., DePristo, M. A., Lindsay, S. J., Zhang, Y., et al. 2011. Variation in genome-wide mutation rates within and between human families. *Nat. Genet.* 43: 712–714.

Cooper, D. N., Stenson, P. D., and Chuzhanova, N. A. 2006. The Human Gene Mutation Database (HGMD) and its exploitation in the study of mutational mechanisms. *Curr. Protoc. Bioinf.* Ch 1: Unit 1.13.

Enard, W., Przeworski, M., Fisher, S. E., Lai, C. S., Wiebe, V., et al. 2002. Molecular evolution of *FOXP2*, a gene involved in speech and language. *Nature* 418: 869–872.

Fu, W. and Akey, J. M. 2013. Selection and adaptation in the human genome. *Annu. Rev. Genomics Hum. Genet.* 14: 467–489.

Fu, W., O'Connor, T. D., Jun, G., et al., NHLBI Exome Sequencing Project, Akey, J. M. 2013. Analysis of 6,515 exomes reveals the recent origin of most human protein-coding variants. *Nature* 493: 216–220.

Futuyma, D. 2013. *Evolution*. Sinauer Associates, Sunderland MA.

Gibson, G. 2009. Decanalization and the origin of disease. *Nat. Rev. Genet.* 10: 134–140.

Haak, W., Forster, P., Bramanti, B., Matsumura, S., Brandt, G., et al. 2005. Ancient DNA from the first European farmers in 7500-year-old Neolithic sites. *Science* 310: 1016–1018.

Hancock, A. M., Witonsky, D. B., Alkorta-Aranburu, G., Beall, C. M., Gebremedhin, A., et al. 2011. Adaptations to climate-mediated selective pressures in humans. *PLoS Genet.* 7: e1001375.

Henn, B. M., Cavalli-Sforza, L. L., and Feldman, M. W. 2012. The great human expansion. *Proc. Natl. Acad. Sci. U.S.A.* 109: 17758–17764.

Hermisson, J. and Pennings, P. S. 2005. Soft sweeps: Molecular population genetics of adaptation from standing genetic variation. *Genetics* 169: 2335–2352.

Holsinger, K. E. and Weir, B. S. 2009. Genetics in geographically structured populations: Defining, estimating and interpreting F(ST). *Nat. Rev. Genet.* 10: 639–650.

International HapMap Consortium. 2005. A haplotype map of the human genome. *Nature* 437: 1299–1320.

International HapMap Consortium. 2007. A second generation human haplotype map of over 3.1 million SNPs. *Nature* 449: 851–861.

Jorde, L. B., Watkins, W. S., Bamshad, M. J., Dixon, M. E., Ricker, C. E., et al. 2000. The distribution of human genetic diversity: A comparison of mitochondrial, autosomal, and Y-chromosome data. *Am. J. Hum. Genet.* 66: 979–988.

Kreitman, M. 2000. Methods to detect selection in populations with applications to the human. *Annu. Rev. Genomics Hum. Genet.* 1: 539–559.

Lachance, J., Vernot, B., Elbers, C. C., Ferwerda, B., Froment, A., et al. 2012. Evolutionary history and adaptation from high-coverage whole-genome sequences of diverse African hunter-gatherers. *Cell* 150: 457–469.

Lesecque, Y., Keightley, P. D., and Eyre-Walker, A. 2012. A resolution of the mutation load paradox in humans. *Genetics* 191:1321–1330.

Li, J. Z., Absher, D. M., Tang, H., Southwick, A. M., Casto, A. M., et al. 2008. Worldwide human relationships inferred from genome-wide patterns of variation. *Science* 319: 1100–1104.

Lynch, M. 2010. Rate, molecular spectrum, and consequences of human mutation. *Proc. Natl. Acad. Sci. U.S.A.* 107: 961–968.

Marigorta, U. M. and Navarro, A. 2013. High trans-ethnic replicability of GWAS results implies common causal variants. *PLoS Genet.* 9: e1003566.

McVicker, G., Gordon, D., Davis, C., and Green, P. 2009. Widespread genomic signatures of natural selection in hominid evolution. *PLoS Genet.* 5: e1000471.

Medawar, P. B. 1952. *An Unsolved Problem of Biology*. London: HK Lewis and Co.

Meyer, M., Kircher, M., Gansauge, M. T., Li, H., Racimo, F., et al. 2012. A high-coverage genome sequence from an archaic Denisovan individual. *Science* 338: 222–226.

Mikkelsen, T. S., Hillier, L. W., Eichler, E. E., Zody, M. C., Jaffe, D. B., et al. 2005. Initial sequence of the chimpanzee genome and comparison with the human genome. *Nature* 437: 69–87

Moreau, C., Bhérer, C., Vézina, H., Jomphe, M., Labuda, D., and Excoffier, L. 2011. Deep human genealogies reveal a selective advantage to be on an expanding wave front. *Science* 334: 1148–1150.

Myers, S., Bottolo, L., Freeman, C., McVean, G., and Donnelly, P. 2005. A fine-scale map of recombination rates and hotspots across the human genome. *Science* 310: 321–324.

Nelson, M. R., Wegmann, D., Ehm, M. G., Kessner, D., St Jean, P., et al. 2012. An abundance of rare functional variants in 202 drug target genes sequenced in 14,002 people. *Science* 337: 100–104.

Ng, P. C. and Henikoff, S. 2003. SIFT: Predicting amino acid changes that affect protein function. *Nucl. Acids Res.* 31: 3812–3814.

Novembre, J. and Di Rienzo, A. 2009. Spatial patterns of variation due to natural selection in humans. *Nat. Rev. Genet.* 10: 745–755.

Pennings, P. S. and Hermisson, J. 2006. Soft sweeps III: The signature of positive selection from recurrent mutation. *PLoS Genet.* 2: e186.

Petrovski, S., Wang, Q., Heinzen, E. L., Allen, A. S., and Goldstein, D. B. 2013. Genic intolerance to functional variation and the interpretation of personal genomes. *PLoS Genet.* 9: e1003709.

Price, A. L., Patterson, N. J., Plenge, R. M., Weinblatt, M. E., Shadick, N. A., Reich, D. 2006. Principal components analysis corrects for stratification in genome-wide association studies. *Nat. Genet.* 38: 904–909.

Pritchard, J. K., Pickrell, J. K., and Coop, G. 2010. The genetics of human adaptation: Hard sweeps, soft sweeps, and polygenic adaptation. *Curr. Biol.* 20: R208–R215.

Pritchard, J. K., Stephens, M., and Donnelly, P. 2000. Inference of population structure using multilocus genotype data. *Genetics* 155: 945–959.

Prüfer, K., Racimo F., Patterson N., Jay F., Sankararaman S., et al. 2014. The complete genome sequence of a Neanderthal from the Altai Mountains. *Nature* 505: 43–49.

Rasmussen, M., Guo, X., Wang, Y., Lohmueller, K. E., Rasmussen, S., et al. 2011. An Aboriginal Australian genome reveals separate human dispersals into Asia. *Science* 334: 94–98.

Reich, D. E., Cargill, M., Bolk, S., Ireland, J., Sabeti, P. C., et al. 2001. Linkage disequilibrium in the human genome. *Nature* 411: 199–204.

Reich, D. E. and Lander, E. S. 2001. On the allelic spectrum of human disease. *Trends Genet*. 17: 502–510.

Roach, J. C., Glusman, G., Smit, A. F. A., Huff, C. D., Hubley, R., et al. 2010. Analysis of genetic inheritance in a family quartet by whole-genome sequencing. *Science* 328: 636–639.

Rosenberg, N., Pritchard, J. K., Weber, J. L., Cann, H. M., Kidd, K. K., et al. 2002. Genetic structure of human populations. *Science* 298: 2381–2385.

Sabeti, P. C., Reich, D. E., Higgins, J. M., Levine, H. Z. P., Richter, D. J., et al. 2002. Detecting recent positive selection in the human genome from haplotype structure. *Nature* 419: 832–837.

Sankararaman, S., Mallick, S., Dannemann, M., Prüfer, K., Kelso, J., et al. 2014. The genomic landscape of Neanderthal ancestry in present-day humans. *Nature* In press.

Scally, A., Dutheil, J. Y., Hillier, L. W., Jordan, G. E., Goodhead, I., et al. 2012. Insights into hominid evolution from the gorilla genome sequence. *Nature* 483: 169–175.

Schuster, S. C., Miller, W., Ratan, A., Tomsho, L. P., Giardine, B., et al. 2010. Complete Khoisan and Bantu genomes from southern Africa. *Nature* 463: 943–947.

Skoglund, P., Malmström, H., Raghavan, M., Storå, J., Hall, P., et al. 2012. Origins and genetic legacy of Neolithic farmers and hunter-gatherers in Europe. *Science* 336: 466–469.

Stoneking, M. and Krause, J. 2011. Learning about human population history from ancient and modern genomes. *Nat. Rev. Genet.* 12: 603–614.

Templeton, A. 2002. Out of Africa again and again. *Nature* 416: 45–51.

Tishkoff, S. A., Reed, F. A., Ranciaro, A., Voight, B. F., Cabbitt, C. C., et al. 2006. Convergent adaptation of human lactase persistence in Africa and Europe. *Nat. Genet.* 39: 31–40.

Vernot, B. and Akey, J. M. 2014. Resurrecting surviving Neandertal lineages from modern human genomes. *Science* In press.

Voight, B. F., Kudaravalli, S., Wen, X., and Pritchard, J. K. 2006. A map of recent positive selection in the human genome. *PLoS Biol.* 4: e72.

Williams, G. C. and Nesse, R. M. 1991. The dawn of Darwinian medicine. *Quart. Rev. Biol.* 66: 1–22.

Wright, S. 1943. Isolation by distance. *Genetics* 28: 114–138.

Zaghloul, N. A. and Katsanis, N. 2011. Zebrafish assays of ciliopathies. *Methods Cell Biol.* 105: 257–272.

4

Normal Variation

W e continue this opening section with a discussion of the genetics of normal human variation, in light of the conceptual models introduced in Chapter 1. Two aspects of humanness that everyone is familiar with are appearance and behavior. Skin color, hair color, and eye color are among the first features we notice upon meeting another person, and it turns out that their genetics is relatively simple: they are oligogenic traits, meaning that they are influenced by a small number of large-effect genes. Height and weight, by contrast, are very close fits to the infinitesimal model and are strongly influenced by the environment. Not enough is known about the genetics of facial features or body shape to discuss them in any depth, but a handful of interesting findings will be discussed here. Debates have raged for decades over how much of intelligence is due to genetic variation, and while genome-wide association studies (GWAS) have supplied definitive evidence that the effect of genes is substantial, individual genes make only tiny contributions. The same can be said for the dimensions of personality—how extroverted or agreeable a person is, or whether they become easily addicted, or have a very spiritual nature—so our discussion here will focus more on aspects of the architecture of human behavior.

While reading this chapter, keep in mind the question of whether and how normality fades into abnormality and disease. On the one hand, most biological features are approximately normally distributed—they have a bell-curve distribution of frequencies. Most of us are somewhat similar, but inevitably some people must be at the extremes. On the other hand, humans are good at categorizing, and we like to place people into groups of otherness. If obesity were defined relative to the small percentage of individuals at the top of the body mass index distribution in a population, few South Asians would be regarded as obese in America, whereas over half of all Americans would be obese in India. How should this realization influence considerations of diabetes risk, given that obesity is one of the major risk factors, yet diabetes prevalence is similar in the two countries? What of attention deficit disorder: is it a discrete condition or just the recognition of adolescents at the tail of hyperactivity? Should we be surprised that there is such wide variation in sexual practices, including orientation, once we recognize that infinitesimal genetic variation has such a widespread influence on cognitive function? Furthermore, even if there is a clear distinction between healthy and unhealthy, are individuals closer to one extreme of normality at most risk of moving over into what is generally regarded as abnormal and unhealthy?

Pigmentation

You might think that the discovery of genes influencing skin pigmentation would be the easiest challenge for a geneticist to solve, but it has turned out to be very difficult. If you simply compared the allele frequencies of people from Norway and Nigeria, you would find tens of thousands of single nucleotide polymorphisms (SNPs) that are, statistically speaking, associated with either light or dark skin but actually have nothing to do with pigmentation. This situation is an example of what is known as the **population stratification** problem: wherever trait and allele frequencies differ between two populations, genetic association studies will find correlations that are not causal. So the identification of alleles that really do influence skin or eye color has taken different routes, borrowing from model organisms and studies of populations with more narrow distributions of color, and seeking biochemical evidence that gives a mechanism to the proposed role of the variation (Sturm 2009).

Normal human eye color ranges from light blue through green to hazel, dark brown, and almost black. It is due to the amount of deposition of two types of pigment, eumelanin and pheomelanin, in melanocyte cells within the iris. Genetic variation in several classes of genes influences pigment deposition: enzymes that catalyze the half-dozen steps in the conversion of tyrosine and dopamine into these pigments account for some of the phenotypic variation, but so too do genes that regulate uptake of the pigment precursors or differentiation of the melanocytes. Investigations in Iceland and Holland initially identified six loci that account for the majority of the variation. Blue versus brown eye color is largely attributable to a single gene, *OCA2*, which encodes a small-molecule transporter. Over 80 *OCA2* mutations give rise to one form of albinism, but the single nucleotide change that leads to most human blue eyes, which has attained high frequency in northern Europeans, is actually a regulatory variant in a highly conserved segment of an intron located in the adjacent gene, 200 kb away (Sturm et al. 2008). An amino acid substitution in *OCA2* itself modifies the penetrance of blue eyes in people with the main regulatory variant. The transition to green eyes requires additional variants in the solute carrier *SLC45A4* and in the tyrosinase enzyme encoded by *TYR*. Combining these three loci, it is possible to predict blue eyes over 90% of the time and brown eyes two-thirds of the time (Sulem et al. 2007; **Figure 4.1**). While most people with green eyes have a typical combination of three genotypes, many others with the same genes have either blue or brown eyes.

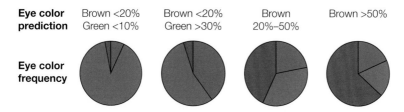

Figure 4.1 Genetic Influences on Eye Color.   Results of genetic prediction of eye color. Given polymorphisms in three genes, it is possible to classify people with respect to their likelihood of having blue, green, or brown eyes. Pie charts show the proportions of Icelanders who actually had the three eye colors (shaded corresponding to the actual color) predicted by their genetics at different prediction likelihoods shown above the charts. (See Sulem et al. 2007 for details.)

The same three loci also contribute to hair color, as does variation at the melanocortin receptor gene *MC1R*, a gene called *IRF4* that encodes a transcription factor involved in interferon response, and the *KITL* ligand involved in multiple aspects of cellular differentiation (Han et al. 2008). Several of these genes have immune functions, raising the possibility that divergence in pigmentation is partially related to adaptation to novel infectious agents as people migrated northward. Furthermore, they are important components of the risk of skin cancer and are also associated with freckling and tanning ability, suggesting that elevated melanoma risk is offset against the improved vitamin D production and calcium absorption in light-skinned people at northern latitudes (Jablonski and Chaplin 2010). Variation at *MC1R* is commonly associated with hair color in many other mammals, including domestic dogs, and intriguingly, the genotype of two sequenced Neanderthal specimens is the same red/blonde type found in many Europeans. Asians and Africans are essentially devoid of the European polymorphisms, and consequently they appear on a list of the few thousand nucleotide sites that have diverged the most in allele frequencies among the major human population groups. Because of population stratification, it is difficult to estimate how much of hair color variation is attributable to these half-dozen identified genes, but a reasonable estimate is around 20%. Nevertheless, as with blue eyes, red hair can be predicted genetically with over 80% accuracy, and brown hair in Europeans has close to two-thirds predictability, but blonde hair is less strongly determined.

Skin color is yet more complex. Another half-dozen candidate pigmentation genes have derived variants that are found at high frequencies in Europeans and are part of slightly extended haplotype blocks that are consistent with positive, but weak, selection (Lao et al. 2007). Another solute carrier, *SLC24A5*, was first identified as the source of the *golden* mutation in zebrafish, and the derived variant significantly correlates with lighter skin in admixed populations (Lamason et al. 2005). There is some overlap between East Asians and Europeans in which haplotypes distinguish both population groups from Africans, but there are also several differences in which loci and which polymorphisms diverge between the two Northern Hemisphere groups. This observation implies that genetic regulation of skin color is complex and has involved parallel divergence from the African source on the two Northern Hemisphere continents. A study of the admixed population of the Cape Verde Islands confirmed the involvement of most of these genes and suggested they account for as much as 35% of skin color on the islands, while unmapped modifiers of small effect explain a similar proportion of the variance (Beleza et al. 2013).

Height

With a heritability estimated at 80%, height is among the most strongly genetically influenced of all human traits. This finding is a little surprising, given that it is not unusual for siblings to differ in height by 20 cm or more and that parents often watch in dismay as their teenage children grow past them. First-generation immigrants from developing countries are usually taller than their parents, confirming that environment also has a strong influence on height. Recall, however, that heritability refers to the variance in a single population, so once corrections are made for generational, gender, and other influences, it turns out that the regression of offspring on parental height is remarkably tight and positive. If you wish to have tall children, pick a tall partner.

Strong genetic influence does not, however, translate into large genetic effect sizes. In fact, the genetic architecture of height is a perfect fit to the infinitesimal model: most of the normal variation is attributable to thousands of common polymorphisms, each with very small effects, while rare variants with larger effects are important at the extremes of the distribution (Weedon and Frayling 2008). An early GWAS of 30,000 people detected just 20 genes that collectively explained no more than 3% of the variance for height in Europeans (Weedon et al. 2008). If those variants were somehow removed from the population, it is doubtful anyone would notice a difference. Yet they are sufficient to categorize people as "at risk of being tall" (or short) quite well, as illustrated in **Figure 4.2**. Not shown on the figure are the error bars for each data point, which are ±7 cm, yet on average there is a strong relationship between genetic risk score (GRS) and phenotype.

The distribution of the number of tall alleles per person is approximately normal. In theory, with 20 genes, the score ranges from 0 to 40, but in practice, almost everyone has between 15 and 30 alleles that are associated with greater height, the average being 22 (since tall alleles are slightly more frequent in the European population). The individuals in the bottom and top deciles (10% bins) of the distribution differ on average by 6 cm. The average effect of substituting one tall for one short allele is thus just 4 mm. Obviously, these 20 genes are not sufficient to say whether someone is going to be the height of a jockey or an NBA basketballer, and in fact there is no reason why a 7-footer cannot be in the bottom tail of the

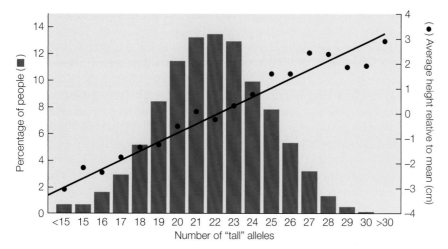

Figure 4.2 Prediction of Height from Genetic Risk Scores. The histogram shows the proportions of individuals with the indicated number of alleles at 20 genes that are associated with taller adult heights, and the points show the average height difference relative to the mean height for people in each category. There is an overall average difference of at least 4 cm between people at the extremes of the genetic risk score. (After Weedon et al. 2008.)

distribution in Figure 4.2. The main reasons are that the vast majority of the genetic variance that influences height remains to be identified, and environment makes a contribution. It is also likely that rare variants of large effect, or nongenetic factors, are particularly influential at the tails of the distribution, and detailed analyses suggest that this is particularly true for the shortest individuals (Chan et al. 2012).

Prediction accuracy can be expressed simply as the square root of the heritability explained by the SNPs in the model, since that is the upper limit of the correlation between genotype and phenotype. Thus, if the standard deviation for height is 7 cm, then the prediction error for any individual, if genotypes explain 10% of the variation, is $7 \times \sqrt{(1-0.1)} = 6.6\,\mathrm{cm}$: the predicted height for most people will be within that range. If half the heritability is captured by SNPs, then accuracy improves to closer to 5 cm, while the best that could be achieved is 3 cm. We can also use this reasoning to estimate how divergent people at the extremes are expected to be. By definition, the top 10% of a normal distribution are around 1.4 standard deviations greater than the mean. With an explained heritability of just 3%, this translates to a deviation of $1.4 \times 7 \times \sqrt{0.03} = 1.7\,\mathrm{cm}$, close to the average for the people with more than 26 "tall" alleles in Figure 4.2. A highly genetically accurate predictor, by contrast, would discriminate the top and bottom deciles of the height distribution by 17 cm (over half a foot).

By the time over 180,000 people had been included in GWAS, the proportion of variation explained for height had exceeded 10% due to the identification of 180 loci (Lango Allen et al. 2010). It is likely that twice this number of loci, with similar substitution effects in the range of 1 or 2 mm, exist. At least 19 loci have been shown to have multiple SNPs that associate independently with height, telling us that where there is one common variant for a trait, it will not be unusual to find a second one. Of the 21 genes affecting height that were first identified by mutations that affect skeletal growth, more than half also have common polymorphisms of small effect, and the genes in the vicinity of the 180 loci that have now been identified are strongly enriched for biological pathways that are plausibly related to growth, as they influence cell growth and division (as well as cancer predisposition), insulin signaling, and bone growth. They also seem to be connected in a network of genes that are likely to function together, since "GRAIL" analysis indicates that they are mentioned together in the scientific literature more commonly than would be expected by chance.

What accounts for the remaining at least 60% of the genetic contribution to height that has not yet been discovered? The vast majority of it is consistent with common variants with average effect sizes smaller than a millimeter. There could be several thousand of these variants, and SNP-based heritability estimation indicates that they are spread fairly uniformly across all chromosomes (Yang et al. 2011). The amount of variance explained per chromosome is more strongly correlated with chromosome length than with number of genes; this observation is most easily explained if regulatory variants are dispersed throughout intergenic regions. There is little evidence that common copy number variations (CNVs) make a strong contribution, but short individuals are mildly enriched for rare and larger deletions (Dauber et al. 2011). Finally, consistent with the soft infinitesimal model, there is now indirect evidence that rare recessive alleles also depress height, since the children of first-cousin marriages tend to be an average of 3 cm shorter than expected. Such consanguinity is expected to increase the likelihood of homozygosity, bringing together rare variant carriers more often than random mating would (McQuillan et al. 2012).

Weight

Weight is as genetically complex as height, with a few extra twists (Loos 2012). It is also arguably of greater biomedical importance, since body mass and fat deposition are so closely related to metabolic disease. The most common clinical measure of body mass is the metric ratio of weight to height squared, which gives rise to the body mass index (BMI) in units of kilograms per square meter. Historically, human adults have had a BMI in the range of 18–22, but nearly half of all Americans now have a BMI greater than 25, and increasing numbers have a BMI greater than 30, which makes them clinically obese. This measure can be deceptive, since muscle mass also increases BMI (an NFL running back who is 5'9" tall and weighs 217 pounds has a BMI of 32, and so by this definition is obese, despite being one of the fittest athletes on the planet). Nevertheless, BMI is generally correlated with percentage of body fat and is an easily measured indicator of weight adjusted for height.

The largest GWAS of BMI, which included almost 250,000 individuals, identified just 32 loci with allelic substitution effects as small as 150 g for an adult of average height (Speliotes et al. 2010). Collectively, these SNPs explain just 1.5% of the variance in BMI, which has a heritability of approximately 50%. In other words, despite the sampling of the equivalent of a small city, only a very small fraction of the genetic variance has been reduced to individual genes. Just as striking is the observation that one-fourth of the explained genetics is due to a single gene, *FTO* (Frayling et al. 2007). Homozygotes for the alternate haplotypes of *FTO* typically differ by as much as 2 kg; the effect is mostly attributable to a difference in growth rate throughout childhood (Elks et al. 2012). The function of the *FTO* locus is still being investigated, but it is thought to involve central nervous system regulation of the balance between food intake and fat deposition.

Just as height differences can be due to growth of the legs or torso, weight differences can arise from differential deposition of fat in different reserves. People typically conform to one of four body types, known as apple, pear, hourglass, or tube shapes, reflecting in part where they are most likely to put on weight, namely around their upper body, hips, or waist (see Figure 12.8). Intriguingly, GWAS for waist-to-hip ratio has uncovered 19 loci, 14 of which are independent of the BMI signals (Heid et al. 2010). Half of these loci are sexually dimorphic, having a greater effect on waist-to-hip ratio in women than in men. The candidate genes associated with them are more likely to be involved directly in lipid metabolism and fat deposition than the BMI loci, many of which are thought to be active in the brain. Central obesity, which mainly involves the deposition of visceral fat in the abdomen, leads to potbellies, cardiovascular disease, and diabetes, whereas total body fat includes subcutaneous fat, the volume of which appears to be affected by yet more distinct loci.

Given the dramatic global increase in obesity over the past two generations, it is interesting to ask how genetic differences may interact with environmental and behavioral choices. Several studies have found that children who have the "heavy" BMI genotype at *FTO* are less likely to put on weight when they are physically active than when they are sedentary (Andreasen et al. 2008). As a consequence, this genotype is one of the few in the genome that is associated not only with a quantitative difference in a trait, but also with its variance: there is much more phenotypic variation among people with the "heavy" BMI genotype than with the "light" BMI genotype (Yang et al. 2012). Furthermore, consumption of sugary beverages increases BMI, but it does so at a greater rate in those with a higher genetic risk of obesity (**Figure 4.3**): based on a sum of the effects of alleles at 32 genes, weighted by their effect size, Qi et al. (2012) found an average increase of up to 1 BMI unit for every 10 risk alleles and replicated the result in three large studies. The diet-dependence is shown as a difference in slope of the curves for each risk score category: the higher the genetic risk, the stronger the impact of the sodas. We also know that dual parental obesity during infancy is a tremendous risk factor for adult obesity, increasing the odds 15-fold over the general population, whereas obese babies are

Figure 4.3 Genotype-By-Diet Interaction for Obesity. This figure summarizes the trends observed in three large adult health studies carried out in the United States. Both sugary beverage consumption (ranging from none to at least one a day), and genetic risk, influence adult body mass index (BMI). However, there is also an interaction effect, such that those in the highest quartile of genetic risk of obesity (top, red line) gain more weight in connection with sugary beverage consumption than those in the lowest quartile of genetic risk. Without the interaction effect, the lines would be parallel. The highest and lowest genetic risk quartiles differ, on average, by 2 BMI units, and the highest and lowest sugary beverage consumption categories by as much as 1 BMI unit, depending on the specifics of the population. (After Qi et al. 2012.)

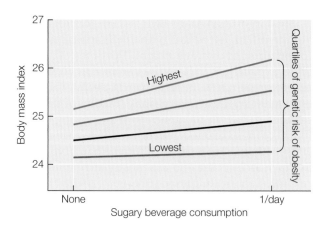

not greatly at risk of adult obesity. This parental effect dwarfs the known genetic risk based on summing the allelic effects at the 32 known loci, strongly implying that genetic influences can be overcome by appropriate behaviors in childhood. Whether or not children are likely to change their dietary and exercise habits as a consequence of knowing their genotype remains to be determined, but it seems unlikely.

Intelligence and Temperament

Genetic analysis of cognitive traits has historically been controversial, since inferences have often been misinterpreted or misused to justify or inhibit social policy. These traits are also notoriously difficult to define, and measures of IQ are typically culturally biased due to the nature of the tests. Although there is general agreement that there is significant resemblance among relatives for measures of intelligence, the possibility that shared environments and educational practices covary with familial differences has resulted in caution in concluding that intelligence has a substantial genetic component. However, recent genome-wide studies confirm that there is substantial SNP-based heritability for both crystallized and fluid IQ (measures that broadly capture acquired knowledge and abstract problem-solving ability, respectively). This conclusion is derived from an analysis of 3511 members of five northern English and Scottish 1920s and 1930s birth cohorts, which showed that as much as half the variance in IQ is associated with common variants (Davies et al. 2011). No single variant has been identified that unambiguously influences intelligence, so again, the model is infinitesimal. The genome-wide measures were only very weakly predictive of IQ in an independent Norwegian cohort.

An interesting extension of this study considered the genetic basis of change (usually decline!) in IQ with age (Deary et al. 2012). By contrasting similar IQ tests for almost 2000 individuals taken 55–70 years apart, it was shown that despite a high general correlation of 0.6 between childhood IQ and retirement IQ, as much as one-fourth of the change in IQ could be accounted for by genotypic differences. So not only do thousands of genes contribute to each person's general intelligence, but there is also a strong genetic influence on how robustly intelligence is retained throughout life. On the other hand, there also seem to be substantial nongenetic components to both aspects of intelligence. Steven Pinker (2011) noted the so-called Flynn effect, that IQ has increased by more than 10 points in the past two generations, more than offsetting the prediction of some geneticists that the accumulation of rare mutations in the expanding human population would result in a progressive decline in human intelligence.

Another aspect of cognitive function that is particularly interesting is personality. There are two major classifications of mood that psychologists recognize: temperament and personality. Cloninger's temperament scale classifies people with respect to novelty seeking, persistence, harm avoidance, and reward dependence (**Figure 4.4A**); females are likely to have higher scores on the latter two measures. The Big Five dimensions of personality—extroversion, agreeableness, openness to experience, neuroticism, and conscientiousness—are captured by an independent questionnaire (versions of which can be found on the Internet for recreational use; **Figure 4.4B**). No replicated significant associations have been identified for either scale in the first series of GWAS, despite a sample size approaching 20,000 and heritability estimates suggesting that more than one-third of the variance in these temperament and personality measures should be genetic (de Moor et al. 2010; Verweij et al. 2012).

Another reason why these findings are surprising is the vast literature of several thousand published papers relating variation in neurotransmitter receptors and transporters with a wide range of behaviors, from age at first sexual encounter to suicide ideation and from dancing ability to thrill-seeking. A temperament GWAS conducted on over 5000 Australian adults was powered to detect any variants that explain 1% or more of the variation in the four temperament traits. Its results effectively falsified a popular hypothesis that behavioral variation is maintained by balancing selection on variants that are advantageous under some circumstances encountered by people in a population and disadvantageous under others. The GWAS data cannot exclude a mutation-selection balance model, which assumes that most of the observed variation is due to rare variants that could have sizable effects, but there are many reasons to doubt that this is the only source of genetic variance. A third possibility, that much of the variation affecting mood is neutral, at least explains why there are no large-effect alleles, but it is difficult to imagine that harm avoidance

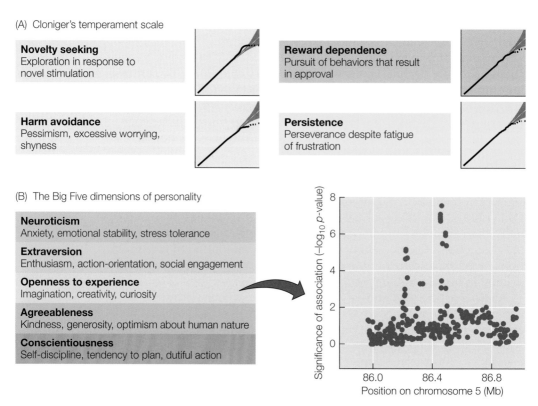

(A) Cloniger's temperament scale

Novelty seeking
Exploration in response to
novel stimulation

Reward dependence
Pursuit of behaviors that result
in approval

Harm avoidance
Pessimism, excessive worrying,
shyness

Persistence
Perseverance despite fatigue
of frustration

(B) The Big Five dimensions of personality

Neuroticism
Anxiety, emotional stability, stress tolerance

Extraversion
Enthusiasm, action-orientation, social engagement

Openness to experience
Imagination, creativity, curiosity

Agreeableness
Kindness, generosity, optimism about human nature

Conscientiousness
Self-discipline, tendency to plan, dutiful action

Figure 4.4 GWAS for Temperament and Personality.   (A) Cloninger's temperament scale recognizes four measures of temperament. The Q-Q plots (see Chapter 6) for each of these measures show no evidence for contributions by common variants, since each line of black points fails to climb above the gray zone. (B) The big five personality scale measures recognize different dimensions of personality. The strongest association with any of these measures observed in GWAS on 17,375 people, with openness, was only marginally significant. Each point in the graph represents the significance and location of a single SNP in a region of chromosome 5 near the *RASA1* gene. (A, after Verweij et al. 2012; B, after de Moor et al. 2010.)

and openness to experience have not affected fitness throughout human history.

A third aspect of cognition that is of general interest is memory performance. Psychologists recognize several different types of memory, including episodic memory (the ability to recall events), working memory and executive function (the ability to hold and manipulate short-term information), and more specific types of pattern recall. People vary tremendously in episodic memory, with some people apparently able to recall what they had for dinner or who won a basketball game on specific dates throughout their lives. Agnosias represent a particularly interesting aspect of episodic memory that also involves the ability to recognize differences in the first place; prosopagnosia, for example, is the inability to recall faces, for which there is wide variability among humans. One gene, *KIBRA*, which is thought to be involved in synaptic plasticity, has been associated with episodic memory in several studies (Milnik et al. 2012), but as with other behavioral traits, its effect size is so small that these results are not always replicated. Working memory is measured by instruments such as the Stroop test that evaluate how distraction (in this case, writing the words red, blue, and green in the wrong colors) influences recall. Deficiencies in working memory are correlated with some psychiatric disease.

Specific Behaviors

It is a little awkward to lump substance abuse, sexual orientation, and spirituality into one category, but each of these topics is of great general interest. There is copious evidence from animal models that single genes can have surprisingly large effects on behavior in natural populations (Anholt and Mackay 2010). Foraging activity is modulated in an almost Mendelian manner by polymorphisms at a cyclic GMP kinase in *Drosophila melanogaster* and a neuropeptide in *Caenorhabditis elegans*. Whether or not voles are monogamous or polygamous differs between mountain and prairie species, largely because of a substitution at a gene that encodes vasopressin. Just as strikingly, the ability of horses to adopt their specific gait has been traced to the activity of a single gene, *DMRT3*, which is active in regulation of neuronal connection between the central nervous system and limbs (Andersson et al. 2012). Is there any evidence for such effects mediating specific human behaviors?

The short answer is no, but there are a few exceptions, one of which relates to alcohol tolerance and abuse. Individuals who experience facial flush and drowsiness following alcohol consumption, which is due to the accumulation of high levels of acetaldehyde, generally have a strongly reduced desire to drink. Variation in the enzymes that catalyze the synthesis of acetaldehyde from alcohol (ADH) and that metabolize acetaldehyde to acetate (ALDH), is associated with the average number of alcoholic drinks an individual consumes per day across populations (**Figure 4.5**). Variants that increase ADH activity and reduce ALDH activity are both found at elevated frequencies in East Asians and probably account for a small proportion of the reduced levels of alcoholism in Japanese, Chinese, and Korean populations (Wang et al. 2012). That they do not explain all of the low alcoholism in East Asians follows from the finding in a Czech study that the variants make a relatively minor (yet significant) impact on alcohol consumption in that country (see Figure 4.5C). Furthermore, a research strategy known as Mendelian randomization (Davey-Smith 2010) supports a causal role of alcohol consumption in increased blood pressure in men, and in a reduced risk of heart disease via its effect on cholesterol levels, since each of these traits is also associated with genetic polymorphism at the relevant ADH or ALDH loci. Twin studies suggest that there is also a strong genetic component to the risk of becoming alcoholic, independent of a clear environmental effect in families where

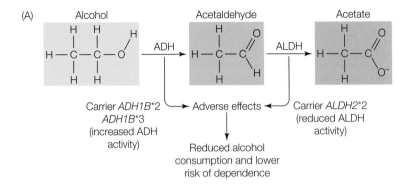

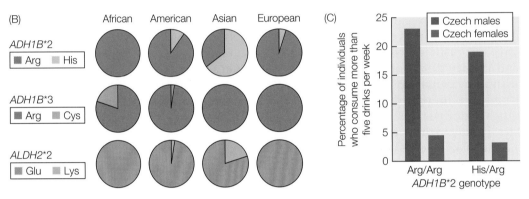

Figure 4.5 **Genetics of Alcohol Consumption Behavior.** (A) Ethanol is converted to acetaldehyde by a complex of seven ADH enzymes, and thence to acetate by ALDH. High levels of acetaldehyde lead to adverse responses that lower the risk of alcohol dependence. (B) The three major polymorphisms in humans that influence acetaldehyde levels are the Arg48His and Arg370Cys alleles of *ADH1* (known as the *ADH1B*2* and *ADH1B*3* polymorphisms respectively), the mutant alleles for both of which have 70 times higher enzyme activity than the wild type allele; and the Glu504Lys substitution allele *ALDH2*2*, which nearly inactivates ALDH. *ADH1B*2* and *ALDH2*2* have elevated frequencies in Asians, while *ADH1B*3* is more common in Africans. (C) In Czech Europeans, carriers of the Arg48His allele show significantly reduced alcohol consumption compared with homozygotes for the Arg allele, with a stronger protective effect in women than in men (and an associated reduction in incidence of binge drinking). (After Wang et al. 2012 and Hubacek et al. 2012.)

drinking is prevalent, but GWAS have not yet shed much light on the contributing genes.

Although less common than it was a generation ago, cigarette smoking continues to account for 30% of cancer deaths and 80% of chronic obstructive pulmonary disease in Americans, and as much as 10% of global mortality. Genetic factors are thought to affect the likelihood of initiation, cessation, and level of tobacco consumption (cigarettes per day), and GWAS have successfully pinpointed two clusters of nicotinic acetylcholine receptors, as well as the CYP2A6 nicotine catabolism enzyme, as key players (Thorgeirsson et al. 2010; Tobacco Genetics Consortium 2010). The *CHRNA5-A3-B4* cluster on chromosome 15 includes three independent SNPs that affect nicotinic acetylcholine receptor activity and/or expression in the frontal cortex, clearly implicating psychological reward in the regulation of smoking behavior. There is some suggestion that these variants may also interact with parental monitoring and/or early life stress to influence the onset of smoking (Xie et al. 2012), but the effects are relatively small, and it will take studies of tens of thousands of people with careful monitoring of environmental variables to tease apart what aspects of behavior are affected. Since the *CHRNA5* locus is also associated with lung cancer (see Chapter 14), it is interesting to ask whether nicotine addiction per se is responsible for the disease, but this question is complicated by a likely pleiotropic role for the protein in the lung epithelium. Studies of the genetics of cannabis and cocaine abuse are also ongoing, but, as with cigarette smoking, they are complicated by cultural and behavioral confounders.

Geneticist Dean Hamer published two provocative books that raised public awareness of the role of genetics in homosexuality (*The Science of Desire*, 1994) and in spirituality (*The God Gene*, 2004). His scientific investigations were performed in the pre-GWAS era, and thus focused in each case on linkage to a single chromosomal region; neither of those studies has been replicated or validated to the degree accepted as providing convincing evidence for association. The idea that there is a heritable component to sexual orientation is no longer controversial, but as with most other complex traits, no variant that explains more than 1% of the observed variance has been identified (Mustanski et al. 2005). There are plenty of studies pointing to biological differences between straight and gay men—for example, in fetal androgen sensitivity and in the morphology of specific regions of the brain—but biological differentiation alone does not imply a genetic etiology. Ongoing studies of sexually dimorphic traits, including facial and body morphology, will shed light on the role of hormones in promoting gender differences in general, but we are a long way from understanding the genetic basis for the coevolution of attractiveness and attraction.

The unfortunate subtitle of Hamer's second book, *How Faith Is Hardwired into Our Genes*, marketed conclusions that are hardly supported by the analyses. First, so long as heritability is less than 100% (let alone 50%), nothing is "hardwired," and second, the book was more about spirituality than faith: it is quite possible for people to have a sense of wonder in relation to creation and purpose without explicitly believing in a god or gods, and there is emphatically no evidence that genes contribute to the adoption of any specific faith (Islam, Christianity, or Hinduism, for example). However, religiosity and spirituality are correlated, and they can be measured fairly robustly from questionnaires; they have been shown to have high heritability and to correlate negatively with alcoholism (Haber et al. 2011). Since mental well-being is a strong predictor of prolonged good health, and by many measures includes a religiosity component, it is not uninteresting to consider the genetics of spirituality. Furthermore, researchers are beginning to consider the genetics of economic decision making (Hatemi and

McDermott 2012), voting behavior (Hatemi et al. 2007), educational attainment (Reitveld et al. 2013), and other phenomena that are to some extent related to mental attributes, but this endeavor is obviously complicated by the pervasive influence of socioeconomic and cultural factors.

Miscellaneous Traits

As background for the final section of this chapter, consider the simple genetic architecture that has been uncovered for many of the traits that most clearly distinguish breeds of dogs (Boyko 2011). The relatively simple approach of documenting regions of the genome that have very different allele frequencies among breeds has led to the discovery of dozens of loci that associate with visible traits. For phenotypes as diverse as body weight, snout length, tail length, coat color, and ear floppiness, just a handful of loci explain a majority of the variation. Just three genes account for most of the differences among breeds in the length, curliness, and wiriness of the hair and the presence of longer facial hair ("furnishings") around the snout (Cadieu et al. 2009), while a single gene, *IGF-1*, which encodes an insulin-like growth factor, accounts for much of the variation in breed size (Sutter et al. 2007). The simple genetic basis of so many canine traits is attributed to strong artificial selection on small breeding populations, while alleles of small and intermediate effect almost certainly contribute to within-breed variation in behavior, morphology, and disease risk.

Alleles of relatively large effect have been discovered for a handful of human traits, but they are the exception. A novel approach to the detection of such variants is to use the rich Web-based survey information provided by participants in personal genome studies, such as the findings reported by 23andme. With sample sizes of between 3000 and 8000 people, that company reported genome-wide associations for 8 of 22 "recreational genomics" traits (Eriksson et al. 2010) that included hair curl (two genes thought to influence the symmetry of hair secretion in the follicle), the ability to smell an unpleasant odor in one's urine after eating asparagus (an olfactory receptor variant), and the *a*utosomal-dominant *c*ompelling *h*elio-*o*pthalmic *o*utburst (ACHOO)/photic sneeze reflex—the tendency of some people to sneeze when they move into bright sunlight (a yet-to-be characterized locus). Intriguingly, self-knowledge of some genotypes affects how individuals assess themselves, as shown for athletic ability, where knowledge of a muscle actin genotype alters whether some people call themselves sprinters. In my own case, these reports correctly predicted 6 of 8 traits that I can self-evaluate, including blue eyes and lactose tolerance, but they got both the probable hair curliness and only average likelihood of male pattern baldness wrong! They also report on the likely response (adverse or no-effect) to 21 drugs. Some of these predictions could be clinically useful—such as those for warfarin sensitivity and Plavix effectiveness—while others are more whimsical and less robust—responses to caffeine and heroin, for example.

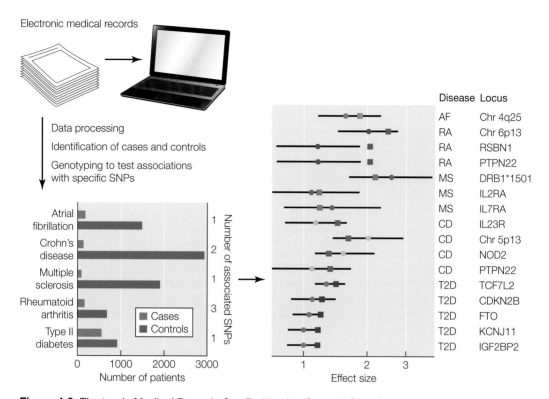

Figure 4.6 **Electronic Medical Records Can Be Used to Support Genetic Research.** Patient data collected by healthcare professionals is digitized and de-identified, then natural language processors are used to extract relevant information that is used to identify cases and controls (individuals who definitely do not have the condition). Ritchie et al. (2010) scanned 9483 patient records in the Vanderbilt University BioVU archive, then reevaluated 21 SNPs for association with five diseases: atrial fibrillation (AR), rheumatoid arthritis (RA), multiple sclerosis (MS), Crohn's disease (CD), and type 2 diabetes (T2D). The forest plot at the bottom right compares the effect size for the 16 indicated SNPs in the literature (squares), with the replication in BioVU (circles) with 95% confidence interval. Eight of fourteen SNPs with an effect size (technically, odds ratio) of greater than 1.25 replicated despite relatively small samples.

The notion that self-reporting can lead to useful genetic discoveries also has important clinical implications. 23andme successfully replicated over 180 disease associations when their database was still relatively small at 20,000 participants (Tung et al. 2011). Similarly, Vanderbilt University has pioneered an approach utilizing electronic medical record entries as a proxy for disease status in their database of almost 10,000 genotyped patients, confirming associations with atrial fibrillation, diabetes, and several other common diseases (**Figure 4.6**; Ritchie et al. 2010). Normal variation meets clinical significance in the example of a GWAS for breast size (Eriksson et al. 2012), in which bra cup size (adjusted for strap size, age, surgical augmentation, and history of pregnancy) reported by over 15,000 women led to the detection of six genes, three of which had been associated previously with breast cancer, while the other three are plausibly related to it. Such findings raise the question of how individuals are likely to respond to learning that they have variants that contribute to cancer or other diseases, a topic that we turn to in the next chapter of this book.

Summary

1. The study of normal human variation provides numerous fascinating models for the genetic analysis of biologically interesting as well as disease-relevant traits.

2. Large genome-wide association studies typically collect anthropomorphic data that has been used to support analysis of the genetics of height, weight, and other simple traits.

3. Blue eyes are most often attributable to homozygosity for a single polymorphism affecting the *OCA2* gene that has attained high frequency in northern Europeans. A small number of other genes further explain much of the difference between blue, green, and brown eyes.

4. Height conforms perfectly to the infinitesimal model of regulation by hundreds of common variants of very small effect, but rare variant contributions are also important for short or very tall stature.

5. Genetic contributions to height are spread throughout the genome approximately in proportion to the length of each chromosome. 180 loci, some of which harbor two or more independent common functional variants, explain just 10% of height.

6. Weight is usually measured as body mass index in kilograms per square meter. It conforms to the infinitesimal model, with less than 1.5% of the phenotypic variance explained by SNPs identified in GWAS of over 200,000 people.

7. Allelic substitution at one locus, *FTO*, leads to an average difference in total weight of 2 kg between homozygous classes, where the heavier genotype is more strongly influenced by behavior, including sugary beverage consumption.

8. Intelligence now appears to have a strong genetic component, but no genes have yet been identified that consistently contribute to it. Decline in IQ with age is also genetically influenced.

9. Temperament and personality have modest heritability, but a very complex genetic basis.

10. Alcoholism is less prevalent in individuals who rapidly metabolize alcohol to acetaldehyde or fail to convert acetaldehyde to acetate, due to adverse effects of this chemical influencing drinking behavior.

11. Cigarette smoking is affected by genetic response to nicotine consumption mediated in part by nicotinic acetylcholine receptors. Other

features of substance abuse, including likelihood of starting or cessation, and number of cigarettes or other drugs consumed per day, are
involved in addiction.

12. Recreational genomics supported by personal genome services has
led to the discovery of variants influencing phenotypes as diverse as
athletic ability and anosmia.

References

Andersson, L. S., Larhammar, M., Memic, F., Wootz, H., Schwochow, D., et al. 2012.
Mutations in *DMRT3* affect locomotion in horses and spinal circuit function in
mice. *Nature* 488: 642–646.

Andreasen, C. H., Stender-Petersen, K. L., Mogensen, M. S., Torekov, S. S., Wegner,
L., et al. 2008. Low physical activity accentuates the effect of the *FTO* rs9939609
polymorphism on body fat accumulation. *Diabetes* 57: 95–101.

Anholt, R. R. H. and Mackay, T. F. C. 2010. *Principles of Behavioral Genetics*. Elsevier,
Oxford UK.

Beleza, S., Johnson, N. A., Candille, S. I., Absher, D. M., Coram, M. A., et al. 2013.
Genetic architecture of skin and eye color in an African-European admixed population. *PLoS Genet.* 9: e1003372.

Boyko, A. R. 2011. The domestic dog: Man's best friend in the genomic era. *Genome
Biol.* 12: 216.

Cadieu, E., Neff, M. W., Quignon, P., Walsh, K., Chase, K., et al. 2009. Coat variation
in the domestic dog is governed by variants in three genes. *Science* 326: 150–153.

Chan, Y., Holmen, O. L., Dauber, A., Vatten, L., Havulinna, A. S., et al. 2012. Common
variants show predicted polygenic effects on height in the tails of the distribution,
except in extremely short individuals. *PLoS Genet.* 7: e1002439.

Dauber, A., Yu, Y., Turchin, M. C., Chiang, C. W., Meng, Y. A., et al. 2011. Genome-
wide association of copy-number variation reveals an association between short
stature and the presence of low-frequency genomic deletions. *Am. J. Hum. Genet.*
89: 751–759.

Davey-Smith, G. 2010. Mendelian randomization for strengthening causal inference
in observational studies: Application to gene × environment interactions. *Persp.
Psychol. Science* 5: 527–545.

Davies, G., Tenesa, A., Payton, A., Yang, J., Harris, S. E., et al. 2011. Genome-wide
association studies establish that human intelligence is highly heritable and polygenic. *Mol. Psychiatry* 16: 996–1005.

de Moor, M. H., Costa, P. T., Terracciano, A., Krueger, R. F., de Gues, E. J., et al. 2010.
Meta-analysis of genome-wide association studies for personality. *Mol. Psychiatry*
17: 337–349.

Deary, I. J., Yang, J., Davies, G., Harris, S. E., Tenesa, A., et al. 2012. Genetic contributions to stability and change in intelligence from childhood to old age. *Nature* 482:
212–215.

Elks, C. E., Loos, R. J. F., Hardy, R., Wills, A. K., Wong, A., et al. 2012. Adult obesity
susceptibility variants are associated with greater childhood weight gain and a
faster tempo of growth: The 1946 British Birth Cohort Study. *Am. J. Clin. Nutr.* 95:
1150–1156.

Eriksson, N., Benton, G. M., Do, C. B., Kiefer, A. K., Mountain, J. L., et al. 2012.
Genetic variants associated with breast size also influence breast cancer risk. *BMC
Med. Genet.* 13: 53.

Eriksson, N., Macpherson, J. M., Tung, J. Y., Hon, L. S., Naughton, B., et al. 2010. Web-based, participant-driven studies yield novel genetic associations for common traits. *PLoS Genet.* 6: e1000993.

Frayling, T. M., Timpson, N. J., Weedon, M. N., Zeggini, E., Freathy, R.M., et al. 2007. A common variant in the *FTO* gene is associated with body mass index and pre-disposes to childhood and adult obesity. *Science* 316: 889–894.

Haber, J. R., Koenig, L. B., and Jacob, T. 2011. Alcoholism, personality, and religion/spirituality: An integrative review. *Curr. Drug Abuse Rev.* 4: 250–260.

Hamer, D. 1994. *The Science of Desire: The Search for the Gay Gene and the Biology of Behavior*. Simon and Schuster, New York.

Hamer, D. 2004. *The God Gene: How Faith Is Hardwired into our Genes*. Doubleday, New York.

Han, J., Kraft, P., Nan, H., Guo, Q., Chen, C., et al. 2008. A genome-wide association study identifies novel alleles associated with hair color and skin pigmentation. *PLoS Genet.* 4:e1000074.

Hatemi, P. K. and McDermott, R. 2012. The genetics of politics: Discovery, challenges, and progress. *Trends Genet.* 28: 525–533.

Hatemi, P. K., Medland, S. E., Morley, K. I., Heath, A. C., and Martin, N. G. 2007. The genetics of voting: An Australian twin study. *Behav. Genet.* 37: 435–438.

Heid, I. M., Jackson, A. U., Randall, J. C., Winkler, T. W., Qi, L., et al. 2010. Meta-analysis identifies 13 new loci associated with waist-hip ratio and reveals sexual dimorphism in the genetic basis of fat distribution. *Nat. Genet.* 42: 949–960.

Hubacek, J. A., Pikhart, H., Peasey, A., Kubinova, R., and Bobak, M. 2012. *ADH1B* polymorphism, alcohol consumption and binge drinking in Slavic Caucasians. Results from the Czech HAPIEE study. *Alcohol Clin. Exp. Res.* 36: 900–905.

Jablonski, N. G. and Chaplin, G. 2010. Human skin pigmentation as an adaptation to UV radiation. *Proc. Natl. Acad. Sci. U.S.A.* 107(Supplement 2): 8962–8968.

Lamason, R. L., Mohideen, M. P. K., Mest, J. R., Wong, A. C., Norton, H. L., et al. 2005. SLC24A5, a putative cation exchanger, affects pigmentation in zebrafish and humans. *Science* 5755: 1782–1786.

Lango Allen, H., Estrada, K., Lettre, G., Berndt, S. I., Weedon, M. N., et al. 2010. Hundreds of variants clustered in genomic loci and biological pathways affect human height. *Nature* 467: 832–838.

Lao, O., de Gruijter, J. M., van Duijn, K., Navarro, A., and Kayser, M. 2007. Signatures of positive selection in genes associated with human skin pigmentation as revealed from analyses of single nucleotide polymorphisms. *Ann. Hum. Genet.* 71: 354–369.

Loos, R. J. F. 2012. Genetic determinants of common obesity and their value in predic-tion. *Best Pract. Res. Clin. Endocrin. Metab.* 26: 211–226.

McQuillan, R., Eklund, N., Pirastu, N., Kuningas, M., McEvoy, B. P., et al. 2012. Evidence of inbreeding depression on human height. *PLoS Genet.* 8: e1002655.

Milnik, A., Heck, A., Vogler, C., Heinze H-J., de Quervain, D. J. F., and Papassotiropoulos, A. 2012. Association of KIBRA with episodic and working memory: A meta-analysis. *Am. J. Med. Genet. B. Neuropsych. Genet.* 159B: 958–969.

Mustanski, B. S., Dupree, M. G., Nievergelt, C. M., Bocklandt, S., Schork, N. J., and Hamer, D. H. 2005. A genomewide scan of male sexual orientation. *Hum. Genet.* 116: 272–278.

Pinker, S. 2011. *The Better Angels of Our Nature: Why Violence has Declined*. Penguin, New York.

Qi, Q., Chu, A. Y., Kang, J. H., Jensen, M. K., Curhan, G. C., et al. 2012. Sugar-sweetened beverages and genetic risk of obesity. *N. Engl. J. Med*. 367: 1387–1396.

Reitveld, C. A., Medland, S. E., Derringer, J., Yang, J., Esko, T., et al. 2013. GWAS of 126,559 individuals identifies genetic variants associated with educational attainment. *Science* 340: 1467–1471.

Ritchie, M. D., Denny, J. C., Crawford, D. C., Ramirez, A. H., Weiner, J. B., et al. 2010. Robust replication of genotype-phenotype associations across multiple diseases in an electronic medical record. *Am. J. Hum. Genet*. 86: 560–572.

Speliotes, E. K., Willer, C. J., Berndt, S. I., Monda, K. L., Thorlelfsson, G., et al. 2010. Association analyses of 249,796 individuals reveal 18 new loci associated with body mass index. *Nat. Genet*. 42: 937–948.

Sturm, R. A. 2009. Molecular genetics of human pigmentation diversity. *Hum. Mol. Genet*. 18: R9–R17.

Sturm, R. A., Duffy, D. L., Zhao, Z. Z., Leite, F., Stark, M., et al. 2008. A single SNP in an evolutionary conserved region within intron 86 of the *HERC2* gene determines human blue-brown eye color. *Am. J. Hum. Genet*. 82: 424–431.

Sulem, P., Gudbjartsson, D. F., Stacey, S. N., Helgason, A., Rafnar, T., et al. 2007. Genetic determinants of hair, eye and skin pigmentation in Europeans. *Nat. Genetics* 39: 1443–1452.

Sutter, N. B., Bustamante, C. D., Chase, K., Gray, M. M., Zhao, K., et al. 2007. A single *IGF1* allele is a major determinant of small size in dogs. *Science* 316: 112–115.

Thorgeirsson, T. E., Gudbjartsson, D. F., Surakka, I., Vink, J. M., Amim, N., et al. 2010. Sequence variants at *CHRNB3-CHRNA6* and *CYP2A6* affect smoking behavior. *Nat. Genet*. 42: 448–453.

Tobacco Genetics Consortium. 2010. Genome-wide meta-analyses identify multiple loci associated with smoking behavior. *Nat. Genet*. 42: 441–447

Tung, J. Y., Do, C. B., and Hinds, D. A. 2011. Efficient replication of over 180 genetic associations with self-reported medical data. *PLoS ONE* 6: e23473.

Verweij, K. J. H., Zietsch, B. P., Medland, S. E., Gordon, S. D., Benyamin, B., et al. 2012. A genome-wide association study of Cloninger's temperament scales: Implications for the evolutionary genetics of personality. *Biol. Psychol*. 85: 306–317.

Wang J-C., Kapoor, M., and Goated, A. M. 2012. The genetics of substance dependence. *Annu. Rev. Genomics Hum. Genet*. 13: 241–261.

Weedon, M. N. and Frayling, T. M. 2008. Reaching new heights: Insights into the genetics of human stature. *Trends Genet*. 24: 595–603.

Weedon, M. N., Lango, H., Lindgren, C. M., Wallace, C., Evans, D. M., et al. 2008. Genome-wide association analysis identifies 20 loci that influence adult height. *Nat. Genet*. 40: 575–583.

Xie, P., Kranzler H. R., Zhang, H., Oslin, D., Anton, R. F., et al. 2012. Childhood adversity increases risk for nicotine dependence and interacts with α5 nicotinic acetylcholine receptor genotype specifically in males. *Neuropsychopharmacology* 37: 669–676.

Yang, J., Loos, R. J. F., Powell, J. E., Medland, S. E., Speliotes, E. K., et al. 2012. *FTO* genotype is associated with phenotypic variability of body mass index. *Nature* 490: 267–272.

Yang, J., Manolio, T. A., Pasquale, L. R., Boerwinkle, E., Caporaso, N., et al. 2011. Genome partitioning of genetic variation for complex traits using common SNPs. *Nat. Genet*. 43: 519–525.

5

Personalized Genomics

Associated with the revolutionary changes in our understanding of human genetics over the past decade is a growing sense that personal genome profiles will soon be a core component of personalized medicine. Babies will have their genomes sequenced at birth and will carry with them through life a listing of all their rare mutations of concern as well as a readout of their polygenic risk for all common diseases. The first steps in this direction have already been taken by providers such as Personalis and Sanford Imagenetics, but it is too early to know whether widespread adoption of personalized medicine will take years or decades, or whether it will be mainly adopted direct-to-consumer, allied with physician care, or follow some other model. Perhaps other types of genomic data will also be a part of the equation, following a model pioneered by Stanford professor Mike Snyder, whose integrated personal omics profile (iPOP), including transcriptomic, proteomic, and metabolomic profiles every few weeks over two years, provided him with early warning of likely onset of diabetes "predicted" by his genome (Chen et al. 2012).

Commercial pioneers in the personal genome realm using only genotype data have not fared so well. The company 23andme, referenced at several places in this book, was forced by the FDA to shut down its disease and trait assessments in December 2013. Navigenics was bought by Life Technologies and is being reconceived as a molecular diagnostics company, and deCODEme has discontinued its consumer genome scans. In this chapter, we explore some of the reasons for these false starts while also discussing the enormous potential of personalized medicine.

Genomic medicine is alive and well in the domains of oncology and pediatrics, as will be discussed in later chapters, and it is poised to inform **predictive health** following Leroy Hood's vision of the four P's of personalized medicine: it should be predictive, preventive, personalized, and participatory (P4Mi.org; Auffray et al. 2009).

It is worth noting that personalized medicine has been practiced for thousands of years, particularly on the Asian continent. In India, for example, Ayurvedic medicine continues to be the major mode of healthcare for

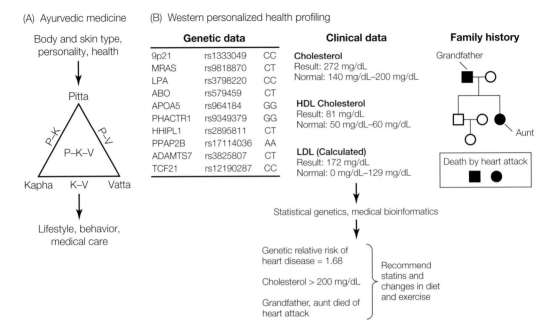

Figure 5.1 Two Approaches to Personalized Medicine. (A) In Ayurvedic medicine, individuals are assigned a prakriti based on the strength of the three doshas, assessed in relation to a series of physical and mental states. (B) An example of Western personalized health profiling, which integrates modern clinical measurements, genomic data, and family history to generate a profile of an individual's health.

hundreds of millions of people (Sethi et al. 2011; Dey and Pahwa 2014), with new hospitals and training centers opening alongside those offering Western-style medicine. The core idea of Ayurvedic medicine is that each one of us has a mixture of three doshas, known as Pitta, Kapha, and Vatta, which allow assignment of individuals to categories called prakritis. These are apparent after one's body build, skin type, eating preferences, and temperament are considered (**Figure 5.1A**); you can get an approximate prakriti assessment at doshaquiz.chopra.com. Each prakriti carries particular health risks—for example, neurological concerns for Vatta and metabolic disease for Kapha individuals—and according to ancient scripts, lifestyle measures can reduce the likelihood of disease progression. In the West, a growing "quantified self" movement has similar objectives. Tens of millions of people now record their daily habits with apps and devices such as RunKeeper, FitBit, and SparkPeople, which provide an unfolding portrait of their health. OneCare.me and mdRevolution, among others, integrate these quantified self data and link them to healthcare providers in an effort to increase patients' personal control over health decisions (**Figure 5.1B**).

The major challenges for personalized genomics are reducing extremely complex information to a palatable profile and demonstrating that this knowledge improves wellness and/or reduces medical costs (Khoury et al. 2007; Green and Guyer 2011). Two factors that are highly motivating for humans are immediacy and certainty, and a personal genomic profile provides neither. There is a danger that geneticists will oversell genomic profiling, just as regulators may overestimate its dangers, but if the right balance is found, it seems inevitable that genomic medicine will revolutionize healthcare.

Concerns over Personal Genome Projects

Whether in the context of wellness or not, **personal genome projects** (**PGPs**), namely efforts to introduce genomic profiling into everyday medical care and development of health behavior plans, face a number of obstacles to adoption. Most of these can be traced to five concerns:

1. *Privacy*. Many people regard their genomic data as deeply private and especially privileged, not something to be shared with others any more than you would share your medical records (Guttmacher and Collins 2003). We don't generally care to have other people know what conditions we have or what medicines are in our closets, so there is a similar presumption that we would protect the privacy of our genome sequence. Furthermore, concerns over genetic discrimination abound, and there is widespread uncertainty over the extent of the protections offered by the Genetic Non-Discrimination Act in the United States (Hudson et al. 2008) or similar statutes in the rest of the world (where they exist). These statutes basically prevent employers and health insurance providers from discriminating or adjusting premiums on the basis of genetic data. There is, however, a gray area regarding life insurance and the expectation that individuals will report information that they are aware of, for example, following participation in a PGP. Finally, it is almost impossible for providers of genomic profiling to give complete assurance that data will never be leaked from secure databases or sold to third-party companies.

2. *Informed consent*. Before anyone participates in any human genetic study, whether for research or for clinical purposes, they must sign a consent form, which acknowledges that they understand the dangers of participation, are well informed about the intent behind their participation, are consenting voluntarily, and have the right to withdraw at any time without penalty (**Box 5.1**). All researchers must complete a certification program that provides basic knowledge about relevant regulations, the moral basis of human studies, and protections laid out in the Helsinki Declaration of the World Medical Association (www.wma.net/en/20activities/10ethics/10helsinki). Minors under the age of 18 provide assent where feasible, but parental or guardian

BOX 5.1
Consent and Responsible Conduct of Research

All human research studies must be conducted with the approval of an institutional review board (IRB) whose responsibility it is to ensure that the participants' rights are protected and that they are appropriately informed about the conduct and purpose of the study. These requirements apply equally to surveys, to clinical research studies, and to basic research involving genetics. Most IRBs at academic centers also require all researchers, whether students, assistants, or principal investigators, to complete training and certification in the responsible conduct of research (RCR). In the United States, the Collaborative Institutional Training Initiative (CITI) provides a series of Internet-based courses that each take several hours to complete, and the National Institutes of Health (NIH) requires evidence that researchers are receiving ongoing RCR training. If a study involves protected medical information, additional knowledge of HIPA (the Health Information Portability Act) or similar regulations must be demonstrated.

Three of the core elements of RCR training are understanding the statutes that apply to human research, appreciating the history of abuses that has encouraged implementation of strict ethical codes, and knowing where to turn to in the event of an incident (e.g., a participant requiring hospitalization) or potential violation of policy (e.g., an undisclosed conflict of interest). Two of the more egregious historical abuses are the Tuskegee syphilis experiments, in which rural African American men were deliberately exposed to the pathogen without their consent, and Operation Whitecoat at Fort Detrick where conscientious objectors were used as volunteers in biological weapons challenge experiments. A more recent case highlighting the need for cultural sensitivity concerns the Havasupai Indians of the Grand Canyon, who consented to genetic research on diabetes, but objected to other research that was subsequently conducted using their genetic data—for example, research on their genetic history, which produced results that conflicted with the tribe's traditional oral history.

For this reason, the consent process has become critically important for all studies. Where the determination of genotypes is involved, the situation is further complicated by the new requirement that researchers make as much data as possible freely available to other researchers so that they may attempt to replicate findings. Inevitably, many researchers will conduct studies that the participants would not have conceived of or which were not covered by historically narrow consent documents that have left of a legacy of uncertainty over what is permissible. These studies set up potential conflicts between the desire of journals and funding agencies to promote open access to data and the need to protect participants' rights.

A further complication arose when it was realized that participants in large genome-wide studies could potentially be identified simply by their allele frequencies for a very large number of polymorphisms. That is to say, if someone had a person's genotypes, they could, in theory, infer whether that person was in the case or control group of a study without needing access to each of the individual genome profiles in the sample, which would be a violation of the participant's privacy. For this reason, the NIH and other agencies restrict access to the raw data to bona fide investigators, who must go through a rigorous application process overseen both by their own IRB and by the Data Access Committee of the NIH database of Genotypes and Phenotypes (dbGaP) before they can download data. It is also the reason why only summary statistics for the most significant SNPs in each study are published, since that amount of information is insufficient to support inference of identity. Because Y chromosomes can often be matched to surnames, some of which in turn have a restricted geographic distribution, they may be removed even from the public databases.

A typical consent document has a dozen or so components. The first is a description of the study in nontechnical language, including some scientific background on the research question and a clear statement of the purpose of the research, as well as justification for why a particular subset of the population (e.g., children, one ethnic group, or people with a medical condition) is excluded or preferentially included. Next, the document must outline the study's expectations of the participant: exactly what will be required (donate a blood sample, fill in a questionnaire, provide medical information), how long it will take, and whether and how participants will be compensated for their participation. Subsequently, there are short statements addressing participant rights, typically including a statement that participants have the right to withdraw from the study at any time without penalty or consequence (and explaining what will happen to their data in that event); statements of how long their biological material will be stored before being discarded; a description of procedures in place to protect privacy and security; a description of the assays that will be performed (genotyping, sequencing, RNA measurement) and the types of genetic analyses that will be allowed (single genes or diseases, or whole-genome studies without restriction); and whether their data will be made freely available to other investigators, before or after publication. The consent document must also contain a concise summary of the risks of participation as well as the procedures participants should take and people they should contact in the case of an adverse event or a perceived violation of the study protocol. Finally, participants are required to sign the document in the presence of a witness, affirming that they have read the document, understand what they are consenting to, and do so voluntarily. Participants under the age of 18 must also obtain the consent of a parent or guardian. Participants are provided with a copy for their records, and another copy is kept by the study team.

consent must also be provided. One of the precepts of informed consent is that the study must be described in nontechnical language understandable by a middle-school student. This requirement presents a particular challenge in genetic research since the basic concepts of DNA and genetic variability are not well understood—even by many healthcare professionals—and practicing geneticists also debate the interpretation of key findings. Furthermore, there is widespread variation among institutions in regard to the policies of the local institutional review board (IRB) that monitors each study. Some prominent researchers have wondered whether informed consent is even possible in the personalized genomic setting, but all agree that careful deliberation and openness are essential (Caulfield et al. 2008).

3. *Incidental findings*. Clinical geneticists are grappling with the implications of results that are incidental to the reasons why a person has sought genetic counseling (Dimmock 2012; Wolf 2013). Until recently, a family seeking advice on carrier status for a disease such as cystic fibrosis, for example, would simply have been tested for the common *CFTR* genotypes. Now, with exome sequencing, we can expect that some people will discover that they harbor a variant that is known to cause an adult-onset disease or that strongly predisposes them toward one. The general principle in such cases is that the information should not be provided unless there is certainty of the prognosis and an intervention exists that could alter the likely course of the disease. However, it is becoming increasingly clear that the penetrance

and expressivity of rare variants is usually much less than 100%
(Cooper et al. 2013). Similarly, interventions such as routine screening for breast or prostate cancer or brain exercises probably make a
difference—but does the benefit outweigh the cost and the anxiety?
Should *BRCA1* or *APOE* breast cancer and Alzheimer's susceptibility
genotypes be reported to PGP participants? Should all participants
be treated equally, even though people are known to vary greatly
in whether they will respond to incidental findings positively or
negatively?

4. *Uncertainty.* Very soon after entering a PGP, many people realize that
   most of the information they will receive is uncertain. This situation
   is slightly ironic because much of the anxiety surrounding the decision to participate in a PGP revolves around participants' fear that
   they will learn things they would rather not know about themselves,
   such as predisposition to dementia or heart disease. In fact, the information provided by common variants is classificatory ("I am in the
   top quartile of risk for diabetes"), but not predictive ("I will have diabetes by the age of 55"). Many of the findings are simply interesting,
   and the term "recreational genomics" is apt, but this does not mean
   that the findings should not influence our behavior. The situation in
   relation to single rare variants is complicated by the fact that many
   of these variants are of unknown significance (e.g., a new mutation
   in a gene in which other mutations are known to cause late-onset
   peripheral neuropathy may or may not promote the disease), that the
   prediction of deleteriousness is based on evolutionary or biochemical
   rather than clinical data, and it is becoming clear that false positive
   attributions of disease causation abound in the literature. Furthermore, different genome analysis algorithms yield quite different
   prioritization of which variants are most likely causal. This situation
   will gradually improve, but participants' uncertainty over the meaning of the results they will receive needs to be overcome.

5. *Availability of counseling.* Ideally, everyone who gets a personal
   genome profile would have someone they could turn to who would
   help them interpret it. That person is currently not likely to be their
   family physician: genetics is not a major part of medical curricula,
   and doctors are simply not trained in contemporary genomics.
   An increasing proportion of healthcare providers are comfortable
   ordering single-gene tests for specific conditions, but very few are
   comfortable with whole-genome data (Bernhardt et al. 2012). Similarly, most genetic counselors were trained in the single-gene era.
   There is a need for a new breed of genomic counselors, but in the
   meantime, people must make do with whatever information they can
   find online or that is provided by the PGP. Such information may be
   accurate and extensive, but without a professional to guide participants through the maze and answer questions, data alone can only
   be of limited utility.

Existing PGP Studies

To date, just a half-dozen large-scale personal genome projects have been discussed in the scientific literature. Some European countries, notably the Netherlands, Finland, Estonia, and the United Kingdom, as well as several Canadian provinces, have extensive public genotyping and sequencing efforts, which in the case of Iceland extends to virtual genome sequences for the majority of the population. However, publications to date have focused on population genetic and epidemiological inference without addressing how the data are being employed in the service of personalized medicine. In this section, I briefly discuss the experiences of a handful of U.S. initiatives.

One of the most highly publicized PGPs is the Harvard Personal Genome Project, led by George Church (www.personalgenomes.org; **Figure 5.2**). Founded in 2005, the hallmark of this initiative has been a commitment to openness: volunteer participants are invited to share their genetic and medical information with the community under an open consent model (Ball et al. 2012). This PGP acknowledges that genomic data may be identifiable and predictive of adverse conditions, and it urges participants to evaluate these risks before enrollment but after an

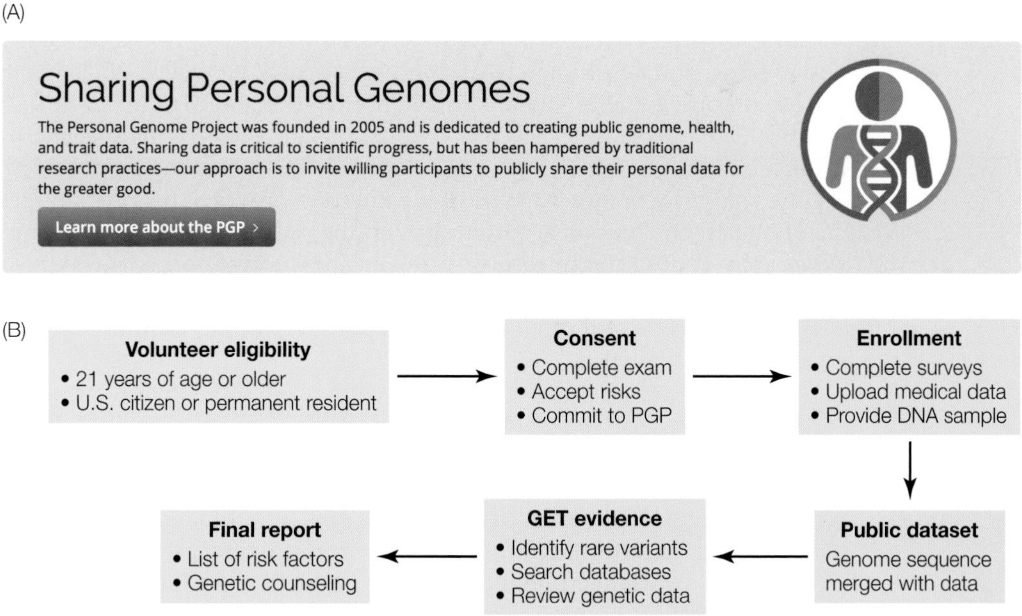

(A)

Sharing Personal Genomes

The Personal Genome Project was founded in 2005 and is dedicated to creating public genome, health, and trait data. Sharing data is critical to scientific progress, but has been hampered by traditional research practices—our approach is to invite willing participants to publicly share their personal data for the greater good.

Learn more about the PGP >

(B)

Figure 5.2 The Harvard Personal Genome Project. (A) This not-for-profit initiative aims to develop tools and resources to promote voluntary participation in personal genome studies. (B) The flow diagram outlines the work flow from participant recruitment to personal genome report. (A, from www.personalgenomes.org; B, after Ball et al. 2012.)

extensive series of tutorials have been taken. A 2014 report (Ball et al. 2014) found that of the first 2294 applications, over 1000 never completed the enrollment exam, and only 1143 fully enrolled in the study. Fewer than 1% of these participants have withdrawn subsequently, but only 16% have shared their information publicly. Only a small minority of the 163 individuals who have had their whole-genome sequences determined (free of charge) have sought feedback through the online mechanisms provided. It seems that the major concern expressed by these individuals is "false positive findings." For example, one participant has a variant in the *SCN5A* gene that is associated with long-QT syndrome in other people but shows no sign of the heart arrhythmia. Because the literature on variant pathology is often inaccurate and our knowledge of the penetrance of deleterious variants is poor, such findings can be a source of unnecessary anxiety. On the other hand, the PGP has generated at least one success story; one participant was found to have a mutation in the *JAK2* gene that leads to high platelet counts and potential thrombosis, but these syptoms are readily controlled by aspirin.

A similar initiative at the Mount Sinai School of Medicine in New York has reported on recruitment of graduate and medical students into a PGP, specifically focusing on attitudes toward participation. Sanderson et al. (2013) employed a decisional conflict scale to evaluate how comfortable 19 students were with receiving their own genome sequence, balancing anxiety over potentially negative findings against the desire to learn potentially useful information. Most started with some anxiety, but after 26 hours of class, the students had moved to more comfort with the process, and all ultimately decided to proceed. However, objectively assessed knowledge of "what's in a genome" was not high, even in this medically biased sample, and few participants interacted with the genetic counselor made available to them. An additional interesting point raised by the study is whether the type of information provided by PGPs might itself be personalized in accordance with the desires of the participants: some may choose not to receive rare variant information, for example, and others may simply focus on domains of disease identified by family risk factors, while still others may be open to active exploration of their own genome sequence.

The largest commercial direct-to-consumer personal genome project was conducted by 23andme (Tung et al. 2011). The company provided hundreds of thousands of clients with common genotypes generated from saliva samples submitted by mail for the cost of $99. It reported on genetic risk for 122 diseases or health concerns (e.g., diabetes, gallstones, macular degeneration), 60 traits (eye color, taste perception, hair curl), 53 Mendelian variants (phenylketonuria, cystic fibrosis, hemachromatosis), and 25 drug responses (warfarin sensitivity, sulfonylurea clearance, Plavix efficacy). Some of its predictions were patently absurd, such as the likelihood of heroin addiction based on one genotype evaluated in a small sample, but others, notably risk of Parkinson's disease, were based on leading-edge research enabled by the company's very large sample.

Despite extensive educational materials and mandatory tutorials before certain genotypes were released (e.g., BRCA breast cancer risk; Francke et al. 2013), online chat comments indicated that some users remained very confused about the relationship between genotypes and their own health. Citing such concerns and the nonresponsiveness of company leadership, the U.S. Food and Drug Administration (FDA) instructed 23andme to stop providing health information to new customers as of December 2013, though the company continues to provide ancestry data. Opinions on this decision are mixed: some hail the protection of naive consumers who may make life decisions based on inconclusive data, while others see it as a major setback for the introduction of personalized medicine (Baudhuin 2014; Green and Farahany 2014). The full text of the FDA's warning letter can be found at www.fda.gov/iceci/enforcementactions/warningletters/2013/ucm376296.htm.

Prenatal Diagnostics

One area in which genomic medicine is having an immediate and significant impact is pediatrics: both prenatal and postnatal diagnostics are rapidly coming to rely on genome sequencing (Wade et al. 2013). In some circumstances, parents may wish to perform genetic testing toward the end of the first trimester, or later, so that they can prepare to care for a child with a disability, arrange for in utero surgery, or choose to abort the pregnancy. For example, both parents might be known carriers of a Mendelian mutation, such as the mutation in the *HEXA* gene that leads to Tay-Sachs disease, a progressive neuropathy that is generally lethal by the age of 4. Or there may be an elevated risk of chromosomal abnormality based on family history or advanced maternal age. The incidence of trisomy approximately triples every 5 years after the maternal age of 30, so that the risk of Down syndrome rises from 1 in 1000 at the age of 30 to 1 in 30 at the age of 45. This rise is most often attributed to the increased likelihood over time that the molecular scaffolds that hold chromosomes suspended in meiosis in oocytes throughout the life of the mother will become disrupted. Genetic testing may also be performed when ultrasound results suggest a physical abnormality or when serum biochemistry is abnormal.

Amniocentesis has been used as a prenatal diagnostic method since the 1930s, and chorionic villus sampling has been used since the 1980s. Both procedures involve sampling of fetal cells, from the amniotic fluid under local anesthesia between weeks 15 and 20 of pregnancy, or from the placental wall by cervical catheter between 10 and 12 weeks, respectively. The risk of miscarriage attributable to these procedures may be as high as 0.5%–1%. The diagnostic techniques applied to the samples include PCR, to amplify and test specific genes, and clinical cytogenetics—namely, karyotype analysis—to observe chromosome numbers. The latter may be supplemented by fluorescence in situ hybridization (FISH) to visualize small regions of a chromosome that are suspected to be deleted, duplicated, or involved in a translocation (Guyot et al. 1988; Hulten et al. 2003). In this technique, DNA corresponding to the region of interest is labeled with fluorescent dyes in such a way that it can be seen when it is hybridized to a chromosome preparation, resulting in extra bands or bands in an abnormal location.

Very recently, it has become possible to perform many of these diagnostic procedures less invasively by sequencing of fetal DNA in maternal peripheral blood, which is routinely sampled throughout pregnancy (Fan et al. 2008). Fetal DNA increases from a small percentage early in pregnancy to as much as 50% of circulating serum DNA during the third trimester. In the United States, at least three commercial options for fetal diagnostics using maternal blood have been available since 2013: Sequenom's MaterniT21 PLUS and Ariosa Diagnostic's Harmony Prenatal Test are primarily for evaluation of trisomy 21, 18, or 13, but they also provide accurate sex determination by detection of Y chromosome DNA. Natera offers prenatal paternity testing, genetic carrier screening, and preimplantation screening and diagnosis (**Figure 5.3**). All three tests utilize sequencing of free circulating DNA in maternal blood, searching for overrepresentation of sequences from the relevant chromosomes. Imbalance in the ratio of alleles relative to the mother's genotype can also be utilized to confirm the identity of the father or to see whether the child carries both copies of a mutation for which the two parents are carriers (Lun et al. 2008). The trisomy 21 and 18 tests are believed to have a **sensitivity** of around 99%, meaning that they will detect almost all cases. They have even higher **specificity**, meaning that the number of false positives is just 1 in 500 or fewer. Both of these measures outperform classical sampling methods, and the tests are much less expensive (though not covered by all health insurance plans). Whole-fetal-genome sequencing from maternal blood has been achieved, but is not yet robust or cheap enough to be offered to expectant parents (Lo et al. 2010; Fan et al. 2012).

Preimplantation diagnosis refers to determination of the genotype or sequence of in vitro fertilized eggs before they are implanted in the mother's womb (Sermon et al. 2004). If cells isolated from ten very early embryos are screened for a rare recessive mutation carried by both parents, for example, two or three would be expected to be homozygous and would not be implanted. It may soon be possible to sequence whole genomes from single cells, which would permit identification of de novo mutations predicted to be deleterious. There is also the prospect that polygenic scores may be determined for traits of interest, such as depression, IQ, or eye color, and once sufficient heritability is explained by these scores, embryos at the two extremes would differ by a standard deviation or more for "risk" (Visscher and Gibson 2013). Though such scores would not be predictive, parents may choose to bias the odds in the direction they choose for specific traits, so "designer babies" are no longer a complete fantasy.

A major obstacle to the use of DNA analysis for prenatal detection of causal mutations is genetic heterogeneity. Although many inborn errors of metabolism can be attributed to one or a few mutations that account for

Genetic carrier screening

Noninvasive prenatal testing

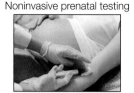

Fetal sex testing

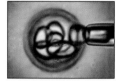

Prenatal paternity testing

Miscarriage testing

Preimplantation genetic
screening

Figure 5.3 Prospects for Prenatal Diagnosis.   Three California companies (Natera, Ariosa, and Sequenom) have become the first to provide analyses of fetal DNA that is circulating in maternal blood to detect abnormal chromosome numbers or inheritance of damaging variants from carrier parents. Extensions include assessment of fetal sex, determination of paternity, and confirmation of miscarriage.

most cases of a disease, the known mutations are just a fraction of all that exist. Almost all the mutations are at low frequencies in the population, so there will always be cases that are due to very rare mutations either in yet-to-be-ascertained genes or at unknown locations in known genes. Of course, de novo mutations that are not present in either parent could be detected only by genome sequencing. Nevertheless, some diagnostic tests target specific known mutations, and panels of genetic markers that are specific for heritable congenital conditions are being developed for this purpose. Procedures described in Chapter 7 and below allow for the capture of all known exons and detection of all polymorphisms or mutations in an individual. Detection of regulatory mutations may eventually follow by means of whole-genome sequencing, but challenges in the interpretation of such data will need to be overcome.

Postnatal Diagnosis

At the time of the passage of the Newborn Screening Saves Lives Act (NBSSL) in 2007, only 15 U.S. states offered government-sponsored newborn screening for a core set of 29 conditions recommended by the

TABLE 5.1 Core postnatal genetic tests recommended by the American College of Medical Genetics

Type of disorder	Disease	Gene	Prevalence
Red blood cells	Sickle-cell anemia	*HB* (coding)	1/400 (African American)
	β-Thalassemia	*HB* (regulatory)	1/50,000
Inborn errors of amino acid metabolism	Tyrosinemia	*FAH/TAT/HPD*	1/100,000
	Argininosuccinic aciduria	*ASL*	1/100,000
	Citrullinenmia	*ASS/SLC25A13*	1/100,000
	Phenylketonuria	*PAH*	1/25,000
	Maple syrup urine disease	*DBT/BCKDH*	1/100,000
	Homocysteinuria	*CBS*	1/100,000
Inborn errors of organic acid metabolism	Glutaric academia type I	*GCDH*	1/75,000
	HMG-lyase deficiency	*HMGCL*	1/100,000
	Isovaleric academia	*IVD*	1/100,000
	3MCC deficiency	*MCCC1,2*	1/75,000
	MM-CoA mutase deficiency	*MUT*	1/75,000
	Methylmalonic aciduria	*MMA A,B,C,D*	1/100,000
	Beta-ketothiolase deficiency	*ACAT1*	1/100,000
	Propionic academia	*PCC A,B*	1/75,000
	Multiple-CoA carboxylase deficiency	*HLCS/BTD*	1/100,000
Inborn errors of fatty acid metabolism	LCHAD	*HADHA*	1/75,000
	MCAD	*ACADM*	1/25,000
	VLCAD	*ACADVL*	1/75,000
	Trifunctional protein deficiency	*HADH A,B*	1/100,000
	Carnitine uptake defect	*OCTN2 (SLC22A5)*	1/100,000
Miscellaneous multisystem diseases	Cystic fibrosis	*CFTR*	1/5000
	Congenital hypothyroidism	*TSHR/TSHB/PAX8*	1/5000
	Biotinidase deficiency	*BTD*	1/75,000
	Congenital adrenal hyperplasia	*CYP21A*	1/25,000
	Classical galactosemia	*GAL E,K1,T*	1/50,000
Screened by other methods	Severe combined immune deficiency		1/50,000
	Congenital deafness		1/5000
	Critical congenital heart defects		1/100

Source: American College of Medical Genetics 2006.

American College of Medical Genetics (**Table 5.1**). Many of these conditions are inborn errors of metabolism that are detected by enzymatic or similar assays on dried blood spots obtained by pricking the baby's heel. In addition, primary immune deficiencies, congenital heart defects, and hearing problems are also routinely screened. The goals of these tests are improved access to healthcare, elimination of health disparities, and better outcomes and quality of early management, including prevention of onset of disease. The NBSSL fostered awareness, education, and research, and today all U.S. states offer an ever-expanding set of postnatal tests. There is considerable variation across jurisdictions in newborn screening policies, however, and some civil rights groups oppose the routine bio-banking of blood spots (Anderson et al. 2011). Similarly, there is wide variation across Europe in the extent and type of newborn screening.

The first widely adopted postnatal test was developed by Robert Guthrie for phenylketonuria, an irreversible neurological condition that can nevertheless be prevented by administration of a diet low in phenylalanine if detected early enough. Guthrie devised a bacterial inhibition assay utilizing dried blood spots, which was later extended to two other inborn errors of metabolism, galactosemia and maple sugar urine disease (Guthrie 1992; Clague and Thomas 2002). A different type of assay was introduced in the 1970s for early detection of congenital hypothyroidism by measurement of thyroxine levels in serum, supplementation of which in tablet form can prevent the symptoms of delayed growth and mental retardation. Mass spectrometry–based measurement of serum metabolites, introduced in the 1990s, brought the potential to expand the range of diagnoses to hundreds of disorders (Wilcken et al. 2003). Current postnatal testing recommendations are listed at the following website: en.wikipedia.org/wiki/List_of_disorders_included_in_newborn_screening_programs and in expanded form at www.hrsa.gov/advisorycommittees/mchbadvisory/heritabledisorders/recommendedpanel/.

Whole-exome and whole-genome sequencing now seem set to be introduced as a routine component of pediatric assessment at some point. Today, however, DNA-based postnatal diagnosis centers on targeted assessments and genome-wide evaluation of copy number variation (Thiessen 2008). The technique of array-based comparative genome hybridization (aCGH) can be used to detect large **copy number variants** (**CNVs**) over several megabases in length. Rather than hybridizing a labeled short region of one chromosome to a chromosome spread as in FISH, the entire genome of the patient is hybridized to a glass or silicon microarray on which hundreds of thousands of probes from across the genome are deposited (**Figure 5.4**). Areas of increased or decreased fluorescence over a stretch of probes from adjacent regions of the chromosome correspond to copy number variations. The resolution of this method is a function of the density and length of the probes, which are now standardized in commercial arrays produced primarily by Illumina and Affymetrix. One study by Shaffer et al. (2007) detected 1049 clinically relevant chromosome abnormalities in 8789 aCGH hybridizations for babies considered at risk, which represents a detection

Figure 5.4 Array-Based Comparative Genome Hybridization. DNA from a test individual (sample) and control DNA from someone without the disease are labeled with two different fluorescent dyes, then hybridized to a chip that contains tens of thousands of arrayed probes, each corresponding to a short chromosome interval. Scanning of the chip reveals locations where multiple adjacent probes either have an excess of test DNA (indicating that the test individual has a CNV duplication) or a deficit of test DNA (indicating a CNV deletion). In both cases, the control DNA shows the opposite trend. More recent technologies, such as the Illumina Infinium CytoSNP-850K BeadChip or Affymetrix CytoScan 750K Array, dispense with the control DNA and simply look for regions of increased or decreased overall fluorescence in the CNV regions relative to the rest of the genome.

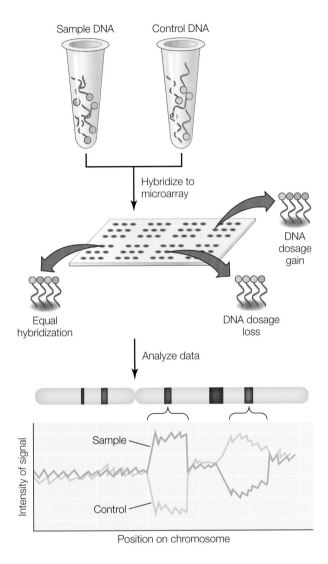

rate of 11%; no more than 5% of the abnormalities discovered were likely to have been detected by other means. The aCGH technique is also used in cancer profiling, as explained in Chapter 14, but in the context of newborn screening, it has proved most useful in screening for intellectual disability, developmental delay, craniofacial dysmorphology defects, and inherited syndromes (Lee et al. 2007; Miller et al. 2010).

A common set of abnormalities that to some extent distinguishes one set of patients from another is termed a **syndrome**. The sporadic manifestation of disease that is restricted to one phenotype is nonsyndromic, or **idiopathic**. Dozens of rare human syndromes are attributed to large chromosomal deletions. Cri du chat syndrome, also known as Lejeune syndrome, usually involves deletion of up to 20% of the short arm of chromosome 5, but localizes to a critical region at 5p15.3 containing two genes, *SEMA5A* and *CTNND2*. Most cases are de novo deletions, but inheritance from unaffected parents, one of whom has a balanced translocation (where the loss of 5p15 is compensated by an additional copy elsewhere in the genome), is occasionally seen. Similar genetics governs the etiology of Wolf-Hirschhorn syndrome, which involves deletion of the tip of chromosome 4 and has a variety of craniofacial and cognitive features. By contrast, velo-cardio-facial syndrome, also known as diGeorge syndrome, results from an intrachromosomal microdeletion of 22q11.2. It presents with a wide range of cardiac, facial, neuromuscular, and immunological abnormalities that are due to haploinsufficiency of various genes in the deleted region. All these conditions can be detected by aCGH, though it is likely that genome sequencing will be utilized in the near future.

Many other congenital abnormalities are attributed to single-gene point mutations. These abnormalities may or may not be syndromic. For example, craniosynostosis, which is premature closure of the skull plates, occurs in approximately 1 in 2000 births. Most cases are nonsyndromic, but the condition is associated with over 180 recognized phenotypes. At least five of these syndromes are due to mutations in the fibroblast growth factor receptor family of genes (*FGFR1, 2*, and *3*); they have been named Crouzon, Apert, Jackson-Weiss, Muenke, and Pfeiffer syndromes after the pediatricians who first described the associated facial or digital abnormalities that are typically observed. Three other syndromes accompanying craniosynostosis, Loeys-Dietz, Shprintzen-Goldberg, and Saethre-Chotzen, are due to mutations in the *TGF-β* receptor genes, *Fibrillin-1*, or *Twist-1*, respectively, and result in abnormalities of the heart, mental function, and symmetry (Kimonis et al. 2007). Postnatal diagnosis by targeted sequencing of previously implicated candidate genes can cost in the vicinity of $30,000, often without detecting any abnormality, so the advent of exome-wide sequencing has enormous potential to change clinical genetic practice in relation to congenital birth defects.

Genome Sequence–Enabled Unbiased Mutation Identification

Genome sequence–enabled gene mapping condenses three steps—linkage mapping, gene identification, and mutation detection—into one procedure. Sequencing, if done to sufficient depth and with appropriate quality control, should identify almost all polymorphisms and mutations without being biased by preconceptions of what gene is a good candidate. Since the "next-generation sequencing" methods explained in Chapter 7 are not infallible, it is essential that all likely pathogenic variants are confirmed by traditional Sanger sequencing. A sobering comparison of different next-generation sequencing platforms also concluded that some types of variants, notably small insertions and deletions, are poorly called and indicated that the bioinformatic challenges that remain to be solved are larger than

(A)

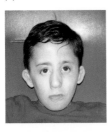

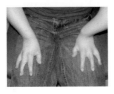

Filter	1 Individual		3 Kindreds	
	Dominant	**Recessive**	**Dominant**	**Recessive**
NS/SS/I	4650	2850	2650	1525
Novel	460	32	8	1
Damaging	228	9	2	0

(B)

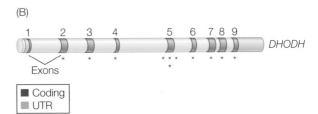

Figure 5.5 Pediatric Diagnostic Sequencing. (A) The causal gene for Miller syndrome was identified by whole-exome sequencing of four affected individuals from three different kindreds (the column labeled "1 Individual" averages the two siblings in the first kindred). NS/SS/I variants (nonsynonymous variants, splice site–disrupting variants, or indels) were identified in each exome. Under a dominant model, just one copy of a variant would be required to cause Miller syndrome, whereas under a recessive model, both alleles would have to have an NS/SS/I mutation (which can be compound heterozygous). Most candidate variants are eliminated as causal if they are already listed in databases of unaffected individuals (such as the Exome Variant Server). Software that predicts whether the variants are likely to be damaging can also be used, though ironically in this case, this filter removed what is thought to be the causal gene, *DHOHD*. (B) Follow-up screening uncovered a total of 11 damaging mutations (indicated by the asterisks) in six kindreds, affecting all nine exons of the gene and including one site disrupted by two different mutations. (After Ng et al. 2010; Photographs courtesy of Michael Bamshad, University of Washington, Department of Pediatrics.)

generally appreciated (Dewey et al. 2014). Nevertheless, the comprehensive annotation of the human genome that is now available, along with an expanding arsenal of analytical tools, ensures that geneticists can readily identify genes that are good candidates for causing the pathology in each case. Much progress has been made simply by sequencing trios of a **proband** (the individual child in whom a condition is found) and both of his or her biological parents.

Proof of principle was first provided by a study of a rare disorder called Miller syndrome, which is characterized by a suite of morphological traits including micrognathia (smaller than normal lower jaw), cleft lip, supernumerary nipples, and limb abnormalities. **Whole-exome resequencing (WER)** was first performed to an average depth of 40X in a kindred that included a proband and a second affected sibling (**Figure 5.5**; Ng et al. 2010). Under a dominant model requiring both affected children to have at least one novel mutation in the same gene—namely, a variant not existing in the databases of polymorphisms—228 candidate genes were identified. Under a recessive model, the mutation needed to be homozygous in each patient, and this simple filter reduced the candidates to just 9 genes. Further WER was performed on two additional proband-parent trios, reducing the candidate genes to 8 under a dominant model and just one (*DHODH*) under the more likely recessive model. The probability of finding four probands with a mutation in the same gene was just 0.000015. Further sequencing of the *DHODH* gene in three more proband-parent trios discovered six additional mutations predicted to be deleterious; each of these new cases was found to be an instance of compound heterozygosity (that is, different mutations inherited from each parent). Intriguingly, an enzyme involved in pyrimidine biosynthesis would not have been regarded as a likely contributor to a craniofacial syndrome a priori, which illustrates the power of an unbiased approach.

A high-profile case of using whole-genome sequencing (WGS) to identify a causal gene that quickly led to successful therapeutic intervention was that of the Beery twins, Alexis and Noah, as reported by Bainbridge et al. (2011). Both had been diagnosed with dopa-responsive dystonia, a defect of dopamine activity, at the age of 5, but were experiencing life-threatening complications that were not responsive to normal drug therapy in their teens. The WGS identified sepiapterin reductase (*SPR*) as the likely causal gene, implicating defects in the biosynthesis of both dopamine and serotonin, and supplementation of the latter quickly alleviated the breathing difficulties and tremors in the teenagers. This case was aided by a strong prior expectation for the role of the gene, but nevertheless demonstrates how powerful WGS may be not only for discovery of the causal mutation, but also for clinical translation into an effective intervention.

Genetic Counseling

Genetic counseling is a profession dedicated to advising families about their options for dealing with inherited disorders and preventing transmission of those disorders to unborn children, and for providing education about inheritance and genetic testing (Resta et al. 2006). Genetic counselors have a specialized graduate degree, allowing them to work in a wide variety of settings, following board certification, as part of a healthcare team. Specialties within the profession include obstetrics, pediatrics, oncology, and adult health. Traditionally, genetic counselors have utilized pedigree analyses and results of cytogenetic tests or a limited repertoire of Mendelian mutation assessments. They work closely with families in which a condition is known to occur, or with individuals thought to be at risk for bearing children with birth defects.

The practice of clinical diagnostics is advancing rapidly with the advent of genome-based mutation detection, and this will likely also impact the manner in which genetic counselors provide support for parents, prospective parents, and patients. Currently there are 19,000 genetic tests reported to be available for over 4000 monogenic conditions, as documented by the

Genetic Testing Registry (GTR; www.ncbi.nlm.nih.gov/gtr) maintained by the National Institutes of Health (NIH; **Figure 5.6**). Clinicians are now invited to report new associations to the ClinVar database, which communicates assertions of clinical significance by simultaneously reporting observed health status and details of the genetic evidence. All new submissions are mapped to the HuRef19 reference human genome coordinates and reported according to Human Genome Variation Society nomenclature standards. There have been over 64,000 submissions, each of which is assigned an accession number that is then linked to other records for the same allele, allowing clinicians to compare symptoms and pleiotropic conditions associated with a site of interest. The GTR site is searchable and also provides access to GeneReviews, which are expert-curated summaries of knowledge with respect to specific conditions, as well as directories of certified clinical geneticists, other consumer resources, and educational materials.

One of the most important roles for genetic counselors is prenatal diagnosis, as is well illustrated by the case of Tay-Sachs disease. The disease is classically autosomal recessive, often due to compound heterozygosity for mutations in the *HEXA* gene, inherited from carrier parents who may be identified from family history or a specific enzyme test. Fetal testing has recently been supplemented by preimplantation diagnosis, and similar tests are available for cystic fibrosis and sickle-cell anemia. Since the frequency of carriers among Ashkenazi Jews is as high as 1 in 30, Tay-Sachs is one of 10 diseases included in a testing service provided by the Committee for Prevention of Genetic Diseases, or Dor Yeshorim, for the worldwide Jewish

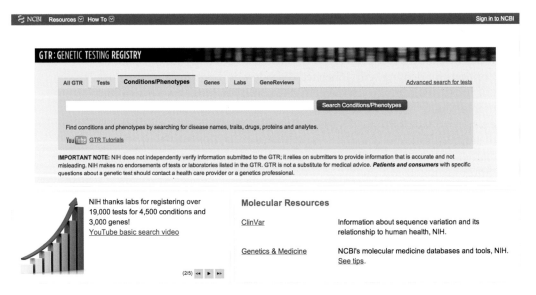

Figure 5.6 The Genetic Testing Registry.   Screenshot of the GTR website at www.ncbi.nlm.nih/gtr website.

community (Ekstein and Katzenstein 2001; Raz 2009). Participants have their DNA tested while in high school, and they may opt to anonymously compare their carrier status with that of a potential partner during the first couple of dates, considering discontinuation of the relationship if both are carriers. Most family planning based on genetic counseling is provided to married couples and is based on clear principles of advice and compassion rather than coercion or direction.

A special circumstance in which adults may seek genetic counseling involves a relatively rare series of cognitive and motor-neuron diseases known as the triplet-repeat-expansion or **trinucleotide-repeat disorders** (Orr and Zoghbi 2007). There are 14 known such diseases, including fragile X syndrome, Huntington's disease, and seven poly-Q class spinocerebellar ataxias. Most of these diseases exhibit the odd phenomenon of genetic anticipation, whereby the onset of disease is earlier and the symptoms more severe in each successive generation. This phenomenon is attributed to expansion of the number of the CAG, CGG, or similar repeats in the coding region of the causal gene: once the number exceeds a pathogenic threshold (typically 35 or more copies, depending on the gene), disease is likely to ensue. It is probable that unequal recombination and/or stuttering of the DNA polymerase leads to expansion of the repeat number in successive generations. Counselors are well aware that there is wide variation in how patients receive the news that they have inherited an expanded allele; some descend into depression, while others choose to live life to the fullest as long as possible. Many people at risk of these diseases refuse testing, citing fear of genetic discrimination by insurers and the high price of the tests (Powell et al. 2010).

Pharmacogenetics

Pharmacogenetics is a rapidly growing field because adverse responses to drugs are a major source of morbidity and because there is wide variation among people in the efficacy of drugs (Urban and Goldstein 2014). Genetic factors can influence the uptake, metabolism, and secretion of a drug (collectively called **pharmacokinetics**) as well as the ways it acts, inside cells or systemically, on intended targets or off-target (collectively called **pharmacodynamics**; Figure 5.7A,B). Most well-established cases of association between genotype and drug response involve drug-metabolizing enzymes and drug transporters, particularly polymorphisms in the cytochrome P450 and ABC transporter gene families. A genetically relatively simple example of pharmacokinetic variation is warfarin resistance and sensitivity, which is sufficiently influenced by *VKOR* and *CYP2C9* polymorphisms that these sites are now tested before dosing with this anticoagulant. Individuals who are sensitive to warfarin may be subject to bleeding events if they start on a dose that is too high. A pharmacodynamics example of a variant that is routinely tested for in Taiwanese clinics is the *HLA-B*1502* allele, which induces a severe skin condition in two-thirds of epileptic patients who have the allele and receive carbamazepine therapy (Chen et al. 2011). It

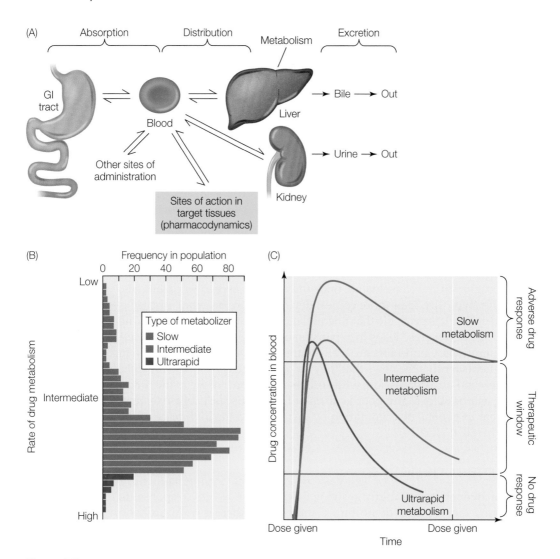

Figure 5.7 Pharmacogenetics. (A) Pharmacokinetics is the study of how drugs are absorbed, distributed, metabolized, and excreted from the body. Once they reach the correct target tissue, their effectiveness is the domain of pharmacodynamics. (B) The metabolism of most drugs is under genetic control, in some cases by a single protein. This situation gives rise to a distribution of rates of metabolism like the one shown here, so it is often possible to recognize ultrarapid, intermediate (sometimes subdivided to recognize extensive as well), and slow metabolizers of any drug. (C) Since the drug concentration must be in the therapeutic range, ultrarapid metabolizers often end up with insufficient bioactive drug, which contributes to the failure of over 25% of prescriptions. Slow metabolizers are unable to digest the drug, which can cause adverse side effects. Adverse reactions to drugs contribute to 6% of all hospitalizations and over 100,000 deaths in the United States each year. (After Bertilsson et al. 2002.)

must be emphasized, however, that most drug responses are genetically complex and that there are currently only two dozen widely approved single-gene tests for either responsiveness or adverse effects. The Pharmacogenomics Knowledge Base (www.PharmGKB.org) provides a convenient portal describing each of these cases as well as much more information.

An example of the use of pharmacogenetics to guide dosing comes from treatment of hepatitis C, a virus that causes acute liver disease in tens of millions of people worldwide. A genotype at the *IL28B* locus, which encodes interferon-λ3, is associated with a doubling of the likelihood of response of patients to the combination of pegylated interferon-α and ribavirin (Ge et al. 2009; Tanaka et al. 2009). The same genotype also predicts spontaneous clearance of the virus. A major side effect of this drug combination is hemolytic anemia, which is further predicted by a variant that reduces the function of the enzyme inosine triphosphatase encoded by *ITPA*, most likely by elevating the ATP pool (Thompson et al. 2010). Thus, these variants, discovered in the context of large clinical trials, provide both information about the mechanism of action of the drugs and clinically useful biomarkers. Furthermore, they help to explain differences in drug response rates among populations, since the frequencies of some of the alleles involved are quite different in Africans, Caucasians, and Asians. It is expected that even larger trials, perhaps facilitated by mobile networks of patients willing to share genotypic information with investigators after drugs come to market, will uncover more opportunities for genetically guided prophylaxis.

Two of the major factors that influence whether pharmaceutical companies will bring a new drug to market, as well as whether insurance companies or governments are willing to cover the costs, are its efficacy and toxicity. Given that the cost of each new drug brought to market is of the order of a billion dollars, there is considerable interest in identifying patients who are most likely to respond to drugs or to experience adverse side effects. For example, some anticancer drugs, such as Herceptin, are extremely expensive and are reserved only for late stages of treatment of *HER2*-positive breast tumors. Establishing that a variant influences drug metabolism, or even a biochemical measure of drug activity, is not necessarily sufficient to establish that the drug has a clinically relevant effect on recovery or other patient outcomes. In general, drugs that have wider therapeutic indices (are effective for a wider range of the patient population) will always be favored over drugs that require companion diagnostics, but genetics is expected to play a role in aiding prescription of drugs that are suitable for only some individuals (**Figure 5.7C**).

One further application of pharmacogenetics may lie in the stratification of apparently homogenous disease groups into subclasses that are likely to respond to different treatments. For example, GWAS have identified dozens of genes that contribute to risk for rheumatoid arthritis, many of which are targets of existing drugs (Okada et al. 2013). If it can be established which variants are most relevant in particular patients, the appropriate drugs may be used in a targeted manner. This argument has also been made in the case of amyotrophic lateral sclerosis (ALS, also known as Lou Gehrig's disease; Gibson and Bromberg 2012) and epilepsy (Urban and Goldstein 2014), in which different rare variants are thought to be responsible for different cases. It is thus possible that genomic medicine may soon converge on simultaneous diagnosis of the cause and personalized therapies for a wide range of diseases.

Integrated Health

It is no doubt foolhardy to attempt to portray the future of personalized genomic medicine since there are so many opportunities yet so many scientific, economic, and behavioral unknowns. Already, though, we can see multiple tiers of its application. Among the most prominent will be genetic diagnosis of rare diseases, primarily in newborn babies with birth defects, in some cases supporting interventions that will make a significant difference in quality of life. Cancer treatment, at least in refractory cases, will be tailored to individual profiles not just of the primary tumor, but in some cases of metastases that have taken on new genetic properties or evolved resistance to standard drugs. For the general relatively healthy population, we will be able to predict late-onset disease associated with highly penetrant variants, bearing in mind that we should expect many false positives, since the genetic background can protect people from diseases that are attributed to the same variant in other people. As the quantified self movement grows and people take more control of their own medical destiny, common genotypes will be used to help categorize risks and focus attention on behaviors by which people can avoid turning susceptibility into reality. At some point in the not too distant future many parents will be confronted with choices involving the use of genetic diagnostics during pregnancy, if not preimplantation, and society will be brought into challenging debates over the suitability of all these new genomic technologies.

Ideally, personalized genomic medicine will be integrated with all other aspects of healthcare. Family history, personal health history, lifestyle, and behavior will be modeled alongside clinical data gathered at regular visits to the doctor (**Figure 5.8**). Internet communities such as Patients Like Me will expand, and some people will choose to share their genetic information with networks of family and friends, and even strangers with similar interests and profiles. A whole new profession of genomic counseling is likely to be born, consisting of people whose job it is to help translate personal genome profiles into everyday health behaviors designed to prolong wellness and avoid disease. Only time will tell whether these changes will reach everyone or just a fraction of society, whether the benefits will outweigh the costs, and whether these technologies will contribute to changing the epidemiological landscape not just of Americans, but of people all over the world.

(A)

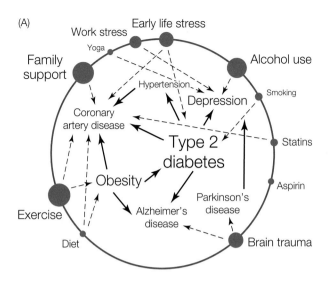

(B)

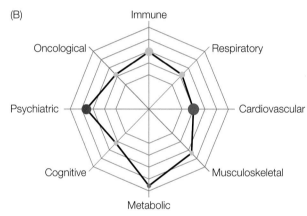

Figure 5.8 Personal Genomic Risk Profiles.   (A) Risk-o-grams. Genetic risk scores are generated for each major disease, and these scores are mathematically integrated, recognizing that risk in one domain of disease can interact with others, resulting in an overall risk represented here by the size of font. Environmental and behavioral factors also feed into the model. (B) This radar plot shows genetic risk scores for disease in eight domains of health along the rays (low risk near center, high risk toward the periphery), with the size of each marker proportional to relevant clinical parameters. Red, discordant for high genetic risk of diabetes but low blood glucose and insulin. Blue, discordant for low genetic risk of hypertension but high blood pressure. Green, concordant for high genetic risk of depression and a combination of social isolation and anxiety. (A, after Ashley et al. 2010; B, after Patel et al. 2013.)

Summary

1. Personalized medicine should allow individuals to take a more proactive role in their own health. Personal genome profiles are likely to become a core component of personalized medicine in the near future.

2. Genes do not act alone, so while personal genome profiles may make it possible to classify individuals with respect to risk, certain prediction of health outcomes will not be possible for common diseases.

3. Five concerns about the introduction of personal genomics relate to privacy, informed consent, incidental findings, uncertainty, and availability of counseling.

4. Amniocentesis and chorionic villus sampling are being replaced by sequencing of fetal DNA from maternal blood. This technique facilitates diagnosis of chromosome abnormalities and will soon be able to reveal targeted mutations.

5. Preimplantation diagnosis is used to evaluate health risks in embryos fertilized in vitro, allowing selection of which embryo to implant.

6. Postnatal diagnosis of dozens of very rare inborn errors of metabolism is routine, but uses mass spectrometry rather than genetic diagnosis. This may change in the near future.

7. Array-based comparative genome hybridization (aCGH) is a technique for detecting copy number variation genome-wide. It is particularly useful in detecting mutations that contribute to intellectual disability and other developmental abnormalities in newborns as well as to cancer.

8. Whole-exome sequencing is an unbiased approach to detection of mutations that are likely to be responsible for rare conditions such as birth defects.

9. The Genetic Testing Registry is a database of rare conditions that can be tested for by genetic diagnosis. Hundreds of companies offer services of this nature.

10. Pharmacokinetics is the study of the uptake, metabolism, and excretion of drugs. Pharmacodynamics is the study of the ways a drug acts, inside cells or systemically, on intended targets or off-target. Both pharmacokinetics and pharmacodynamics are influenced by genotypes.

11. Personalized medicine will require integration of genomic profiling with family history and clinical data and will be facilitated by online networks of physicians, peers, and genomic counselors.

References

American College of Medical Genetics. 2006. Newborn screening: Towards a uniform screening panel and system. *Genet. Med.* 8(5)Suppl: S12–S252.

Anderson, R., Rothwell, E., and Botkin, J. R. 2011. Newborn screening: Ethical, legal, and social implications. *Annu. Rev. Nurs Res.* 29: 113–132.

Ashley, E. A., Butte, A. J., Wheeler, M. T., Chen, R., Klein, T. E., et al. 2010. Clinical assessment incorporating a personal genome. *Lancet* 375:1525–1535.

Auffray, C., Chen, Z., and Hood, L. 2009. Systems medicine: The future of medical genomics and healthcare. *Genome Med.* 1: 2.

Bainbridge, M. N., Wiszienewski, W., Murdock, D.R., Friedman, J., Gonzaga-Jauregui, C., et al. 2011. Whole genome sequencing for optimized patient management. *Sci. Transl. Med.* 3: 87re3.

Ball, M. P., Bobe, J. R., Chou, M. F., Clegg, T., Estep, P. W., et al. 2014. Harvard Personal Genome Project: Lessons from participatory public research. *Genome Med.* 6: 10.

Ball, M. P., Thakuria, J. V., Zaranek, J. W., Clegg, T., Rosenbaum, A. M., et al. 2012. A public resource facilitating clinical use of genomes. *Proc. Natl. Acad. Sci. U.S.A.* 109: 11920–11927.

Baudhuin, L. M. 2014. The FDA and 23andMe: Violating the First Amendment or protecting the rights of consumers? *Clin. Chem.* 60: 835–837.

Bernhardt, B. A., Zayac, C., Gordon, E. S., Wawak, L., Pyeritz, R. E., and Gollust, S. E. 2012. Incorporating direct-to-consumer genomic information into patient care: attitudes and experiences of primary care physicians. *Per. Med.* 9: 683–692.

Bertilsson, L., Dahl, M-J., Dalén, P., and Al-Shurbaji, A. 2002. Molecular genetics of CYP2D6: Clinical relevance with focus on psychotropic drugs. *Br. J. Clin. Pharmacol.* 53: 111–122.

Biesecker, B. B., Boehnke, M., Calzone, K., Markel, D. S., Garber, J. E., et al. 1993. Genetic counseling for families with inherited susceptibility to breast and ovarian cancer. *JAMA* 269: 1970–1974.

Caulfield, T., McGuire, A. L., Cho, M., Buchanan, J. A., Burgess, M. M., et al. 2008. Research ethics recommendations for whole-genome research: consensus statement. *PLoS Biol.* 6: e73.

Chen, P., Lin, J. J., Lu, C. S., Ong, C. T., Hsieh, P. F., et al. Taiwan SJS Consortium. 2011. Carbamazepine-induced toxic effects and *HLA-B*1502* screening in Taiwan. *N. Engl. J. Med.* 364: 1126–1133.

Chen, R., Mias, G. I., Li-Pook-Than, J., Jiang, L., Lam, H. Y. K., et al. 2012. Personal omics profiling reveals dynamic molecular and medical phenotypes. *Cell* 148: 1293–1307.

Clague, A. and Thomas, A. 2002. Neonatal biochemical screening for disease. *Clin. Chim. Acta* 315: 99–110.

Cooper, D. N., Krawczak, M., Polychronakos, C., Tyler-Smith, C., and Kehrer-Sawatzki, H. 2013. Where genotype is not predictive of phenotype: towards an understanding of the molecular basis of reduced penetrance in human inherited disease. *Hum. Genet.* 132: 1077–1130.

Dewey, F. E., Grove, M. E., Pan, C., Goldstein, B. A., Bernstein, J. A., et al. 2014. Clinical interpretation and implications of whole-genome sequencing. *JAMA* 311: 1035–1044.

Dey, S. and Pahwa, P. 2014. *Prakriti* and its associations with metabolism, chronic diseases, and genotypes: Possibilities of newborn screening and a lifetime of personalized prevention. *J. Ayurveda Integr. Med.* 5: 15–24.

Dimmock, D. 2012. A personal perspective on returning secondary results of clinical genome sequencing. *Genome Med.* 4: 54.

Ekstein, J. and Katzenstein, H. 2001. The Dor Yeshorim story: Community-based carrier screening for Tay-Sachs disease. *Adv. Genet.* 44: 297–310.

Fan, H. C., Blumenfeld, Y. J., Chitkara, U., Hudgins, L., and Quake, S. R. 2008. Noninvasive diagnosis of fetal aneuploidy by shotgun sequencing DNA from maternal blood. *Proc. Natl. Acad. Sci. U.S.A.* 105: 16266–16271.

Fan, H. C., Gu, W., Wang, J., Blumenfeld, Y. J., El-Sayed, Y. Y., and Quake, S. R. 2012. Non-invasive prenatal measurement of the fetal genome. *Nature* 487: 320–324.

Francke, U., Dijamco, C., Kiefer, A. K., Eriksson, N., Moiseff, B., et al. 2013. Dealing with the unexpected: Consumer responses to direct-access BRCA mutation testing. *Peer J.* 1: e8.

Ge, D., Fellay, J., Thompson, A. J., Simon, J. S., Shianna, K. V., et al. 2009. Genetic variation in IL28B predicts hepatitis C treatment-induced viral clearance. *Nature* 461: 399–401.

Gibson, S. B. and Bromberg, M. B. 2012. Amyotrophic lateral sclerosis: Drug therapy from the bench to the bedside. *Semin. Neurol.* 32: 173–178.

Green, E., Guyer, M. S., and National Institute Human Genome Research. 2011. Charting a course for genomic medicine from base pairs to bedside. *Nature* 470: 204–213.

Green, R. C. and Farahany, N. A. 2014. The FDA is over-cautious on consumer genomics. *Nature* 505: 286–287.

Guthrie, R. 1992. The origin of newborn screening. *Screening* 1: 5–15.

Guttmacher, A. E. and Collins, F. S. 2003. Ethical, legal, and social implications of genomic medicine. *N. Engl. J. Med.* 349: 562–569.

Guyot, B., Bazin, A., Sole, Y., Julien, C., Daffos, F., and Forestier, F. 1988. Prenatal diagnosis with biotinylated chromosome specific probes. *Prenat. Diagn.* 8: 485–493.

Hudson, K. L., Holohan, M. K., and Collins, F. S. 2008. Keeping pace with the times— The Genetic Information Nondiscrimination Act of 2008. *N. Engl. J. Med.* 358: 2661–2663.

Hulten, M. A., Dhanjal, S., and Pertl, B. 2003. Rapid and simple prenatal diagnosis of common chromosome disorders: Advantages and disadvantages of the molecular methods FISH and QF-PCR. *Reproduction* 126: 279–297.

Khoury, M. J., Gwinn, M., Yoon, P. W., Dowling, N., Moore, C. A., and Bradley, L. 2007. The continuum of translation research in genomic medicine: How can we accelerate the appropriate integration of human genome discoveries into health care and disease prevention? *Genet. Med.* 9: 665–674.

Kimonis, V., Gold, J., Hoffman, T. L., Panchal, J., and Boyadjiev, S. A. 2007. Genetics of craniosynostosis. *Sem. Ped. Neurol.* 14: 150–161.

Lee, C., Iafrate, A. J., and Brothman, A. R. 2007. Copy number variations and clinical cytogenetic diagnosis of constitutional disorders. *Nat. Genet.* 39: S48-S54.

Limdi, N. A. and Veenstra, D. L. 2008. Warfarin pharmacogenetics. *Pharmacotherapy* 28: 1084–1097.

Lo, Y. M., Chan, K. C., Sun, H., Chen, E. Z., Jiang, P., et al. 2010. Maternal plasma DNA sequencing reveals the genome-wide genetic and mutational profile of the fetus. *Sci. Transl. Med.* 2: 61ra91.

Lun, F. M., Tsui, N. B., Chan, K. C., Leung, T. Y., Lau, T. K., et al. 2008. Noninvasive prenatal diagnosis of monogenic diseases by digital size selection and relative mutation dosage on DNA in maternal plasma. *Proc. Natl. Acad. Sci. U.S.A.* 105: 19920–19925.

Miller, D. T., Adam, M. P., Araadhya, S., Biesecker, L. G., Brothman, A. R., et al. 2010. Consensus statement: Chromosomal microarray is a first-tier clinical diagnostic test for individuals with developmental disabilities or congenital abnormalities. *Am. J. Hum. Genet.* 86: 749–764.

Ng, S. B., Buckingham, K. J., Lee, C., Bigham, A. W., Tabor, H. K., et al. 2010. Exome sequencing identifies the cause of a Mendelian disorder. *Nat. Genet.* 42: 30–35.

Okada, Y., Wu, D., Trynka, G., Raj, T., Terao, C., et al. 2013. Genetics of rheumatoid arthritis contributes to biology and drug discovery. *Nature* 506: 376–381.

Orr, H. T. and Zoghbi, H. Y. 2007. Trinucleotide repeat disorders. *Annu. Rev. Neurosci.* 30: 575–621.

Pagon, R. A., Adam, M. P., Bird, T. D., Dolan, C. R., Fong, C-T., et al., eds. 1993–2014. *GeneReviews*. University of Washington, Seattle. Available online at www.ncbi.nlm.nih.gov/books/NBK1116/

Patel, C., Sivadas, A., Tabassum, R., Preeprem, T., Zhau, J., et al. 2013. Whole genome sequencing in support of wellness and health maintenance. *Genome Med.* 5: 58.

Powell, A., Chandrasekharan, S., and Cook-Deegan, R. 2010. Spinocerebellar ataxia: Patient and health professional perspectives on whether and how patents affect access to clinical genetic testing. *Genet. Med*.12: S83-S110.

Raz, A. E. 2009. Can population-based carrier screening be left to the community? *J. Genet. Couns.* 18: 114–118.

Resta, R., Biesecker, B. B., Bennett, R. L., Blum, S., Hahn Estabrooks, S., et al. 2006. A new definition of genetic counseling: National Society of Genetic Counselor's Definition Task Force report. *J. Genet Couns.* 15: 77–83.

Sanderson, S. C., Linderman, M. D., Kasarskis, A., Bashir, A., Diaz, G. A., et al. 2013. Informed decision-making among students analyzing their personal genomes on a whole genome sequencing course: A longitudinal cohort study. *Genome Med.* 5: 113.

Sermon, K., van Steirteghem, A., and Liebaers, I. 2004. Preimplantation genetic diagnosis. *Lancet* 363: 1633–1641.

Sethi, T. P., Prasher, B., and Mukerji, M. 2011. Ayurgenomics: a new way of threading moleculare variability for stratified medicine. *ACS Chem. Biol.* 6: 875–880.

Shaffer, L. G., Bejjani, B. A., Torchia, B., Kirkpatrick, S., Coppinger, J., and Ballif, B. C. 2007. The identification of microdeletion syndromes and other chromosome abnormalities: Cytogenetic methods of the past, new technologies for the future. *Am. J. Med. Genet.* 145C: 335–345.

Tanaka, Y., Nishida, N., Sugiyama, M., Kurosaki, M., Matsuura, K., et al. 2009. Genome-wide association of IL28B with response to pegylated interferon-alpha and ribavirin therapy for chronic hepatitis C. *Nat. Genet.* 41: 1105–1109.

Theisen, A. 2008. Microarray-based comparative genomic hybridization (aCGH). *Nat. Education* 1: 45.

Thompson, A. J., Fellay, J., Patel, K., Tillmann, H. L., Naggie, S., et al. 2010. Variants in the *ITPA* gene protect against ribavirin-induced hemolytic anemia and decrease the need for ribavirin dose reduction. *Gastroenterology* 139: 1181–1189.

Tung, J. Y., Do, C. B., Hinds, D. A., Kiefer, A. K., Macpherson, J. M., et al. 2011. Efficient replication of over 180 genetic associations with self-reported medical data. *PLoS ONE* 6: e23473.

Urban, T. J. and Goldstein, D. B. 2014. Pharmacogenetics at 50: Genomic personalization comes of age. *Sci. Transl. Med.* 6: 1–9.

Visscher, P. M. and Gibson, G. 2013. What if we had whole-genome sequence data for millions of individuals? *Genome Med.* 5: 80.

Wade, C. H., Tarini, B. A., and Wilfond, B. S. 2013. Growing up in the genomic era: Implications of whole-genome sequencing for children, families, pediatric practice. *Annu. Rev. Genom. Hum. Genet.* 14: 535–555.

Wilcken, B. W., Wiley, V., Hammond, J., and Carpenter, K. 2003. Screening newborns for inborn errors of metabolism by tandem mass spectrometry. *N. Engl. J. Med.* 348: 2304–2312.

PART II

Tools

6

Genome-Wide Association Studies

The first transformational research strategy to emerge in the wake of the Human Genome Project was genome-wide association studies, or GWAS. The name says it all: association (correlation) of genotypes with traits or disease status is performed throughout the genome. Gene-based association studies have been performed for some time, particularly in model organisms, but the audacious extension from just hundreds of variants at a single gene to hundreds of thousands of variants genome-wide took major advances in theory, technology, computational power, study management, and financing. It is estimated that in the first 5 years of GWAS, from 2007 to 2011, approximately $250 million was spent on these studies worldwide (Visscher et al. 2012). The result has been thousands of robustly validated genetic loci associated with hundreds of diseases and phenotypes, dwarfing the output from the prior two decades of linkage mapping and opening up new areas of biological inquiry that also present novel clinical possibilities.

First GWAS

The conceptual impetus for GWAS was laid in a short paper by Neil Risch and Kathleen Merikangas entitled "The Future of Genetic Studies of Complex Human Disease," published in *Science* in 1996. They considered the case of multiplicative genetic effects and showed that linkage studies in families, which were the standard mode of genetic mapping at the time, were only powerful enough to identify alleles that increase risk considerably more than twofold. This estimate assumed that data from fewer than 1000 families would be available for most diseases. The authors' calculations showed that such data would support identification of alleles with genotype relative risks (GRR; see Chapter 1) as small as 1.5, simply by comparing the ratios of transmission of the two alleles from heterozygous parents to affected offspring. Presciently, they set a genome-wide significance threshold of $p < 5 \times 10^{-8}$ for the proposed family-based association testing (assuming testing of 10 variants for each of 100,000 genes, rather than the 50 for each of 20,000 that is common today, but this was before the

Figure 6.1 Results of the Initial WTCCC Study in 2007. (A) Manhattan plots for five ▶
of the seven diseases in the study. Each plot shows the significance of the association
at each of 360,000 SNP sites ordered by position along the chromosomes, indicated as
the negative logarithm (base 10) of the p-value: highly significant associations generate
peaks (shown in green). (B) Fine-scale association mapping of two of the peaks shows
that for each highly significant SNP, there tends to be a cluster of related associations
due to linkage disequilibrium in the vicinity. Dotted vertical lines show the locations of
recombination hotspots that set the limits of the haplotype block. Black points are actually genotyped, and gray ones are imputed statistically. Tracks beneath each plot show
the locations of candidate genes in the vicinity and regions of sequence conservation
across mammals. (After WTCCC 2007.)

completion of the first human genome sequence!). The obstacle to genetic
mapping was thus not statistical, but technological—namely, in finding
ways to score hundreds of thousands of markers in thousands of individuals with and without a disease or trait.

It took a decade for the technology to catch up. The first positive association was actually reported in 2005, using data for 100,000 SNPs genotyped
in 96 unrelated Caucasian individuals with the eye disease age-related
macular degeneration and just 50 healthy controls (Klein et al. 2005). The
researchers found a block of SNPs in the complement factor H gene, *CFH*,
that defined a haplotype that confers fourfold increased risk to heterozygotes and over sevenfold risk to homozygotes. Two similar follow-up
studies a year later found that homozygotes for another locus, *HTR1*,
have up to 11-fold increased risk for the so-called wet form of the disease (DeWan et al. 2006; Yang et al. 2006). These studies also carried
out additional experiments, including fine mapping of the likely causal
polymorphisms by sequencing the entire genes. They also demonstrated
altered gene expression as a function of the polymorphism, setting a high
standard for future research, while clearly proving the capability of GWAS
to quickly map genetic factors outside the context of family studies.

The landmark study that ushered in the GWAS era for large-scale studies
of many diseases emerged from the Wellcome Trust Case Control Consortium (WTCCC) in June 2007. They reported results for seven diseases, each
with 2000 cases compared with a healthy control sample of 3000 English
people (**Figure 6.1**). Even at the liberal significance threshold of $p < 10^{-5}$, they
were able to report only 23 associations. No genes were found for either
hypertension or bipolar disorder, but one was found for coronary artery
disease, three for type 2 diabetes, four for rheumatoid arthritis, seven for
type 1 diabetes (one of which was shared with rheumatoid arthritis), and
nine for Crohn's disease. The largest effect size was an 18-fold increase in
risk for type 1 diabetes for homozygotes at a particular site in the HLA
complex, but 21 of the significant associations had per-allele risk increases
of less than twofold, the smallest being less than 1.3, corresponding to just
a 30% elevation in the odds that a heterozygote would have the disease.
Soberingly, the study failed to replicate five previously well-established
associations at candidate genes.

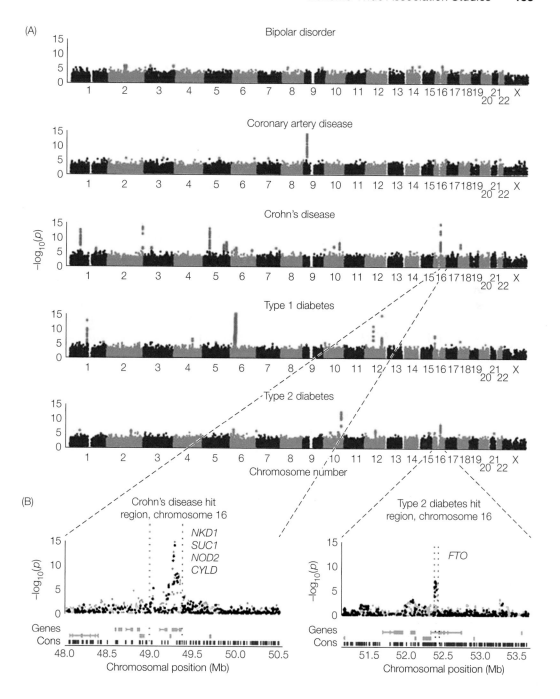

(A)

Bipolar disorder

Coronary artery disease

Crohn's disease

Type 1 diabetes

Type 2 diabetes

Chromosome number

(B)

Crohn's disease hit region, chromosome 16

NKD1
SUC1
NOD2
CYLD

Type 2 diabetes hit region, chromosome 16

FTO

Genes
Cons

Chromosomal position (Mb)

The WTCCC, in this single publication, nevertheless established the standards and expectations that would underlie the rapid adoption of GWAS for all common and many relatively rare diseases. Five and a half years later, over 1500 publications had appeared, reporting almost 9000 associations. These associations are catalogued and constantly updated at www.genome.gov/gwastudies in a freely searchable format that provides researchers with between a handful and over 100 genes for specific diseases and traits (Hindorff et al. 2009). Traits range from cytokine or metabolite levels to working memory performance, and diseases from various forms of dementia to very specific tumor types (though intriguingly, the most common source of morbidity in humans, lower back pain, is not yet represented).

Basic Methodology

The basic idea of GWAS is to perform between 100,000 and 5 million hypothesis tests simultaneously on a large sample of unrelated individuals. Each test is an evaluation of whether one or the other allele at a genetic marker is overrepresented in either people with a disease (**cases**) or people without it (**controls**), or among individuals with high or low values of a quantitative trait. The distribution of test statistics is compared with the distribution expected by chance; that is, relative to the null hypothesis of no association at any marker. The results are plotted as the negative logarithm of the p-value (NLP, on the y axis) against location in the genome (on the x axis) in "Manhattan plots" (see Figure 6.1). Any SNPs that show stronger association than expected are regarded as markers of candidate genes: most will not be the actual causal variant, but they will "tag" nearby variants that could be the true source of the association. Correlation does not necessarily imply causation, so ultimately, functional assays are usually required to provide additional evidence that one or more genes in the vicinity of the tagged variant do actually contribute to variation in the trait. **Box 6.1** provides a simple overview of the whole process in the context of a tiny study of a Mendelian disease.

A popular open-source software platform for performing GWAS is PLINK (Purcell et al. 2007; pngu.mgh.harvard.edu/~purcell/plink), which was developed by Shaun Purcell at Harvard University. It facilitates a wide variety of hypothesis testing options. For the most part, disease associations are evaluated by chi-square or related tests, and continuous trait associations are evaluated by linear regression (Clarke et al. 2011). The simplest case-control association test is an **allelic trend test**, which contrasts the frequencies of the two alleles of a SNP in the cases and in the controls (**Figure 6.2**). **Genotypic trend tests**, which compare the three genotype classes, assume that heterozygotes have risk intermediate to the two homozygotes. A **Cochran-Armitage trend test** is commonly used for this purpose. A straight three-way chi-square (χ^2) test that does not order the genotypes can also be performed, while modifications allow fitting of dominant and recessive models. Most authors generate p-values that can be compared with standard theoretical expectations or evaluated empirically by comparison with permutations of the genotypes against phenotypes. Some researchers prefer Bayesian approaches, which

BOX 6.1
Summary of an Association Study

Because the concepts described in the text may be daunting, especially since they depend on quite a few statistical concepts that may be unfamiliar to students, this box provides a simple outline of the strategy underlying GWAS. In essence, we are searching for highly significant correlations, which are called GWAS signals, or "hits."

Genome-wide association studies start with the gathering of a large number of samples, usually from at least 1000 individuals with a disease and a matched set of healthy individuals who serve as controls. After the participant provides informed consent under the auspices of an IRB (institutional review board; see Box 5.1), peripheral blood or saliva samples are taken, from which DNA is extracted. Each sample is then sent to a genomics facility, which hybridizes it to a genotyping chip so that hundreds of thousands of polymorphic genotypes can be scored. Comprehensive arrays with millions of SNPs, such as those made by Illumina or Affymetrix (see Figure 6.10), are expensive, so if funds are limited, options exist for using a more focused array that either targets specific sets of genes or covers most common variants. The genotypes are automatically identified ("called") by software associated with the arrays. Once this has been done, quality control

metrics are applied, and some manual screening may be required. Bioinformatic methods are then used to "impute" millions of other SNPs (including the ones not represented in the arrays that were used; see Figure 6.7), and then statistical genetics is used to perform the association study. Finally, biologists attempt to interpret the findings, identifying the most likely gene and/or variant that is responsible for the association. The workflow is to (1) collect samples, (2) extract genomic DNA, (3) hybridize DNA to the array, (4) call the genotypes, (5) impute additional SNPs, (6) perform statistical analyses, and (7) interpret the findings.

The basic principles underlying an association study are illustrated below for a set of seven hypothetical SNPs from ten individuals, five with a disease (cases) and five without (controls see figure).

If we look carefully at the relationships between the SNPs and disease status, we will see that for the first three SNPs, there is no relationship stronger than we would expect by chance. For SNP rs1, three of the cases have an A and two a T, while for the controls, it is the other way around, but given that there are five of each type of allele, we would expect a slight deviation from a 50:50 ratio. For rs3,

	SNP							
Individual	rs1	rs2	rs3	rs4	rs5	rs6	rs7	Status
ID001	A	T	C	C	A	G	A	Case
ID002	A	C	T	C	A	G	A	Case
ID003	A	C	C	T	G	–	A	Control
ID004	T	T	C	T	A	G	A	Control
ID005	A	T	C	T	G	G	A	Control
ID006	A	C	C	C	A	–	A	Case
ID007	T	C	T	T	G	–	A	Control
ID008	T	T	C	C	A	G	A	Case
ID009	T	C	T	C	A	–	G	Case
ID010	T	T	T	T	G	–	A	Control

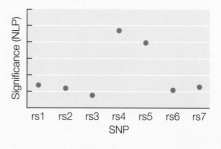

(Continued)

BOX 6.1 (continued)

three Cs are in cases and three in controls, and similarly, two Ts are in cases and two are in controls, so there is no association at all. By contrast, for rs4, there is perfect segregation: all five Ts are in all five controls, and all five Cs are in all five cases, so it appears that C is associated with the disease. The next SNP, rs5, is almost the same, since it is in LD with rs4, but one control individual has an A instead of a G, so the evidence of association is slightly weaker. This difference is clear in the Manhattan plot on the right, since the negative log p-value (NLP) for rs5 is slightly smaller than the peak value at rs4. There is an indel polymorphism at rs6, but it is not associated with the disease. The final SNP is a rare variant, and although the one instance of a G is in a case, this is not enough to give a significant test statistic, as many individuals would be required to establish an association. It is theoretically possible that the G "causes" the disease in this person, but there would have to be other causes in the other four cases.

In reality, with thousands of individuals of each geno-type, highly significant associations can be observed even if there is a difference in allele frequencies

between cases and controls of only a few percent-age points. The situation in our mock example is what we might expect of a Mendelian trait with complete penetrance: all the affected individuals have one genotype in common (C at rs4). Similarly, for continuous traits, we might expect enrichment of one genotype in individuals with either high or low values of the trait. If the effect is large, then genotyp-ing individuals at the extremes of the trait distribution can be an efficient way to perform association stud-ies, since the enrichment will be strong and it may be possible to use a small sample.

The power of a GWAS is a function of the number of individuals of each class that are sampled, the effect size of the allele, the frequency of the allele, and the variation within each genotype class. Typically, a few thousand cases are required to see effects in which the allele increases the risk by more than 1.5, whereas hundreds of thousands of individuals may be needed to consistently see an effect that increases the risk by 1.1 (just 10%). A GWAS power calculator can be found at pngu.mgh.harvard.edu/~purcell/gpc/

generate Bayes factors summarizing the probability of association given a set of prior expectations.

Q-Q plots are a convenient way to visualize the distribution of test statistics by plotting the observed versus expected range of values on the y and x axes, respectively (**Figure 6.3**). Either the chi-square value or the NLP can be used. NLP are easy to think about, since larger val-ues equate to smaller p and higher statistical confidence: an NLP of 2 is $p = 0.01$, while an NLP of 8 is $p = 10^{-8}$. On a Q-Q plot, there is some expected sampling variance for large test statistics, so a gray area indi-cating the expected range of values under the null hypothesis is plotted. The observed curve typically lies slightly above this region throughout the range and deviates more toward higher-than-expected values as NLP (or chi-square) increases. The deviation is due to either an excess of weak associations or technical artifacts, most notably population stratification, which must be corrected for. Points lying well above the curve of the gray area with very small p-values (large NLP) correspond to GWAS "hits."

Individual sample number	Genotype at a particular SNP	Disease status
S0012323	*AA*	Case
S0012324	*AA*	Case
S0012543	*AG*	Case
S0012666	*GG*	Case
S0012687	*AG*	Case
S0034301	*GG*	Control
S0034310	*GG*	Control
S0034533	*AA*	Control
S0034564	*AG*	Control
S0034662	*GG*	Control
:	:	:

Quality control:
- 98% genotype calles in 98% of individuals
- Impute more than 2.5 million genotypes from 1000 Genomes Project
- Check for Hardy-Weinberg equilibrium
- Control for population structure

Allelic trend

Status	*A*	*G*
Case	8680	31,320
Control	8000	32,000

Chi-square test: $p = 1.9 \times 10^{-17}$
Odds ratio (*G:A*) = 0.90

Genotypic trend

Status	*AA*	*AG*	*GG*
Case	940	6800	12,260
Control	800	6400	12,800

Cochran-Armitage trend test: $p = 3.3 \times 10^{-9}$
Odds ratio (*GG:AA*) = 0.815

Figure 6.2 Allelic and Genotypic Trend Tests.   Once samples have been genotyped, a series of quality control steps is applied to the data, and many more SNPs may be imputed. From these data, counts of individuals in the case and control groups are determined, from which it is possible to calculate the frequencies of the two alleles at any SNP in the cases and controls as well. (In this example, there are two As in AA homozygotes and 1 in heterozygotes; you should confirm that $[A/(A + G)]^2 = [AA/(AA + AG + GG)]$ in both sets of individuals.) The allelic trend test is simply a chi-square test with 1 degree of freedom on the 2×2 contingency table shown at the bottom left. A straight genotypic trend test is a chi-square test with 2 degrees of freedom on a 3×2 contingency table (bottom right), but this test must be adjusted for the expectation that the heterozygotes are intermediate under an additive model, which is usually done with Cochran-Armitage trend test using 1 degree of freedom. Further adjustments for non-additivity can increase power. The odds ratios for the G allele relative to the A, and for the GG homozygotes relative to the AA homozygotes, are computed by formulating the ratio of the diagonal products (e.g., $8000 \times 31,320/8680 \times 32,000 = 0.90$). In this case, each G allele protects individuals against the disease, lowering their risk by approximately 1.1-fold. Note that although the allelic trend test may have a smaller p-value, the genotypic trend test is usually preferred since it incorporates more biological information.

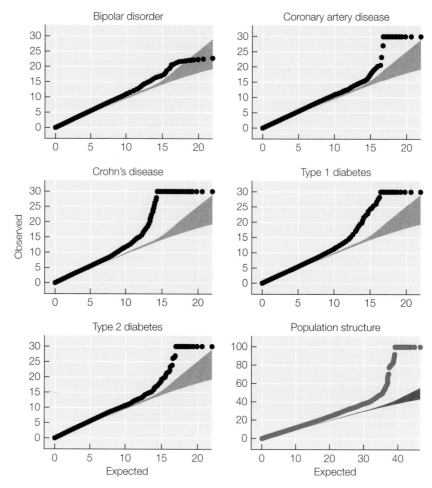

Figure 6.3 Q-Q Plots. These plots from a study performed in the United Kingdom show the overall distribution of test statistics (here Cochran-Armitage χ^2 values for the individuals with the indicated diseases, and population structure in the controls) and the expected values in the absence of any association. The highly significant values stretch above the expected region, truncated at 30 (they would extend well above the plot, otherwise). In each case, the observed values begin to depart subtly from the expected even at small χ^2 values, possibly indicating an excess of alleles of very small effect. For bipolar disorder, no SNPs are significant genome-wide, as none of them exceed the high significance threshold. The population structure curve at bottom right is on a different scale, but clearly shows how much subtle deviation in allele frequencies among populations exists even within the United Kingdom. (After WTCCC 2007.)

A critical question is what constitutes genome-wide significance—that is, what statistical threshold must be exceeded in order to conclude that a SNP contributes to a disease or trait. The multiple comparison problem that arises when hundreds of thousands of variants are tested one after the other must be carefully considered. Generally, statisticians are comfortable with a false positive rate, α, of 5%, which means that they are prepared to incorrectly reject the null hypothesis once in every 20 trials. If I perform a single test, the chance that I will do this is just 1 in 20, so $p = 0.05$. If I perform 20 tests, then, the chances are that one of them will yield a false positive at this level of significance. If I perform 100 tests, then 5 of them will produce p-values of less than 0.05 by chance and generally be false positives. Correspondingly, if we perform 100,000 tests, then 5000 of these "type 1 errors" will occur. Therefore, it is essential to make an adjustment. This is classically done with a Bonferroni correction, by which the study-wide significance is adjusted to α divided by the number of tests. Assuming 1 million tests, the genome-wide significance level is $0.05/10^6 = 5 \times 10^{-8}$, which conservatively says that any p-value smaller than this is expected to occur only once in every 20 random permutations of the entire study.

Now that GWAS has become feasible, this standard of evidence is generally applied to all association studies, even candidate gene studies in which only 50 SNPs are evaluated, since in theory the investigators should have surveyed the entire genome if they desired an unbiased result. If a genotyping chip has just 100,000 markers, we might argue that the threshold should be 5×10^{-7}, or if it has 2.5 million markers (as the most current ones do), then perhaps it should be 2×10^{-8}. The reason why 5×10^{-8} is preferred is because extensive investigations of the amount of linkage disequilibrium (LD) in European and Asian genomes suggest that it takes approximately 1 million SNPs to capture most of the variation in the genome (Dudbridge and Gusnanto 2008). This is thus the number of independent tests it takes to provide genome-wide coverage. Investigators now typically add an imputation step (described later in this chapter) that allows them to assess the most probable genotypes at untyped SNPs, but the GWAS threshold does not change, even though five times as many statistical tests are evaluated than with raw data.

In practice, GWAS is often performed in two or more phases. The gold standard for significance is $p < 5 \times 10^{-8}$ with replication, preferably on an independent cohort but possibly in a second set of individuals from the same cohort. It is not required that both phases exceed the GWAS threshold independently, but together they must, and they should show the effect in the same direction, such that the combined p-value is smaller than that of either phase alone. A common strategy is to analyze the first thousand or so samples, choose the several hundred most strongly associated SNPs (say, with $p < 10^{-4}$), and genotype these with a custom, less expensive genotyping array in a larger sample to identify those that are truly genome-wide significant across the combined sample of several thousand cases and matched controls.

Meta-analysis refers to the practice of combining p-values from multiple studies, whereas **mega-analysis** is doing the joint analysis on all the individuals from multiple studies (which is usually logistically very difficult to do). The results of a meta-analysis can be visualized in the form of a forest plot (**Figure 6.4**, right). The **point estimate** and 95% confidence interval for the effect of each allele is plotted on the x axis, with each study aligned one above the other. The meta-analysis p-value can be thought of as an average p-value weighted by the study sizes and boosted by the combined sample size. Note that there can be quite a range of effect size estimates, and in some studies the effect size will not be significantly different from zero (a GRR of 1, dashed vertical line). This is because GWAS is not powered to detect small effects in the range of 10% to 20% (GRR 1.1–1.2) with regularity, but if the effect is consistently in the same direction in multiple studies, then it can usually be regarded as true. The final effect size estimate obtained by analyzing multiple studies is always a better estimate than that obtained in the first study, which tends to be inflated by the "winner's curse." This inflation occurs because in order to cross the stringent GWAS threshold, SNPs tend to get the benefit of sampling variance in addition to the true effect. Subsequent studies can be biased in the other direction by random sampling, but according to the central limit theorem in statistics, the combined analysis of many studies will converge on the best estimate.

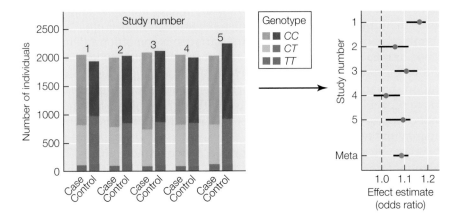

Figure 6.4 Winner's Curse and Forest Plots. The gold standard for GWAS requires independent replication of the findings obtained in the initial study. The bar graphs show the observed genotype frequencies in cases and controls for a series of studies of the same disease, indicating how the estimates include a stochastic component. These then give rise to the odds ratios reported in the forest plot on the right. Typically, the observed odds ratio is greatest in the first study, since the significant SNPs get the benefit of sampling variance in the same direction as the true allele effect. If four follow-up studies are conducted on different samples or cohorts, it is not unusual for one of them (study 4) to be marginally significant and the others to have intermediate significance. Meta-analysis of the statistics converges on the best estimate of the true odds ratio (the brown estimate), which is also more accurate than each of the individual contributing studies.

Population Structure

It is very easy to be misled by GWAS if there is hidden population structure in the sample (**Figure 6.5**). As a trivial and ridiculous example, suppose you were looking for genes that influence religious affiliation in a mixed sample of Americans of Irish and Iranian descent. Any alleles that are at higher frequencies in the Irish-Americans will naturally tend to be more common in Catholics than in Muslims, and hence will give a positive association test result. However, this does not mean that they cause the difference in religious affiliation. They are simply correlated with it because of population structure and environmental and cultural effects. Wherever there is a difference in allele frequency as well as in the prevalence of a disease or

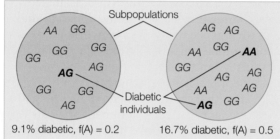

Blue subpopulation

	AA	AG	GG
Case	80	640	1280
Control	800	6400	12,800
Case/control	0.1	0.1	0.1

Red subpopulation

	AA	AG	GG
Case	200	400	200
Control	1000	2000	1000
Case/control	0.2	0.2	0.2

Total population

	AA	AG	GG
Case	280	1040	1480
Control	1800	8400	13,800
Case/control	0.155	0.124	0.107

Odds ratio $(A:G) = 1.2$ $\qquad$ $p = 10^{-8}$

Figure 6.5 Population Structure. If a population consists of two subpopulations, which may be unrecognized, then despite the absence of any association in either subpopulation, there may appear to be a spurious correlation between genotype and disease risk. In this hypothetical example, 9.1% of the individuals in the blue population have diabetes irrespective of genotype, and there is a minor (A) allele frequency of 0.2; in the red population, 16.7% of the individuals have diabetes, and the A allele has a frequency of 0.5. In the total population (yellow), the A alleles are more likely to be observed in people with diabetes, which gives an odds ratio of 1.2 and genome-wide significance, yet the odds ratios in the two sub-populations are 1.0.

trait between two groups, false associations can arise, so those differences must be accounted for.

The most straightforward adjustment for such differences is known as **genomic control** (**GC**; Devlin and Roeder 1999; Devlin et al. 2004). This adjustment is made by downweighting all p-values by a factor (called λ_{gc}) such that the overall distribution of nulls is close to random expectations; any small p-values that remain after the adjustment may be true positives. If the population structure is known, then efforts to ensure that the two or more ethnicities and environmental biases are equally distributed in the cases and controls should ensure that λ_{gc} is close to 1 and there is little bias. This may mean removing individuals from the study, but the loss in sample size is offset by the gain in homogeneity of sample allele frequencies genome-wide. Minimization of population structure is also one reason why GWAS is often restricted to single population groups, such as Caucasians, or even more stringently restricted by ensuring that both sets of grandparents of each participant were also Caucasian. However, as studies have become very large, it has also become apparent that even the very small amount of divergence that occurs between, for example, people in the north and south of the Netherlands, or in the east and west of Iceland, can inflate test statistics and needs to be controlled for.

A more nuanced approach is to measure the population structure itself and include estimators that capture it in the statistical model. Data reduction methods known as principal component analysis (PCA) or multi-dimensional scaling (MDS) are applied to the genotypes to identify population structure that may not be obvious or to give a more objective measure than self-reports. These methods were introduced in Chapter 3, where it was shown that the first two or three principal components (PCs) can be used to reliably distinguish people according to their African, Asian, or European ancestry as well as to detect finer-scale population structure (see Figure 3.3B). Inclusion of the PC values as continuous covariates, along with the genotypes, in a logistic regression model with case versus control status as the dependent variable has been shown to effectively control for population structure without the need for the uniform λ_{gc} adjustment (Price et al. 2006). Similarly, so-called mixture models can include the population structure adjustment for continuous traits (Zhu et al. 2002; Yu et al. 2005). Analogous procedures can also be used to correct for small levels of relatedness among individuals in the sample.

So far, we have discussed association studies that are performed on unrelated individuals drawn from the population at large. GWAS can also be performed in the context of families, in which situation population structure is not an issue. This design utilizes family-based association tests for enrichment of alleles in sibling pairs affected by a disease or, more commonly, transmission disequilibrium tests (TDTs). A TDT asks whether either of the alleles found in a heterozygous parent is more commonly transmitted than the other allele to one affected offspring per family, and there is no need to study controls (Spielman et al. 1993). Thus, for a common variant, if 1000 affected children are sampled along with both of their biological parents, 500 of whom are heterozygous, then under random assortment, the children of those heterozygotes should carry a total of 250 transmitted copies of each allele (that is, half the alleles). If in fact 300 transmissions of one allele are observed, it can be concluded that it is significantly associated with the disease. This is a powerful approach, but it is more difficult to perform, since obtaining biological samples from parents can present logistic difficulties.

Fine Mapping

In order for GWAS to work, there must be strong linkage disequilibrium between the tagging SNPs included in the genotyping panel and the causal variants. It turns out that humans are almost ideal from this perspective, because even spacing of 500,000 SNPs ensures that there is at least one SNP approximately every 10 kb throughout the genome. Since haplotype blocks of SNPs that are in strong LD tend to be between 20 and 100 kb long (recall Figure 3.4), each of these blocks is represented, on average, by 5–10 SNPs. Consequently, there is a strong probability that common causal polymorphisms located within haplotype blocks will be captured by at least one SNP. Most are captured by multiple SNPs, which form a tight cluster on a fine-scale association map (**Figure 6.6**).

The initial aim of the International HapMap Project was to generate a fine-scale map of haplotype blocks. These can be browsed at the hapmap. ncbi.nlm.nih.gov website. For the most part, haplotype blocks are similar among human populations. Consequently, any disease-associated SNP that arose prior to the spread of humans out of Africa should be tagged by common polymorphisms in its haplotype block in any population. The limits of these blocks are established by hotspots of recombination, which are marked by the black peaks in the recombination rate profile that is superimposed on the fine-scale association maps in Figure 6.6. It is estimated that over 90% of all common polymorphisms are tagged by (that is, in linkage disequilibrium with) the SNPs on the two major commercial genotyping platforms (Illumina and Affymetrix, described later in this chapter), at least in Europeans and Asians. Since there is considerably more polymorphism, and LD is considerably less extensive, in Africans, GWAS is less efficient in Africans and requires additional genotyping depth. When successful, however, cross-population studies can provide higher-resolution mapping of the causal site(s) because the extent of the LD interval is reduced.

Once an association peak has been detected, there are two approaches to further fine-scale mapping of the likely causal variants. One is to "impute" additional variants onto the map, and the other is to sequence the region and then use a custom genotyping chip to saturate the region with dense coverage in a larger sample of individuals. **Imputation** is the process of estimating the most likely genotype at untyped SNPs based on the extent of LD in those SNPs in a reference population, as shown

Figure 6.6 Complex Association. These fine-scale association maps show an example of two independent clusters of associations with height in the vicinity of the growth hormone receptor gene *GHSR* at chromosomal location 3q26.31. The lead SNP, rs572169, is in linkage disequilibrium with 25 other variants in a 200 kb window bounded by two hotspots of recombination (the tallest black vertical spikes, which show recombination rates). After fitting the rs572169 SNP in the regression, the significance of the red SNPs was reduced to NLP < 1, but another set of 30 or so blue SNPs in the adjacent interval covering the *FNDC3B* gene remain highly significant, indicating that they capture an an independent association. (After Lango Allen et al. 2010.)

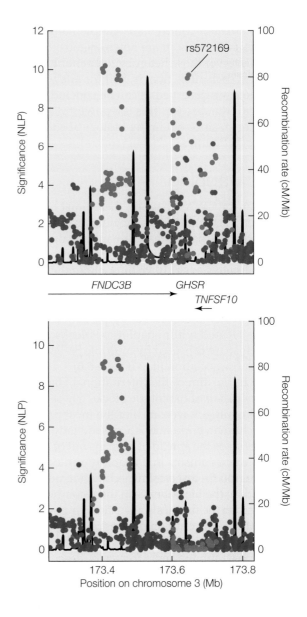

in **Figure 6.7**. Once this is done, association tests are performed just as for directly typed SNPs, and in some cases these tests detect association signals that are even stronger than the observed ones. Open-source software for imputation includes Beagle and Impute-2, which adopt slightly different strategies, the details of which are beyond the scope of this text (Browning and Browning 2009; Howie et al. 2009). The reference set of genotypes is usually from the 1000 Genomes Project (www.1000genomes.org), which now includes a catalog of fully sequenced genomes from

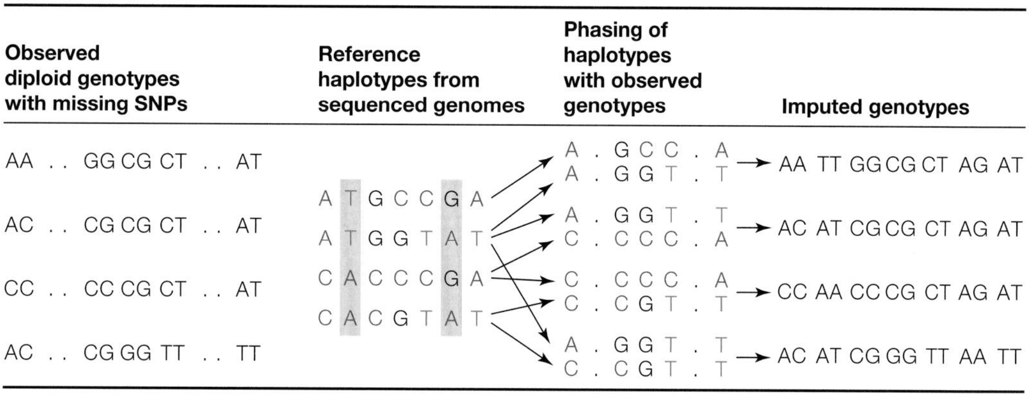

Observed diploid genotypes with missing SNPs	Reference haplotypes from sequenced genomes	Phasing of haplotypes with observed genotypes	Imputed genotypes
AA . . GG CG CT . . AT		A . G C C . A A . G G T . T	AA TT GG CG CT AG AT
AC . . CG CG CT . . AT	A T G C C G A A T G G T A T	A . G G T . T C . C C C . A	AC AT CG CG CT AG AT
CC . . CC CG CT . . AT	C A C C C G A C A C G T A T	C . C C C . A C . C G T . T	CC AA CC CG CT AG AT
AC . . CG GG TT . . TT		A . G G T . T C . C G T . T	AC AT CG GG TT AA TT

Figure 6.7 Phasing and Imputation. Genotype data is generally ascertained as a series of independent genotypes (AA, AC, CC, etc.) rather than as haplotypes. The first step in imputation is to phase the haplotypes; that is, to estimate the series of haploid genotypes on each of the two contributing chromosomes observed in a diploid individual. This step is straightforward for pairs of homozygotes or for a homozygote and heterozygote pair (AA GC could only be a combination of AG and AC haplotypes), but double heterozygotes cannot be phased directly (CG CT could either be CC and GT or CT and GC). Maximum likelihood methods are used to optimize the fit of the observed genotypes to the frequency and identity of haplotypes in a reference population. Once this is done, the identity of missing genotypes (. .) in the observed sequence is "imputed" by reference to the sequences in the reference panel (the genotypes under the gray bars). These imputed genotypes are then used for association studies.

diverse populations. Alternatively, investigators can sequence the candidate region themselves in a subset of their sample, identifying novel variants that can then be imputed into the larger sample. For high confidence, newly identified SNPs should then be directly typed with a targeted assay to confirm the accuracy of the imputation.

Custom genotyping methods are described later in this chapter, but mention should be made of two resources that have seen widespread usage in the context of metabolic and immune disease. These resources are the "metabochip" (Voight et al. 2012) and "immunochip" (Parkes et al. 2013), respectively, each of which covers several hundred genes that experts feel are likely to be associated with common diseases in the two domains, including all known SNPs with a minor allele frequency as small as 1%. For a fraction of the cost of whole-genome genotyping, these chips support detailed interrogation of the best candidates in hundreds of thousands of individuals from most human populations. They are particularly well suited for the detection of associations with so-called Goldilocks variants, those with modest effect sizes in the range of 1.2- to 1.5-fold increased risk but frequencies of less than 2%. Such variants are for the most part not included on the standard genotyping arrays, and they are difficult to impute, for statistical reasons.

Causal Inference

Whenever one SNP at a locus is associated with a disease or trait, it is reasonable to assume that additional SNPs in the same gene may also be involved. In theory, there are two alternative interpretations of associations in a haplotype block: either that all the variants contribute a small amount and, combined, they yield a strong signal; or that one SNP is fully responsible while the others in the block are associated only statistically due to their LD with the causal SNP. In practice, the parsimony principle is usually adopted, and it is assumed that a single SNP is responsible for most, if not all, of the effect. Another possibility is that one SNP is responsible for association with a particular trait, while a different SNP is responsible for association with another trait: in this case, the locus would be pleiotropic, but the individual causal variants would be unique.

A procedure known as **conditional association** is used to distinguish the effects of different variants at a single locus. The idea is to fit the SNP with the smallest p-value as a term in the linear regression model, then evaluate each of the other SNPs conditionally on the already discovered genotype. This procedure is essentially the same as including any other covariate, such as gender or ethnicity, in the model. By taking into account the known effect of the first discovered SNP, we can ask whether the other SNPs can explain even more of the trait. Similarly, for continuous traits, instead of fitting both SNPs jointly into a single model, we can run a second model in which each SNP is evaluated for its association with the "residual" variance left after fitting the effect of the first SNP. The result of either procedure is that SNPs that are in only partial LD with the main causal variant are sometimes seen to remain highly significant even after accounting for the main effect. This finding implies that these variants contribute secondary associations—namely, that they are also contributing to the disease trait. Figure 6.6 shows the example of one of the 19 (out of 180) height-associated genes identified by the GIANT consortium that have multiple associations with height (Lango Allen et al. 2010). Such a situation in which a complex series of effects within a locus is observed is often referred to as **allelic heterogeneity**.

A slightly different situation, known as **synthetic association** (Figure 6.8), is observed when the main effect of the common variant identified in a GWAS is actually found to be due to multiple associations with rare variants that have much larger effects and happen to be in LD with the common variant (Dickson et al. 2010). This phenomenon is difficult to detect because the individual rare variants do not affect enough people to produce a GWAS signal on their own, and they become apparent only if the investigator deliberately scans for them. Furthermore, the rare variants can plausibly act over several megabases of DNA and still be in LD with the common tagging SNP, which makes their identification even more difficult. Without full sequence coverage of the region, it is not possible to systematically scan for synthetic associations. There is considerable debate as to how often common variant effects are actually due to LD with rare variants: some

Haplotype							BMI
C	C	G	A	A	C		21.4
C	C	G	A	A	C		30.3
C	C	G	A	A	C		28.2
C	C	G	A	A	C		23.8
C	C	G	A	A	C		33.7
C	C	G	A	A	C		27.9
C	C	G	A	A	C		22.0
C	C	G	A	A	C		24.4
C	C	G	A	A	C		30.8
C	C	G	A	A	C		29.2
C	C	G	A	A	C		27.7
C	C	G	A	A	C		25.1
G	T	A	A	T	C		21.4
G	T	A	A	T	C		30.3
G	T	A	G	T	C		**40.6**
G	T	A	A	T	C		23.8
G	T	A	A	T	C		33.7
G	T	A	A	T	C		27.9
G	T	A	A	T	C		22.0
G	T	A	A	T	C		24.4
G	T	A	A	T	C		30.8
G	T	A	A	T	G		**36.2**
G	T	A	A	T	C		27.7
G	T	A	A	T	C		25.1

Figure 6.8 Synthetic Association. Synthetic association occurs where the apparent association of a disease or trait with a common variant or haplotype is actually driven by one or a small number of rare variants that happen to be enriched on that haplotype. In this hypothetical example, the actual distribution of body mass index (BMI) phenotypes is the same for the blue and orange haplotypes, yet the mean is higher for the orange haplotype due to the strong effects of two rare (green) variants. Without those variants, the orange haplotype actually shows a slightly reduced effect on BMI.

Blue haplotype mean BMI = 27.0
Orange haplotype mean BMI = 28.7
Haplotype with green allele mean BMI = 38.4
Orange without green mean BMI = 26.7

authors claim that such instances are likely to be prevalent, while others doubt on theoretical grounds that they can explain more than a small fraction of common effects (Wray et al. 2011).

Ultimately, the inference of causation depends on experimental verification that a particular SNP is causal. Statistical evidence alone supports correlation and may or may not imply causation. Examples of studies that demonstrate causal mechanisms will be discussed in the context of specific diseases in Part III of this book. Methods used for this purpose include assays of protein function (enzyme activity, binding assays), cellular activity (overexpression or knockdown of the gene in cell culture followed by physiological assays), gene expression (differential activation of a reporter gene), and targeted alteration of the suspected nucleotide (by homologous recombination or TALEN and CRISP-R/Cas9 mediated genome engineering; Segal and Meckler 2013). In classical genetics, the gold standard for proof of function is reversion analysis, in which the mutation is converted back to the normal state and the phenotype shown to revert as well. Obviously such manipulations are not possible in living humans, but model organisms and cell lines provide ample opportunity for less direct, but no less compelling, demonstration of function. Even though the effect of a single SNP on disease susceptibility may be modest, its effect on gene activity can be quite large, and in many cases it is sufficient to show that the gene itself is related to the trait without needing to go so far as to show how the single nucleotide acts mechanistically.

Interaction Effects

GWAS is explicitly designed to capture common variant contributions to narrow sense heritability; that is, it usually starts with an assumption of additivity. Under broad sense heritability models, the genetic architecture of traits should also be affected by interaction effects, either between two or more genotypes (G × G interaction, or epistasis) or between genotypes and an environmental variable (G × E interaction). For the most part, however, GWAS studies have not provided strong support for departures from additivity. There are a number of possible reasons for this: (1) the allele effects may be truly independent of one another and environmental variables, (2) the effects may be relatively small compared with the main effects and statistical power is lower (since there is an even larger multiple comparison burden to adjust for), (3) interaction effects may be heterogeneous at the level of individuals and so do not appear as strong average effects, or (4) it may be particularly difficult to identify the correct environmental variable or alternative genotype to model. Some solid theory argues against G × G interaction between common variants being a major source of genetic variance in general (Hill et al. 2008). On the other hand, model organism geneticists typically do find it when they look for it specifically in controlled crosses. Since environmental effects are often confounded with population structure, they may be discarded artificially during the analysis, and since the interaction increases variability, it can reduce statistical power, which also makes it hard to detect G × E interaction.

Nevertheless, many authors continue to argue that interaction effects are likely to be ubiquitous. New statistical methods may be required to detect epistasis genome-wide (Cantor et al. 2010; Moore et al. 2006), and studies oriented to testing genes in specific physiological or biochemical pathways may reduce the burden of proof. A study using graphics processor units (GPUs) to evaluate over 10^{17} pairwise G × G interactions influencing the abundance of 7400 transcripts in 846 peripheral blood samples found 501 significant epistatic effects influencing 238 genes, with significant replication in an independent cohort (Hemani et al. 2014). Although the combined contribution of these interactions to the overall variance in this study was small, the impact at the level of particular transcripts in specific individuals can be large. Similarly, concerted efforts are under way to monitor G × E interactions, notably through the GENEVA Consortium of 16 ongoing studies using environmental covariates (Cornelis et al. 2010). To date, only a handful of robust environmental modifications of genotypic effects on disease have been reported, and G × E interaction effects were found to make only a modest contribution to the genetic regulation of transcript abundance in peripheral blood (Idaghdour et al. 2009).

Polygenic Risk and SNP-Based Heritability

There are, in general terms, three major reasons for conducting GWAS. The first is to find the specific genes and variants that contribute to disease risk or phenotypic variance. The second is to understand more about the genetic architecture of a trait: how many genes contribute, are they mostly rare or common, and do they interact with one another and with the environment? The third is to develop genetic predictors that may have clinical utility by identifying at risk individuals and focusing resources or suggesting appropriate therapies. For the first objective, it is essential that strict standards of evidence—namely, genome-wide significance thresholds—be adopted. For the second and third, however, it is not necessary that every SNP have genome-wide significance. Much can be learned by relaxing the threshold and thereby gaining from the inclusion of more variants than those with very robust evidence of association, while accepting that some of these variants will be false positives. There is always a trade-off between losing false negatives and including false positives. This trade-off is particularly apparent in GWAS, since the effect sizes are so small that the studies are not powered to achieve significance for the majority of causal variants, which typically hide among the thousands of tests with p-values in the range of 10^{-3} to 10^{-8}.

Some intuition into the genetic architecture of traits can be gained by extrapolating from initial discoveries (Park et al. 2010). Given the effect sizes observed and the size and hence statistical power of a study, it is possible to derive a density distribution of likely effect sizes that would explain the observed heritability. For example, So et al. (2011) surveyed GWAS findings for ten common diseases and found that the median amount of variance explained by individual discovered SNPs was just 0.25%, and that collectively, the discovered SNPs explained slightly less than 10% of the variance in liability for each disease. Nevertheless, given the power of the studies, it could be concluded that samples of hundreds of thousands of cases would substantially increase the amount of variance explained. This conclusion has been borne out for height and BMI, and ever larger studies, even for psychiatric diseases that initially found no hits, are now reporting more than 50 small-effect variants. Simplistic extrapolation to a median effect size of just 0.1% suggests that it would take 500 variants to explain all the genetic contribution to a disease with a heritability of 50%. **Figure 6.9A** illustrates the confidence intervals around current estimates of the number of common variants, and the proportion of the variance they would explain, for five diseases (Ripke et al. 2013).

The International Schizophrenia Consortium (2009) introduced the notion of using polygenic risk scores that include all variants at nominal significance levels. They identified 74,062 independent common SNPs genome-wide and took the 37,655 of these that showed $p < 0.5$ in a sample of 2176 males with schizophrenia and 1642 male controls. From these data, they created a "risk score" based on how many of the variants a man has, weighted by the effect size, and showed that the same SNPs in women

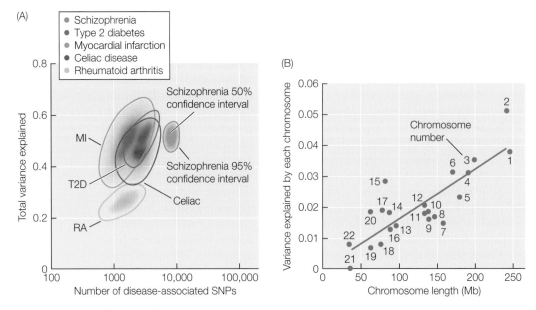

Figure 6.9 Multivariate Association. (A) Extrapolation of observed allele effect sizes led Ripke et al. (2013) to estimate the total number of common variants that explain the observed heritability for five diseases: rheumatoid arthritis (RA), celiac disease, type 2 diabetes (T2D), myocardial infarction (MI), and schizophrenia. The core colored area represents the 50% confidence interval of each estimate, and the outer circles show the 95% confidence intervals. For example, over 8300 SNPs are predicted to contribute to half the liability for schizophrenia. (B) Genomic partitioning allows estimation of the relative contributions of subsets of loci—in this case, of each chromosome—to the total variance for height. There is a highly significant linear regression of variance explained on chromosome length. (B, after Yang et al. 2011.)

produced a score that was highly correlated ($p \sim 10^{-18}$) with schizophrenia in the 1146 women with the disease from the same study. Furthermore, a score derived from all the cases, male and female, was also very strongly associated with the disease in two other studies of Caucasians, as were scores derived from fewer SNPs at less liberal thresholds, such as $p < 0.1$. Even though the amount of variance explained was just 3% in each replication, showing that the scores are not predictive, the approach nevertheless demonstrated that there is hidden heritability throughout the genome. Intriguingly, the schizophrenia scores are also associated with bipolar depression risk, albeit to a lesser degree, but not with metabolic or immune diseases, suggesting polygenic association across multiple neurological disorders (Lee et al. 2012). Polygenic evaluation actually has a long history in animal breeding, where sophisticated Bayesian "genomic selection" methods have been used to select sires or dams for breeding purposes for several years.

Genome-wide SNP-based heritability estimation has since been formalized with the introduction of the GCTA software (www.complex-traitgenomics.com/software/gcta/) for joint assessment of all variants simultaneously (Yang et al. 2011). Whereas the above approach simply adds together the effects of thousands of SNPs identified individually, GCTA evaluates them simultaneously, as if a single model included tens of thousands of genotypes evaluated together. For technical and statistical reasons, that is not what is actually done, and instead a more accurate approximation is performed. The idea is actually to simply regress genotypic similarity on phenotypic similarity, which is the essence of heritability (see Figure 1.10). Genotypic similarity is computed from the full matrix of genotypes: individuals who share more alleles are, in a sense, closer relatives (but the analysis explicitly excludes pairs known to be closer than the equivalent of third cousins). Phenotypic similarity is computed from the distance between the phenotypes or from a threshold liability model for disease. With enough data—at least several thousand individuals' worth—it is possible to show that common variants actually capture as much as half of the phenotypic variance for some traits. In the case of height, further subtle adjustments implied that almost all the heritability can be attributed to common variants, even though the identities of the vast majority of them are unknown. In fact, this approach also allows partitioning of the combined effects of subsets of SNPs, such as those expressed in a relevant tissue, or by chromosome (**Figure 6.9B**).

Given a set of genotypes that contribute to a trait, it is possible to generate genotypic predictors. Predictors built with thousands of variants still explain only a fraction of the variance, however, because they include false positives that add noise, because the effect size estimates are usually very imprecise, and because of incomplete LD between causal variants and the tagging SNPs included in the score. Predictors based on just dozens of variants capture only a minority of the variance: intuitively, how well can you predict something if you have only enough data to explain 10% of the differences among individuals? These issues will be discussed in Part III, but suffice it to say that part of the solution is that the goal is not so much prediction as classification. Individuals who have risk scores at either tail of the population distribution do typically differ by as much as an order of magnitude in risk. This does not mean that they are guaranteed not to get the disease, or that they will eventually do so, but it does suggest that genetics can help identify which conditions each person is relatively susceptible to.

The simplest risk scores are summations of the number of alleles a person has that increase risk (Dudbridge 2013). Such **allelic sum scores** can be modified by weighting the effect size of each allele: variants that have twice the magnitude of effect as the others are in essence tallied twice. Weighted allelic sums can be further modified to account for the prevalence of the allele in the population, and the results, in turn, should be adjusted for differences in allele frequencies among populations. Heterozygotes in a population in which the risk allele is the minor allele may be at higher risk than the population average, whereas they will be relatively protected if the risk allele is the major one. Another approach to risk evaluation is the formulation of odds ratios that are multiplied together with the baseline lifetime probability of disease for the reference population (Morgan et al. 2010). Yet more methods use machine learning to weight combinations of alleles that statistically interact and so generate a score that explains more of the variance (Kruppa et al. 2012). This is an active area of enquiry that is sure to become more and more relevant as GWAS discovers increasing numbers of loci that contribute to complex phenotypes.

Genotyping Technologies

There are currently two major commercial platforms that provide genome-wide genotyping services (Ragoussis 2009). Costs range from $100 to $800 per sample depending on the depth of coverage and the specific array, though large consortia gain cost savings while small users need to factor in additional fees for access to equipment. Nevertheless, at a fraction of a cent per genotype, and with accuracies exceeding 99.5%, the technology is quite remarkable.

Illumina (www.illumina.com/applications/genotyping.ilmn) actually offers several families of arrays that are built on the Infinium II assay genotyping technology shown in **Figure 6.10A**. The Omni family of arrays can produce anywhere from 730,000 (OmniExpress) to 4.3 million (Omni5) markers. These arrays include expert-curated content derived from the 1000 Genomes and International HapMap projects, providing sufficient coverage to facilitate imputation in European or Asian populations or higher-resolution direct genotyping for any population. They can also be custom designed to include project-specific content as requested by a user. The Core family of arrays uses a more focused and cost-effective set of bead chips built around 250,000 common tagging polymorphisms, with or without the inclusion of another 200,000 exonic variants at minor allele frequencies as small as 1%. Some of the markers on the arrays also support analysis of structural variation (CNV) and insertion-deletion polymorphisms (indels) as well as mitochondrial variation.

The Infinium II assay is a single-base extension reaction (Gunderson et al. 2006; see Figure 6.10A). Whole genomic DNA is amplified in a PCR-free manner overnight, then enzymatically fragmented and hybridized to an array that contains multiple copies of each of hundreds of thousands of beads, each coated in an oligonucleotide probe that is specific for a particular SNP. Each 50mer oligonucleotide probe is designed to terminate at the base adjacent to the SNP, so that when a mixture of modified nucleotides is added along with DNA polymerase, just one base is incorporated that is complementary to the SNP adjacent to where the probe has hybridized to the genomic DNA fragment. This base is then detected with a specially designed dye-labeled antibody that fluoresces at a characteristic color when a laser scans across the bead. The wavelengths of the two nucleotides that can incorporate at any position differ, and the relative intensity of each signal tells us which nucleotide is present in the fragment. A limitation of this assay is that it cannot distinguish substitutions of complementary bases (A/T or G/C), so these must be tagged by a proxy SNP in high LD. GenomeStudio software provided by Illumina interprets the signals from its iScan reader, using information across all the individuals sampled in a study (each on their own array), to help call the genotypes accurately.

Affymetrix (www.affymetrix.com/estore) provides a variety of whole-genome and custom genotyping platforms, which for human whole-genome analysis currently use either the Genome Wide Human SNP Array 6.0 (Kennedy et al. 2003) or the newer Axiom arrays. The Human SNP array has over 900,000 SNPs and almost 950,000 probes specifically for copy number variation (CNV). Genomic DNA is fragmented and ligated to a short adapter that facilitates whole-genome amplification by PCR that includes end labeling. These fragments are then hybridized to a chip that contains almost 4 million short oligonucleotide probes, with each variant represented by both alternate sequences at different places on the chip. (The principle is illustrated in the context of RNA profiling in Figure 8.2C.) DNA anneals to short oligonucleotides with such high specificity that if there is a mismatch between the probe and the amplified genomic DNA fragment, there will be less hybridization, and less signal, than if there is a match. Thus, comparison of signal intensity between the oligonucleotides representing the two alleles provides an indication of the genotype. Similarly, the intensity of the signals from CNV probes is proportional to the number of copies of the allele in the genome and provides robust information on the CNV status. The technology used to create gene chips with millions of oligonucleotides is described in Chapter 8 in relation to microarray-based gene expression profiling. The Axiom array illustrated in **Figure 6.10B**, by contrast, depend on amplification of the signal associated with specific ligation of a reporter oligonucleotide whose terminal base matches the variant of the genomic DNA fragment hybridized to a probe.

Both companies also provide custom genotyping services that may be used for targeted replication studies designed to evaluate just those variants of interest discovered in the initial phase of a GWAS. Illumina uses a ligation-based "Golden Gate" genotyping assay, also deploying its bead technology, while Affymetrix offers an Axiom service for bespoke arrays based on well-validated genotype assays. In addition, several other companies, notably Sequenom, Fluidigm, and Raindance, offer very different technologies for typing small numbers of SNPs, from a few dozen up to several thousand. It is possible that as sequencing costs are reduced, whole-genome genotyping will become obsolete, but targeted genotyping and sequencing will remain a staple component of follow-up and validation studies for the foreseeable future.

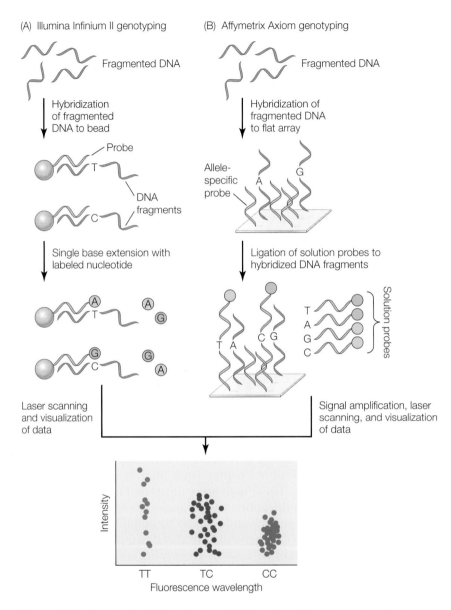

(A) Illumina Infinium II genotyping

Fragmented DNA

Hybridization
of fragmented
DNA to bead

Probe

T

DNA
fragments

C

Single base extension with
labeled nucleotide

A

A
G

T

G

G
A

C

Laser scanning
and visualization
of data

(B) Affymetrix Axiom genotyping

Fragmented DNA

Hybridization of
fragmented DNA
to flat array

Allele-
specific
probe

A

G

Ligation of solution probes to
hybridized DNA fragments

T
A
G
C

Solution probes

T A

C G

Signal amplification, laser
scanning, and visualization
of data

Intensity

Fluorescence wavelength

TT   TC   CC

Figure 6.10 Genotyping Technologies. (A) Illumina's Infinium II assay is a single-base extension protocol. Genomic DNA is fragmented and hybridized to beads coated in a probe oligonucleotide that terminates at the base adjacent to the SNP. The assay is designed so that one of two colored nucleotides is incorporated at each polymorphism. The ratios of intensities across all the samples in a batch are deconvoluted to generate clouds of points corresponding to each genotype (in the graph shown here, minor homozygotes are red, heterozygotes green, and major homozygotes blue). (B) The Affymetrix Axiom assay also relies on capture of fragmented genomic DNA, but visualization of the polymorphism follows from amplification of the signal associated with a short oligonucleotide designed to ligate to the template according to the identity of the SNP. This platform, too, calls genotypes called from the clouds of points from multiple samples.

Summary

1. GWAS is an approach to identification of polymorphisms that are correlated (associated) with phenotypes. It samples genetic variation throughout the genome in an unbiased manner.

2. Due to the multiple comparison problem, a strict statistical threshold of 5×10^{-8} must be exceeded in order to conclude that a SNP contributes to a disease or trait.

3. Studies with fewer than 5000 cases and controls are generally only sufficiently well powered to detect a handful of variants that have effect sizes greater than approximately 1.2-fold.

4. This leaves most of the variants that make smaller contributions undetected, giving rise to the so-called "missing heritability" problem. However, it is now recognized that most of the variation is "hidden" below the strict threshold as false negatives, rather than "missing."

5. Very large studies with several hundred thousand cases have discovered in the vicinity of 100 or more loci that contribute to height, body mass, and several diseases. Collectively, though, these variants rarely explain more than one-fourth of the genetic component of the variation, so most of the loci remain to be discovered.

6. Most associations are first detected by tagging SNPs present on genotyping arrays that are in high linkage disequilibrium with the actual causal variants.

7. Imputation may be used for fine-scale mapping of variation that is likely to be causal without actually genotyping the SNPs directly. This technique uses data from the International HapMap Project and 1000 Genomes Project to make a best estimate of all the genotypes in a region based on LD in a reference population.

8. Common causal variants are generally thought to be embedded in haplotype blocks, the limits of which are defined by hotspots for recombination. If the variant predates the separation of populations, the signal of disease association should be detectable in most derived populations.

9. A fraction of the loci that associate with traits have multiple causal variants that result in secondary associations and allelic heterogeneity.

10. Synthetic association is the situation in which what appears to be the effect of a common polymorphism is actually due to LD between that common SNP and multiple rare variants of large effect.

11. Correlation or association does not necessarily mean causation, so experimental assays are needed to confirm that the gene, and eventually the polymorphism, contributes to the disease.

12. Whole-genome genotyping arrays are generally provided by two companies, Illumina and Affymetrix, using bead array and gene chip technologies, respectively.

References

Browning, B. L. and Browning, S. R. 2009. A unified approach to genotype imputation and haplotype phase inference for large data sets of trios and unrelated individuals. *Am. J. Hum. Genet*. 84: 210–223.

Cantor, R. M., Lange, K., and Sinsheimer, J. S. 2010. Prioritizing GWAS results: A review of statistical methods and recommendations for their application. *Am. J. Hum. Genet*. 86: 6–22.

Clarke, G. M., Anderson, C. A., Pettersson, F. H., Cardon, L. R., Morris, A. P., and Zondervan, K. T. 2011. Basic statistical analysis in genetic case-control studies. *Nat. Protocols* 6: 121–133.

Cornelis, M. C., Agrawal, A., Cole, J. W., Hansel, N. N., Barnes, K. C., et al. 2010. The gene, environment association studies consortium (GENEVA): Maximizing the knowledge obtained from GWAS by collaboration across studies of multiple conditions. *Genet. Epidemiol*. 34: 364–372.

Devlin, B., Bacanu, S., and Roeder, K. 2004. Genomic control to the extreme. *Nat. Genet*. 36: 1129–1130.

Devlin, B. and Roeder, K. 1999. Genomic control for association studies. *Biometrics* 55: 997–1004.

DeWan, A., Liu, M , Hartman, S., Zhang, S. S., Liu, D. T. L., et al. 2006. *HTRA1* promoter polymorphism in wet age-related macular degeneration. *Science* 314: 989–992.

Dickson, S. P., Wang, K., Krantz, I., Hakonarson, H., and Goldstein, D. B. 2010. Rare variants create synthetic genome-wide associations. *PLoS Biol*. 8: e1000294.

Dudbridge, F. 2013. Power and predictive accuracy of polygenic risk scores. *PLoS Genet*. 9: e1003348.

Dudbridge, F. and Gusnanto, A. 2008. Estimation of significance thresholds for genomewide association scans. *Genet. Epidemiol*. 32: 227–234.

Gunderson, K. L., Steemers, F. J., Ren, H., Ng, P., Zhou, L., et al. 2006. Whole-genome genotyping. *Methods Enzymol*. 410: 359–376.

Hemani, G., Shakhbazov, K., Westra, H-J., Esko, T., Henders, A. K., et al. 2014. Detection and replication of epistasis influencing transcription in humans. *Nature* 508: 249–253.

Hill, W. G., Goddard, M. E., and Visscher, P. M. 2008. Data and theory point to mainly additive genetic variance for complex traits. *PLoS Genet*. 4: e1000008.

Hindorff, L. A., Sethupathy, P., Junkins, H. A., Ramos, E. M., Mehta, J. P., et al. 2009. Potential etiologic and functional implications of genome-wide association loci for human diseases and traits. *Proc. Natl. Acad. Sci. U.S.A*. 106: 9362–9367.

Howie, B. N., Donnelly, P., and Marchini, J. 2009. A flexible and accurate genotype imputation method for the next generation of genome-wide association studies. *PLoS Genet*. 5: e1000529.

Idaghdour, Y., Czika, W., Shianna, K. V., Lee, S. H., Visscher, P. M., et al. 2010. Geographical genomics of human leukocyte gene expression variation in southern Morocco. *Nat. Genet*. 42: 62–67.

International Schizophrenia Consortium, Purcell, S. M., Wray, N. M., Stone, J. L., Visscher, P. M., et al. 2009. Common polygenic variation contributes to risk of schizophrenia and bipolar disorder. *Nature* 460:748–752.

Kennedy, G. C., Matsuzaki, H., Dong, S., Liu, W., Huang, J., et al. 2003. Large-scale genotyping of complex DNA. *Nat. Biotech*. 21: 1233–1237.

Klein, R. J., Zeiss, C., Chew, E. Y., Tsai, J. Y., Sackler, R. S., et al. 2005. Complement factor H polymorphism in age-related macular degeneration. *Science* 308: 385–389.

Kruppa, J., Ziegler, A., and König, I. R. 2012. Risk estimation and risk prediction using machine-learning methods. *Hum. Genet.* 131: 1639–1654.

Lango Allen, H., Estrada, K., Lettre, G., Berndt, S., Weedon, M. N., et al. for the GIANT Consortium 2010. Hundreds of variants clustered in genomic loci and biological pathways affect human height. *Nature* 467: 832–838.

Lee, S. H., DeCandia, T. R., Ripke, S., Yang, J., PGC-SCZ, ISC, MGS, et al. 2012. Estimating the proportion of variation in susceptibility to schizophrenia captured by common SNPs. *Nat. Genet.* 44: 247–250.

Moore, J. H., Gilbert, J. C., Tsai, C-T., Chiang, F-T., Holden, T., et al. 2006. A flexible computational framework for detecting, characterizing, and interpreting statistical patterns of epistasis in genetic studies of human disease susceptibility. *J. Theor. Biol.* 41: 52–61.

Morgan, A. A., Chen, R., and Butte, A. J. 2010. Likelihood ratios for genome medicine. *Genome Med.* 2: 30.

Park, J-H., Wacholder, S., Gail, M. H., Peters, U., Jacobs, K., et al. 2010. Estimation of effect size distribution from genome-wide association studies: Implications for future discoveries and risk-prediction. *Nat. Genet.* 42: 570–575.

Parkes, M., Cortes, A., van Heel, D. A., and Brown, M. A. 2013. Genetic insights into common pathways and complex relationships among immune-mediated diseases. *Nat. Rev. Genet.* 14: 661–673.

Price, A. L., Patterson, N. J., Plenge, R. M., Weinblatt, M. E., Shadick, N. A., et al. 2006. Principal components analysis corrects for stratification in genome-wide association studies. *Nat. Genet.* 38: 904–909.

Purcell, S., Neale, B., Todd-Brown, K., Thomas, L., Ferreira, M. A. R., et al. 2007. PLINK: A toolset for whole-genome association and population-based linkage analysis. *Am. J. Hum. Genet.* 81: 559–575.

Ragoussis, J. 2009. Genotyping technologies for genetic research. *Annu. Rev. Genom. Hum. Genet.* 10: 117–133.

Ripke, S., O'Dushlaine, C., Chambert, K., Moran, J. L., Kähler, A. K., et al. 2013. Genome-wide association analysis identified 13 new risk loci for schizophrenia. *Nat. Genet.* 45: 1150–1159.

Risch, N. and Merikangas, K. 1996. The future of genetic studies of complex human diseases. *Science* 273: 1516–1517.

Segal, D. J. and Meckler, J. F. 2013. Genome engineering at the dawn of the golden age. *Annu. Rev. Genomics Hum. Genet.* 14: 135–158.

So, H. C., Gui, A. H., Cherny, S. S., and Sham, P. C. 2011. Evaluating the heritability explained by known susceptibility variants: A survey of ten complex diseases. *Genet. Epidemiol.* 35: 310–317.

Spielman, R. S., McGinnis, R. E., and Ewens, W. J. 1993. Transmission test for linkage disequilibrium: The insulin gene region and insulin-dependent diabetes mellitus (IDDM). *Am. J. Hum. Genet.* 52: 506–516.

Visscher, P. M., Brown, M. A., McCarthy, M. I., and Yang, J. 2012. Five years of GWAS discovery. *Am. J. Hum. Genet.* 90: 7–24.

Voight, B. F., Kang, H. M., Ding, J., Palmer, C. D., Sidore, C., et al. 2012. The metabochip, a custom genotyping array for genetic studies of metabolic, cardiovascular, and anthropometric traits. *PLoS Genet.* 8: e1002793.

Wellcome Trust Case Control Consortium (WTCCC). 2007. Genome-wide association study of 14,000 cases of seven common diseases and 3,000 shared controls. *Nature* 447: 661–678.

Wray, N. R., Purcell, S. M., and Visscher, P. M. 2011. Synthetic associations created by rare variants do not explain most GWAS results. *PLoS Biol.* 9: e1000579.

Yang, J., Lee, S. H., Goddard, M. E., Visscher, P. M. 2011. GCTA: A tool for genome-wide complex trait analysis. *Am. J. Hum. Genet.* 88: 76–82.

Yang, J., Manolio, T. A., Pasquale, L. R., Boerwinkle, E., Caporaso, N., et al. 2011. Genome partitioning of genetic variation for complex traits using common SNPs. *Nat. Genet.* 43: 519–525.

Yang, Z., Camp, N. J., Sun, H., Tong, Z., Gibbs, D., et al. 2006. A variant of the *HTRA1* gene increases susceptibility to age-related macular degeneration. *Science* 314: 992–993.

Yu, J., Pressoir, G., Briggs, W. H., Vroh, Bi I., Yamasaki, M., et al. 2005. A unified mixed-model method for association mapping that accounts for multiple levels of relatedness. *Nat. Genet.* 38: 203–308.

Zhu, X., Zhang, S., Zhao, H., and Cooper, R. S. 2002. Association mapping, using a mixture model for complex traits. *Genet. Epidemiol.* 23: 181–196.

7

Whole-Genome and
Whole-Exome Sequencing

S tarting around 2010, the cost of DNA sequencing had dropped to the point where it became feasible for researchers to begin to incorporate genome-scale sequencing into their genetic analyses. Despite the success of genotype-based genetic analysis, there were four main rationales for pushing the technology to the next level. First, it was considered likely that rare variants are responsible for congenital abnormalities and other highly penetrant diseases, and these variants are not captured by genotyping approaches. Second, it was becoming apparent that GWAS may be missing moderately common variants of modest effect, the so-called Goldilocks variants in the frequency range of 1%–5% that are difficult to accurately impute onto known haplotypes. Third, the only way to capture the full range of regulatory and noncoding variation across the frequency range is complete genome sequencing, which, combined with RNA sequencing, would facilitate systems genomics analyses as discussed in Chapter 10. Fourth, advances in cancer biology were increasing the pressure to contrast normal and tumor genomes to fully catalog the changes that accompany oncogenesis and provide therapeutic targets.

By the spring of 2014, large sequencing centers had achieved whole-genome sequencing for well under $5000, and Illumina had just released its new HiSeq X Ten hardware, which promises to deliver $1000 genomes. Many predict that new technologies may bring the cost closer to $100 or less within a decade. At that point, the expenses associated with data archiving and analysis may be greater than the actual cost of sequence generation, and indeed, much attention is being focused on methods for storing and accessing the information from thousands of genomes. Just for perspective, consider what it would cost to print out a single human genome on paper: assuming a 10 point font and single spacing, which produces 6000 letters (6 kb) per page, and one cent per page for ink and paper, its 3 Gb would consume $5000! Thankfully, electronic data are much cheaper to store, particularly in the cloud.

It is unlikely that anyone's complete genome will ever be sequenced with existing technologies, since there will always be regions of heterochromatin and refractory regions of unusual nucleotide content, not to mention highly repetitive elements, that are not included in the analyses. Nevertheless, the term "**whole-genome sequencing**" (**WGS**) has a clear connotation that the vast majority of the euchromatin, including intergenic regions, introns, and regulatory regions, is sequenced. At around $500 per exome, "**whole-exome sequencing**" (**WES**, or **WER** for "**whole-exome resequencing**") equally clearly refers to efforts to sequence only the coding regions of the majority of the annotated protein coding genes and select RNA genes. Because of the way targeted capture of exon sequences is performed, WES usually also incorporates some regulatory regions and introns adjacent to exons. A third type of genome-scale sequencing is "**targeted genome sequencing**" (**TGS**), in which the researcher specifically targets genes or gene regions of interest, usually between 1 Mb and 30 Mb, and sequences them to high depth. The protocol is very similar to WES, except that the oligonucleotide probes that are used to pull down, or capture, the less than 1% of the genome that is of interest are specifically designed for a purpose other than unbiased discovery.

In this chapter, we will start with a discussion of the technology for genome-scale sequencing, then contrast some applications of WES and WGS, before considering some of the emerging new technologies.

Sequencing Technology and Strategy

There are currently three major companies that provide platforms for genome-scale sequencing: Illumina, Life Technologies, and Complete Genomics (Shendure and Ji 2008; Metzker 2010). These platforms are all quite different from the traditional Sanger sequencing that dominated analyses throughout the 1990s and 2000s (and is still used for verification purposes). They are highly automated and can generate gigabases of data in a single day, feasibly leading to a lag time of just 72 hours between tissue biopsy or blood sample and presentation of a patient or client with a fully annotated genome.

Illumina uses single-base extension "sequencing by synthesis" technology that was initially developed by a company called Solexa. It currently produces four automated sequencers, the MiSeq, HiSeq 2000, HiSeq 2500, and HiSeq X Ten, in increasing order of performance. These essentially allow users to analyze data from WES and TGS through to a single genome in a day, or to perform multiple (now hundreds of) whole-genome analyses in 10 days. Sample preparation involves fragmentation of genomic DNA, either by ultrasonic shearing (TruSeq) or by a novel transposon insertion protocol (Nextera), followed by library preparation, cluster generation, and sequencing, all accompanied by careful quality control steps. Typically the objective is to obtain paired reads of 100 bp—though this number is increasing with improved chemistry—from either end of each of 300 million DNA fragments per lane of a HiSeq, or 30 million per MiSeq run. The technology works by bridging single-stranded DNA fragments to an array surface, locally amplifying them by PCR to generate a cluster, then sequencing from one end by adding one nucleotide at a time (**Figure 7.1A**). Each of the four nucleotides is attached to a different fluorescent dye, and

(A) Illumina's sequencing by synthesis

(B) Life Technologies' semiconductor sequencing

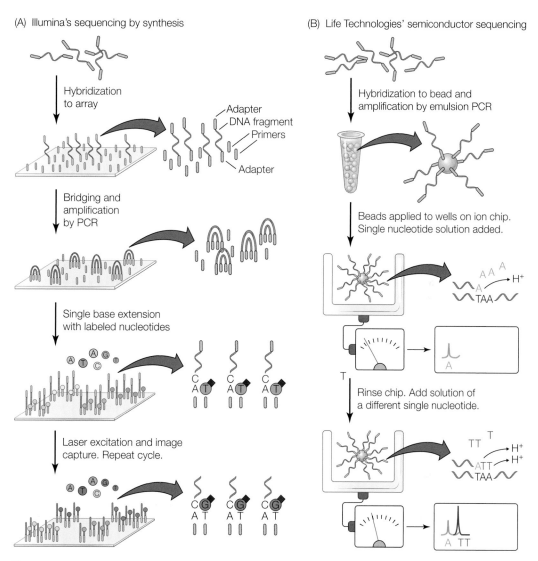

Figure 7.1 Next-Generation Sequencing Technologies.   (A) Illumina's sequencing by synthesis involves generation of cell-free clones of short sequence fragments on a solid surface by bridge-PCR, followed by single-base extension reactions. After each of 100 cycles, the fluorescence of the incorporated dye associated with A,T,G,C is read from a photograph of up to 300 million individual amplicons; sequential alignment of these images enables the sequence to be assembled. (B) Life Technologies' IonTorrent version of pyrosequencing involves measurement of the amount of current produced as nucleotides are incorporated after each addition of either A, T, C, or G in hundreds of cycles, all measured in parallel on millions of microsensors associated with a PCR pool of single DNA fragments.

a photograph of the entire array surface with 300 million parallel reactions thus identifies which nucleotide was added to which cluster in each cycle. Since each reaction is actually performed on a local pool of clones (clustered amplicons), the color is derived from thousands of incorporations, and the error rate is remarkably low. After 100 cycles, the sequencing reaction is reinitiated from the alternate primer at other end of the fragment, generating a "paired-end read."

Life Technologies uses a modification of the pyrosequencing technology initially developed by 454/Roche. In semiconductor (also called PostLight) sequencing, incorporation of each nucleotide is accompanied by production of a small current of protons (Merriman et al. 2012). These currents are recorded on an array of millions of silicon microsensors. Rather than incorporating a single nucleotide at a time, these reactions continue to incorporate nucleotides so long as the stretch of nucleotides on the template is complementary to the particular one of the four nucleotides that was added to the reaction. As each nucleotide is added, the diphosphate that is cleaved from the nucleotide triphosphate during the polymerization step is used to generate an ion current, the strength of which is proportional to the number of nucleotides incorporated (**Figure 7.1B**). Thus, if the template reads TAATCT, then as nucleotides are added in the cycling order A,C,G,T, the program reads a small current (A), then no current, no current again, then twice the current (TT), then a small current (A), then no current, then a small one (G), no current, a small one (A), and so forth. The IonTorrent machine generates tens of millions of short reads up to 500 bp in length from one end of each of tens of millions of fragments to be sequenced, while the ProtonTorrent is the scaled-up version capable of whole-genome sequencing.

Complete Genomics (now owned by BGI-Shenzhen, the Beijing Genome Institute) uses in-house technology and performs all sequencing at its own facility, providing only whole-genome and cancer-genome sequencing services. Its technology has incorporated a novel twist that allows direct **phasing** of the genome: namely, direct assessment of full haplotype structure along a chromosome (Peters et al. 2012). It does this by diluting the template so much that only one-thirtieth of the fragmented genome is included in each of 384 parallel reactions. It is thus very unlikely that the two molecules corresponding to both alleles will be included in any one reaction, so each of the 384 sequences essentially provide snapshots of just a fraction of a haploid genome—but these snapshots are combined to infer the complete diploid sequence. The sequencing technology is based on a complicated series of ligation reactions (cPAL, combinatorial probe-anchored ligation) to concatenated DNA nanoballs (DNBs) that generate hundreds of millions of paired 35 bp reads, each derived from 500 bp fragments. Proprietary software is then used to assemble the reads into a whole-genome sequence, which is reported to investigators in a file that annotates SNPs, copy number variations (CNVs), mobile and repetitive elements, and other structural features.

A standard workflow for assembling complete genomes from raw next-generation sequence data is shown in **Figure 7.2**. It is based on the Genome

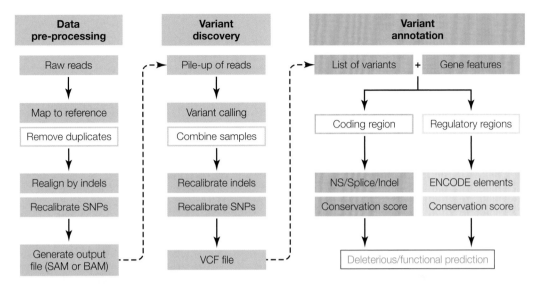

Figure 7.2 The GATK Pipeline for Genome Sequencing. Data preprocessing consists of alignment of short reads to a reference genome, including elimination of duplicate reads, realignment to adjust for indels and sequence mismatches, and generation of an output file (usually SAM or BAM). Variant discovery involves identifying all locations where the read does not match the reference (this step can be done one sample at a time, or by jointly calling variants from multiple genomes), assignment of quality scores to SNPs and indels, and generation of an output file (usually VCF). Variant annotation then involves determining the location of the SNP or indel relative to coding regions and preliminary assessment of whether it is likely to be functional or deleterious. (See www.broadinstitute.org/gatk/about#typical-workflows for more details.)

Analysis Tool Kit (GATK) pipeline developed at the Broad Institute (www.broadinstitute.org/gatk; McKenna et al. 2010). Short reads are first filtered to pass basic read quality metrics, then aligned against the reference human genome (**Figure 7.3A**) using a rapid alignment algorithm such as the Burrows-Wheeler Aligner (BWA; Li and Durbin 2009). The current standard reference genome is HuRef19 (the nineteenth release of the most accurate assembly of the initial human genome sequenced by the International Human Genome Sequencing Consortium in the early 2000s). Paired-end reads increase accuracy both by ensuring that the alignment is over a longer region and by enforcing the requirement that the reads be in the correct orientation and at the expected spacing given the fragment size. Paired-end reads therefore help to overcome gaps that are due to repetitive sequences that are masked, and a local realignment step optimizes the accuracy of the alignments. All the reads are piled up and reported in tab-delimited text format as a sequence alignment/map (SAM) file or in the compressed BAM binary format: see the open-source SAMtools website at samtools.sourceforge.net (Li et al. 2000; **Figure 7.3B**).

(A) Alignment

```
Coor      1 2 3 4 5 6 7 8 9 0 1 2 3 4   5 6 7 8 9 0 1 2 3 4 5 6 7 8 9 0 1 2 3 4 5 6 7 8 9 0 1 2 3 4 5
ref       A G C A T G T T A G A T A A * * G A T A G C T G T G C T A G T A G G C A G T C A G C G C C A T

+r001/1             T T A G A T A A A G G A T A * C T G
+r002           a a a A G A T A A * G G A T A
+r003                         A T A G C T . . . . . . . . . . . . . . . T C A G C
−r001/2                                                             C A G C G C C A T
```

(B) SAM format

```
@HD VN: 1.5  SO: coordinate
@SQ SN: ref  LN: 45
r001  163  ref   7  30  8M2I4M1D3M   = 37   39  T T A G A T A A A G G A T A C T G   @C C F F D A D H J > ; J C A C <
r002    0  ref   9  30  3S6M1P1I4M   *  0    0  A A A A G A T A A G G A T A   @B C F F I J I B / ? >  B  D
r003    0  ref  16  30  6M14N5M      *  0    0  A T A G C T T C A G C   @C C F F D B B B J >
r001   83  ref  37  30  9M           =  7  −39  C A G C G G C A T   @B @ H H D F E ?
```

(B) VCF format

#CHROM	POS	ID	REF	ALT	QUAL	FILTER	INFO	FORMAT	Sample1
2	4370	rs6057	G	A	29	.	NS=2; DP=13; AF=0.5	GT:GQ:DP:HQ	0|1:48:1:52,51
2	7330	.	T	A	3	Q10	NS=5; DP=12; AF=0.017	GT:GQ:DP:HQ	0|0:46:3:58,50
2	110696	rs6055	A	G	67	PASS	NS=2; DP=10; AF=0.333; AA=T	GT:GQ:DP:HQ	0|1:21:6:23,27
2	130237	.	T	.	47	.	NS=2; DP=16; AA=T	GT:GQ:DP:HQ	0|0:54:7:56,60

Figure 7.3 SAM and VCF Formats. (A) The intuitive alignment of a set of short sequence reads to a reference sequence—in this case, a 45 bp reference with three reads, one of which (r001) is paired (that is, has reads from both ends). Read r001 includes a 2 bp insertion and a 1 base deletion, and read r002 is also polymorphic at the insertion site. Read r003 contains 14 nucleotides that do not align. (B) The SAM format starts with two header rows (@HD and @SQ), then reports each read with eleven columns showing the read name, a FLAG that summarizes various bits of information, the name of the reference sequence, the base number of the leftmost sequence, a map quality score, the CIGAR string (a representation of sequential matches, mismatches, insertions, and deletions), signifier of a matched read, location of the matched read, length of the template, the observed sequence, and ASCI-coded single-base quality information. (See samtools.sourceforge.net/SAMv1.pdf for more details.) (C) The VCF format summarizes variants called in an individual genome. These four rows show evidence for four SNPs, indicating the chromosome and position; the rsID if one exists; reference and alternative alleles; quality of the base call and whether a filter is passed; information on (respectively) the number of samples with data, the total depth of reads, the alternative allele frequency, and ancestral allele identity; and the format of the individual sample data presented in the final set of columns (only one sample is shown, but many may be appended), in this case showing the genotype, the quality of the read in that individual, the read depth, and the quality of the inferred haplotype. (See en.wikipedia.org/wiki/Variant_Call_Format for more details.)

Next, the reads must be screened to identify (in colloquial usage, to "call") variants, and each variant must be annotated: Does it fall within an exon or intron? Does it encode an amino acid change, and is that change likely to affect protein function? Is it a potential regulatory site in known functional motifs? Does it map to an already identified SNP that has an rsID? The typical output format for a whole-genome annotation is a VCF (variant call format; **Figure 7.3C**) file. There are numerous tools for performing variant annotation and for visualizing variants. These tools include the Broad Institute's Integrated Genomics Viewer (IGV), the Human Genome Browser at UCSC, and multiple other freely available tools with distinct useful features, two examples of which are VAAST (the variant annotation, analysis, and search tool) and ANNOVAR. Genome annotation is a rapidly changing landscape, and there is little point in trying to summarize the options here: each lab has its preferences based on the aims and design of a study. It is important, however, to emphasize that an annotation is only as good as the assembled genome that it uses as a reference, and there is wide variation among tools in their definitions of exons, introns, and key genomic features. Consequently, important findings need to be cross-referenced with other genome viewers where possible. At the current state of development of genome-scale sequencing, users must be prepared for a steep learning curve accompanied by much frustration, and they must be willing to engage with other users for advice and feedback on a regular basis, but the payoff is well worth the patient effort.

Whole-Exome Sequencing

Whole-exome sequencing refers to the strategy of sequencing all the exons of an individual to an average depth of at least 30X, but preferably 50X or more. Whole-exome sequencing is performed primarily to identify rare coding variants that are likely to disrupt protein function (Bamshad et al. 2011). It is most effective when performed on trios of an affected individual (proband) and that individual's two biological parents. The exon sequences are captured by fragmenting total genomic DNA and hybridizing it either to a solid array or a solution of magnetic beads that are coated in oligo-nucleotides that have been specifically designed to capture (colloquially, "pull down") the vast majority of annotated exons in the human genome. Commercial kits, such as Agilent's SureSelect or Illumina's Exome Enrichment, simplify the process, allowing a sequencing library to be generated with just a few days of effort. They result in nearly 90% enrichment of exons and several hundred bases immediately adjacent to them (thus including small introns and some promoter sequences). This method reduces the number of reads that are required to cover the exome by two orders of magnitude relative to the whole genome, since the exons constitute just 1% of the DNA, approaching 30 Mb of sequence.

The requirement for at least 30X coverage may seem excessive, but there are two reasons for it. First, because capture and sequencing includes a stochastic component, even if the median is 30X, there will always be a small fraction, around 10%, of the exons that are read to only 20X or less, and this depth drops even further with lower coverage. Second, the probability that both alleles of a heterozygote are observed in n sequences is just $(1 - 1/2^{n-1})$. With 2 reads, there is a 50:50 chance that both are the same allele, so the heterozygote would not be called. With 5 reads, there is a 1 in 16 chance that they are all the same allele; with 10 reads, it is 1 in 512; and even with 15 reads, it is approximately 1 in 15,000, which establishes the absolute minimum required to be sure of observing at least one copy of each of the approximately 5000 exonic variants (up to two-thirds of which are nonsynonymous) in a typical human genome. Given that it is desirable to have multiple reads of each allele to be confident, the need for deep coverage should be obvious.

Assuming that a rare congenital disorder is primarily attributable to mutation(s) in a single gene, there are essentially five possible scenarios that can arise from exome sequencing of trios:

1. The proband is homozygous for a deleterious mutation that is carried in the heterozygous state by both parents: this would be a classical Mendelian recessive mutation.

2. The proband is compound heterozygous for two different deleterious mutations, each of which is carried by one of the two parents, that when combined result in recessive loss of function.

3. The proband inherits a dominant mutation that was transmitted from the affected parent, or is incompletely penetrant so that the disease was not observed in the parent who carried it.

4. The proband has a de novo mutation that causes disease in either a dominant or a sex-linked manner and thus was not present in either parent.

5. The mutation is regulatory and is not captured by the exome-sequencing strategy.

Although it is difficult to be sure how often trio exome sequencing is successful in identifying a causal mutation, since negative results tend not to be reported, initial observations suggest that the success rate may be somewhere between 20% and 50%. For example, one study of 12 children with undiagnosed medical conditions such as developmental delays, intellectual disabilities, or craniofacial abnormalities, found the likely causal mutations in 6 cases, most of which were attributed to de novo events (Need et al. 2012). The technique may fail because the mutations are noncoding, because they are not thought to be deleterious, or because the condition has a more complex genetic basis involving disruption of multiple genes. It is also quite likely that causal mutations are often missed because we are unable to read whether candidate mutations are deleterious or not. Furthermore, it is critical to recognize that false positive findings are likely to be common: one study of the exomes of 104 unrelated healthy adults found that each person carried an average of almost 3 (range 0–7) mutations in suspected Mendelian disease-causing genes that were predicted to be deleterious, but that one-fourth of those mutations either were common in the population at large or were probably mis-annotated (Bell et al. 2011).

The first two successful demonstrations that WES can identify the genetic basis of a previously unknown Mendelian condition were published in November 2009 and January 2010. The first was a study of one child subsequently diagnosed with congenital chloride diarrhea due to a mutation in the SLC26A3 chloride exchange protein, which was confirmed in four other children (Choi et al. 2009). The second was a study of four children with Miller syndrome, a disease in which the function of the gene product, dihydroorotate dehydrogenase (DHODH), which is involved in pyrimidine biosynthesis, was plausibly related to the facial phenotypes observed in affected children, since inhibition of the same pathway with the drug leflunomide produces a similar spectrum of defects in mice (Ng et al. 2010). This study was discussed in some detail in Chapter 5, but here we will recap the filters that the researchers applied to reduce the search space of candidate mutations (see Figure 5.5). First, they focused on nonsynonymous variants, splice-site variants, and indels (NS/SS/I variants) because these changes are much more likely to be disruptive than nucleotide changes that do not alter the protein sequence; second, they removed common variants because only novel or very rare mutations are likely to contribute to extremely rare conditions; third, they extended the comparison to additional children from different families.

Both of these studies are notable for the fact that the variant detection was performed simply by sequencing a small number of affected individuals, without utilizing information from close relatives in the primary analyses. Where possible, the extra step of confirming that the mutation only segregates with the disease in family members is desirable as well. WES now commonly employs trios, since the parental information increases the ability to screen genotypes that arise de novo from those that are transmitted from parents, who may show subclinical phenotypes if not the disease itself. Since the inference that a gene is causal in promoting disease generally requires identification of multiple individuals with mutations in the same gene (though not necessarily the same mutation), extended families also provide a convenient framework, at least for initial discovery. As with classical linkage analyses, family-based studies utilize genotypes linked to the candidate mutation to confirm that the region is inherited identical by descent in all individuals in the kindred who have the disease. That is to say, it is assumed that the mutation arose once in a common ancestor who has passed the same portion of the affected chromosome to all affected family members. Pedigree mapping, described in more detail in **Box 7.1**, is an important part of this process. Sequencing of the most distant family members should be most efficient, since they will share fewer genotypes by chance, reducing the search space for the putative causal variants. In fact, one effective strategy is to perform initial linkage analysis, and then simply genotype strong candidate variants identified in the sequence of a single proband that also lie within linkage peaks across the pedigree, as was done to find a gene for the dominant Mendelian condition metachondromatosis (Sobreira et al. 2010).

BOX 7.1
Pedigree Mapping

Arguably the most important research strategy in human genetics over the past half century has been linkage mapping of disease-associated loci in family pedigrees, or **pedigree mapping** for short (Freimer and Sabatti 2004). Pedigrees are descriptions of the relationships between members of an extended family. Mapping of affected individuals onto pedigrees allows inference of the mode of transmission of a trait: namely, whether it is autosomal or sex-linked, recessive or dominant, and whether it is likely to have a single-gene cause.

Addition of genetic markers to the pedigree facilitates linkage mapping (Botstein et al. 1980). If it is assumed that a causal mutation arose once in one of the founders of the pedigree, then all affected individuals should share not only that mutation, but also a suite of genetic markers in the region of the mutation. These individuals are said to be **identical by descent** (**IBD**) for the region. The extent of linkage disequilibrium between markers in families is much greater than in the general population, since there have been only a few generations for recombination to occur within the family. Consequently, tracking markers as far apart as 10 centimorgans (cM, several hundred Mb) is sufficient to detect linkage between a marker and the trait or disease. Further sleuthing is then used to refine the linkage signal to identify the gene and mutation.

The rules for drawing pedigrees are fairly straightforward. Females are drawn as circles, males as squares. Affected individuals are black shaded, unaffected individuals are unshaded, and ambiguous status can be indicated by lighter shading or a question mark. Deceased individuals are shown with an oblique strikethrough. A horizontal line joins partners who conceive offspring, and a vertical line drops down either to a single child or to multiple children joined by a common horizontal bar. Birth and death dates or other information can be added if they are relevant.

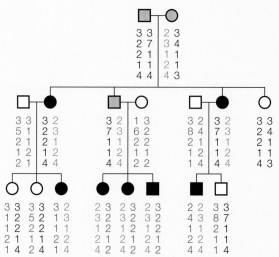

The pedigree shown here is one of those that was used to map autosomal dominant long-QT syndrome (one of the major hereditary causes of arrhythmia and sudden cardiac arrest; Curran et al. 1995); genotype information has been added in the form of a vertical string of numbers. Although there are multiple haplotypes made up by different strings of these genotypes, it can be seen that all the affected (black shaded) individuals are heterozygous for a single haplotype. The likelihood of this pattern occurring by chance can be compared with its likelihood under a genetic model attributing the disease to that region. This comparison gives rise to a **LOD**, or **logarithm of the odds**, score. To a broad approximation, a LOD score of greater than 3 is suggestive of linkage, and a LOD score of greater than 5 provides strong evidence of linkage.

Consider the three affected children in one nuclear family. The chance that one of these children has the same homozygous haplotype derived from two carrier parents is 1 in 4; the chance that two of the children have it is 1 in 16. Additional information can be gleaned from unaffected siblings, who are required not to have the haplotype, and from more distant family members, whose relatedness must be adjusted for during the likelihood computations. With thousands of linkage blocks throughout the genome, odds of even 1 in 256 are not compelling, since several regions are likely to show this level of commonality by chance. Consequently, pedigree mapping generally requires multiple families whose calculations are combined. Open-source software packages for estimating relatedness in pedigrees and performing linkage analyses are listed at link-age.rockefeller.edu/soft; they include the popular packages FISHER/MENDEL, GENEHUNTER, LINK-AGE, MERLIN, and SOLAR.

For complex traits affected by multiple genes as well as gene-by-environment interactions, pedigree mapping is generally only sufficient to highlight a small number of regions of the genome that are likely to harbor contributing genes. More refined analyses are possible in the case of Mendelian genes and traits, especially if the same gene causes the condition in all or most of the pedigrees. Even if different mutations are responsible, so that individuals are not IBD across pedigrees, a linkage signal in the same locus can be generated. Typical linkage intervals cover several centimorgans and dozens to hundreds of genes, any one (or more) of which could cause the disease. Fine-scale mapping generally requires molecular cloning of the interval and sequencing to identify likely deleterious mutations.

The alternative strategy is to sequence a larger number of unrelated individuals who have a disease, or who are at the extremes for the distribution of a physiological or biochemical trait (Cirulli and Goldstein 2010). If there is genetic heterogeneity, as will often be the case, then only a fraction of these individuals will have mutations in the same gene. Yet if multiple individuals are discovered to share rare variants, **statistical burden tests** can be performed to evaluate the likelihood that the observed enrichment is more than expected by chance. Thus, a rare allele that is at a frequency of 0.1% in the general population, but increases risk of disease 20-fold over a background disease prevalence of 1 in 1000, would be expected to be seen in 2% of cases. WES of 500 cases may discover 10 copies, whereas only 1 would be expected in 500 controls. A similar argument applies to a set of very rare mutations that collectively have these frequencies, but it turns out that genome-wide significance in this situation is a complex function of the underlying genetic model (Ladouceur et al. 2012; Zuk et al. 2014). A reasonable estimate provided by MacArthur et al. (2014) in their set of guidelines for implicating rare variants in disease is $p < 1.7 \times 10^{-6}$ (0.05/30,000), implying that it will likely require exome sequences of tens of thousands of cases to establish the significance of variants in any individual gene. A convenient online tool for computing the power of WES, which also provides a tutorial and review of these methods, can be found at bioinformatics.org/spower/.

One of the key concepts for interpreting the results of WES studies is the **recurrence risk**, which is the probability that the disease recurs in another family member, given that a proband has already been identified (Risch 1990). For many rare diseases, the sibling relative risk, λ_S, defined as the ratio of the probability that the disease is observed in siblings of a proband to the probability that it is observed in the general population, may be as high as 10. This risk must be reconciled with the genotype relative risk (GRR), which is the ratio of the disease probability in people with the risk genotype to the disease probability in everyone else who does not have the genotype, but the relationship is a complex function of allele frequencies (Rybicki and Elston 2000). Genetic epidemiologists recognize that even for rare diseases such as epilepsy or autoimmunity, rare variants may make a substantial contribution, but they are unlikely to work alone. Multiple rare variants may work together, in combination with their common background and environmental risk factors, reducing the penetrance of effects. Thus, despite the appeal of WES and its potential to contribute to an understanding of the etiology of non-Mendelian diseases, there are complex statistical considerations involved that are just beginning to be addressed.

Copy Number and De Novo Variation

Two specific applications of whole-exome sequencing are the detection of copy number variation and the detection of de novo mutations. In fact, much of the initial impetus for rare variant studies came from studies of schizophrenia and autism that identified specific rare CNV enrichment in people with these conditions. With appropriate normalization to control for technical artifacts specific to the batch of samples that are processed together, variation in GC content, and sampling variance, CoNIFER (copy number inference from exome reads; sourceforge.net/projects/conifer/; Krumm et al. 2012) and XHMM software (exome hidden Markov model; atgu.mgh.harvard.edu/xhmm; Fromer et al. 2012) support the detection of exons that have underrepresented read depth, and so are CNV deletions, or have overrepresented read depth, and are likely CNV duplications. The former should not harbor individual sequence variation, while the latter may have skewed variant ratios of 1:2 instead of 1:1. A survey of over 1000 control samples found that most individuals have at least one rare CNV deletion and one rare CNV duplication. Itsara et al. (2010) used whole-genome genotyping to observe nine de novo CNV events, each at least 30 kb in length, among 772 children, a finding that implies a de novo CNV mutation rate approaching 0.01. Given the observed distribution of CNVs in human populations, this rate implies quite strong purifying selection against new CNVs in general.

Full discussion of de novo variant detection in psychiatric disease is deferred to Chapter 15; suffice it to say here that comprehensive analyses of several hundred trios show that most healthy individuals have at least one protein-coding mutation that is predicted to be deleterious. Individuals with a disease are not expected to have significant enrichment in de novo mutations throughout their genome; rather, these mutations are likely to be enriched in that subset of the proteome that contributes to their specific disease. Consequently, the strongest evidence that de novo mutations collectively contribute to psychiatric disorders is the finding that they preferentially affect proteins that are known to interact within neurons and hence are likely to be involved in neuropathogenesis.

Whether or not this is also the case for other complex diseases remains to be seen, but it is important to note that de novo mutations have been unambiguously associated with a number of very rare conditions (Veltman and Brunner 2012). For example, 12 of 13 patients with Schinzel-Giedion syndrome, which causes developmental abnormalities, have a new mutation in a short region of the *SETBP1* gene, while six different subunits of the SWI/SNF complex have been found mutated in children with the similarly complex Coffin-Siris syndrome. The likelihood that de novo mutations will lead to disease seems to be influenced by a number of factors, including the mutational target size (the number and length of genes and of the functionally critical stretches of the proteins they encode), the level of redundancy through duplication or functional complementation, CpG density in the vicinity of the genes, and factors and processes that influence the mutation rate. Among the latter, paternal age is emerging as a major factor, since the mutation load increases as men age. A study in Iceland found that the per-nucleotide mutation rate in 30-year-old men results in approximately 1.2×10^{-8} transmissions of new mutations to offspring per generation, but of the order of 2 mutations are added to each spermatocyte per year, resulting in a doubling of the total mutational load every 15 years (Kong et al. 2012).

Whole-Genome Sequencing

Inevitably, WGS will soon become a staple component of genetic analysis of both rare and common diseases. As sequencing costs drop, it will become less expensive to generate a whole-genome sequence than to incorporate the exon capture step into a WES protocol. Furthermore, WGS provides information on the entire genome, including regulatory regions, noncoding RNA genes, and gene deserts that nevertheless contain functional elements.

The first person to have a high-quality whole-genome sequence determined was J. Craig Venter (Levy et al. 2007; Ng et al. 2008), whose company, Celera, sequenced his genome at the same time the International Human Genome Sequencing Consortium was drafting the reference human genome (which was deliberately stitched together from multiple anonymous donors). Venter's diploid genome is typical of a Caucasian individual, containing over 4 million variable sites, 78% of which are SNPs that account for 26% of all variant bases (since indels and CNVs cover multiple nucleotides). Almost half his genes are heterozygous for nonsynonymous substitutions, the total number of which, approximately 7781 (**Figure 7.4**), is similar to that observed in the second high-quality genome that was published, that of the co-discoverer of the structure of DNA, James D. Watson (Wheeler et al. 2008). Between 15% and 20% of these variants are rare, almost half are homozygous, and as many as 10% are predicted to be deleterious to the function of the protein. Slightly more than 100 of Venter's genes have premature stop codons, but these genes tend to be members of gene families or to be duplicated genes. Similarly, several hundred of his genes harbor indels in the coding regions, but these tend to be the length of codons, or multiples thereof, and to be located toward the 3′ ends of the genes. It is nevertheless clear that all of us carry mutations that should

Figure 7.4 Nonsynonymous Variants in J. Craig Venter's Genome.   (A) The two pie charts show the relative proportions of homozygous and heterozygous variants (left) and of common, rare, and novel variants (right) inferred from Venter's complete genome sequence. Note that whereas only 8% of homozygous variants in general are predicted to be deleterious, 22% of all rare variants are. (B) A few examples of Venter's homozygous common variants known to be associated with traits (but note that while Venter does have blue eyes, this does not mean he will necessarily develop the two diseases).

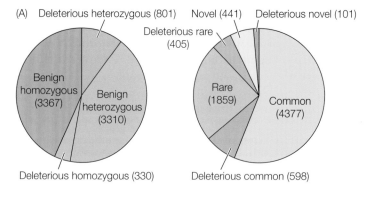

(A) Deleterious heterozygous (801)   Novel (441)   Deleterious novel (101)

Deleterious rare (405)

Benign homozygous (3367)

Benign heterozygous (3310)

Rare (1859)

Common (4377)

Deleterious homozygous (330)

Deleterious common (598)

(B)

Gene	Position	Trait
OCA2	W305R	Blue eyes
BDNF	V74M	Late-onset Parkinson's disease
SOD2	A16V	Prostate cancer

disrupt protein function, and that several of these mutations are likely to be homozygous.

Subsequent studies have reinforced each of these conclusions. Most genomes carry more than 100,000 novel variants that are private to the individual or, presumably, the individual's immediate family members, providing a vast pool of variability that is almost impossible to evaluate for functional significance. The sobering result of one WGS study of ten people with hemophilia was that the casual variant in the factor VIII gene was identified in only six of those people, most likely because the other four had small inversions that were not easy to detect by short-read sequencing (Pelak et al. 2010). Furthermore, 6% of the X chromosome variants were called heterozygous in males (not including another 5% in the pseudo-autosomal region), which was explained by the realization that many genes are recently duplicated and/or contained within variable copy number duplications. Many of these genotypes are not in Hardy-Weinberg equilibrium, since there is an excess of apparent heterozygosity, but again, the technical difficulties in reading WGS data are highlighted. Another analysis of twelve whole genomes found that two different sequencing platforms gave discordant calls for half of the small indels (Dewey et al. 2014). It also highlighted disagreements among professional analysts regarding which variants were likely to be pathogenic, though the analysts did agree on prescribing follow-up tests relating to the most highly implicated mutations. At this stage of development of the field, it must be recognized that different algorithms in the hands of different analysts can yield quite different inferences.

One further issue that is crucial for several aspects of interpretation of variant function is the phasing of haplotypes. Wherever an individual is heterozygous for two closely linked polymorphisms—say, A/G and T/C—it is impossible from the genotype calls alone to know whether the two haplotypes are A-T and G-C or A-C and G-T. If the A and T both affect amino acids, then in the first scenario, there would be one wild-type copy of the gene and one double-mutant copy, whereas in the second, the individual would be compound heterozygous: both copies would be abnormal. This question can be resolved if the two polymorphisms are read together from the same molecule, or if the parental sequences are available and transmission of the two haplotypes can be inferred. Additionally, most of the algorithms that are used for genotype imputation employ a haplotype phase estimation step that fits a maximum likelihood model using data from genome-wide LD to estimate phase with reasonable accuracy.

Some of the most fascinating results to emerge from initial WGS studies relate to human evolution. The sequencing of four Bushmen hunter-gatherers from the Kalahari region of southern Africa found over 1.3 million novel variants among them, and it demonstrated that these individuals were, on average, more different from one another than two typical European or Asian people (Schuster et al. 2010). Whole-genome sequencing of a further fifteen hunter-gatherers from Tanzania and Cameroon found signatures suggestive of the basis for the short stature of central African pygmies, as well as of archaic admixture with a possible relative of Neanderthals (Lachance et al. 2012). Whole-genome sequences derived from DNA contained in a 100-year-old lock of hair from an Aboriginal Australian provided strong support for an early wave of modern human dispersal into East Asia more than 60,000 years ago, predating the more recent dispersal that gave rise to most modern Asians (Rasmussen et al. 2011). Each of these studies also has implications not only for discovering disease associations in different human populations, but also for understanding the evolution of human traits and disease susceptibility.

Another important application of WGS is the detection of somatic changes within individuals. This technology can detect structural changes, such as the transposition of mobile elements, as well as de novo mutations that are tissue specific, in healthy individuals. For example, almost 1% of people are mosaic for large segmental deletions or duplications in their peripheral blood that are present in only some of their cells; this proportion increases with age and is associated with leukemia risk in a small proportion of those individuals. As we will see in Chapter 14, the combination of whole-genome and transcriptome sequencing is beginning to revolutionize individual cancer treatment as it identifies the novel fusion genes, point mutations, and altered copy numbers that may drive oncogenesis (Meyerson et al. 2010).

Inference of Deleteriousness

No matter the context, the inference that rare variants influence a disease depends on additional evidence beyond their statistical enrichment in individuals with that disease. Given that individual genomes harbor dozens of homozygous rare coding variants and hundreds of heterozygous ones, it is necessary to devise filters that will increase confidence that a particular nucleotide change is damaging. The dbNSFP database, which summarizes scores of predicted deleteriousness for all 76 million potential nonsynonymous sites in the human genome, is available at sites.google.com/site/jpopgen/dbNSFP (Liu et al. 2011). It summarizes a variety of prediction scores, all of which are based on the evolutionary and structural principles discussed in this section.

Biochemical or biophysical assays provide the strongest evidence of deleteriousness. In some cases, enzymatic assays allow direct measurement of protein defects in vitro, and in others, the mutation can be engineered into a microbial genome for which a suitable assay exists. Human cell cultures provide another convenient platform for evaluating the effect of overexpression or knockdown of gene function. Animal models, from *Drosophila* to rodents, are very important for establishing that a mutation not only disrupts protein function, but also produces a phenotype. Fruit flies are not always directly relevant to human cases, and mice can be expensive and time-consuming to work with, but both have a distinguished history in genetic analysis. Zebrafish provide a convenient, relatively high-throughput model so long as a tissue-specific assay is available (Niederriter et al. 2013). A common three-step procedure for modeling loss-of-function mutations is to (1) disrupt the normal gene function of zebrafish embryos with a **morpholino**, which is a modified oligonucleotide that binds to the mRNA (preferably at a splice junction) and prevents translation; (2) rescue the function in some of the embryos with mRNA made from the wild-type human gene, restoring normal development or function to the fish; and (3) detect the loss-of-function phenotype by injecting mRNA from the human gene encoding the putative deleterious mutation into other fish embryos (**Figure 7.5A**). Analogous procedures that do not require knockdown can be used to demonstrate gain of function (**Figure 7.5B**). In some cases, clinical intervention may require an immediate decision based on data that can be generated with a few weeks of molecular genetic testing in zebrafish. Ideally, though, positive results from this transient assay are then confirmed in transgenic mice if time permits.

Computational assessment of the likelihood that a variant is damaging is commonly pursued with a series of algorithms that, for the most part, measure whether the nucleotide affects a highly conserved residue (Kumar et al. 2011). It is now clear that the human genome has an excess of very low frequency variants as a result of the recent human population expansion combined with relatively weak purifying selection, as we saw in Chapter 3. Highly conserved sites are much more likely to be affected by very rare nonsynonymous substitutions than are less constrained ones (Fu et al. 2013). This observation makes sense in light of the theory of natural selection, since variants that disrupt functionally important sites are most likely to be deleterious, and hence are likely to be selected against, which prevents them from entering the gene pool. Most are lost within a few generations of their appearance in an individual, particularly if they cause loss of fertility or have a debilitating effect on quality of life.

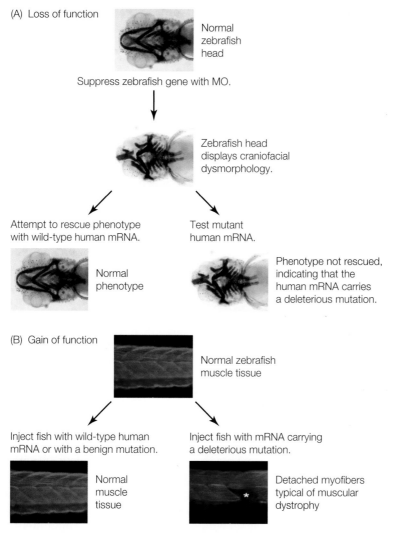

(A) Loss of function

Normal zebrafish head

Suppress zebrafish gene with MO.

Zebrafish head displays craniofacial dysmorphology.

Attempt to rescue phenotype with wild-type human mRNA.

Test mutant human mRNA.

Normal phenotype

Phenotype not rescued, indicating that the human mRNA carries a deleterious mutation.

(B) Gain of function

Normal zebrafish muscle tissue

Inject fish with wild-type human mRNA or with a benign mutation.

Inject fish with mRNA carrying a deleterious mutation.

Normal muscle tissue

Detached myofibers typical of muscular dystrophy

Figure 7.5 **Use of Zebrafish to Show Allelic Function.** (A) Loss of function is demonstrated in three steps: First, the wild-type zebrafish gene is knocked down with a morpholino (MO) and generates a phenotype related to the disease caused by the human mutation in that gene (e.g., craniofacial dysmorphology). Second, the phenotype is rescued by the normal human mRNA. Third, failure of the mutant human mRNA to rescue the phenotype implies that the allele is deficient in activity. (B) By contrast, gain of function can be demonstrated if injection of the mutant human mRNA generates a novel phenotype (e.g., muscular dystrophy), whereas the normal human mRNA has no effect, or a much milder effect. (After Niederriter et al. 2013.)

The two most popular software programs that are used to infer deleteriousness are SIFT and PolyPhen2 (Ng and Henikoff 2003; Adzhubei et al. 2010; **Figure 7.6**). SIFT (sorting intolerant from tolerant) compares a human protein sequence with the sequences of similar proteins (usually

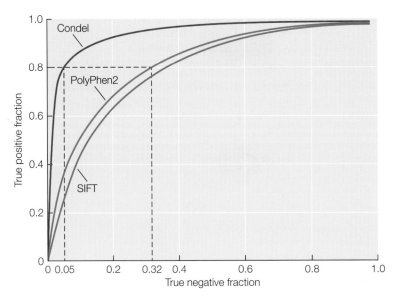

Figure 7.6 Comparative Performance of Deleteriousness Assessment Scores.
This Receiver Operating Curve shows the relationship between the proportion of true
positives detected by a score and the corresponding true negative inclusion rate. The
three curves are for SIFT, PolyPhen2, and Condel. For example, a PolyPhen2 score that
identifies 80% of variants that are known to be deleterious unfortunately also identifies
32% of benign variants as deleterious, whereas Condel is much more specific, only
including 5% of the true negatives. (After González-Pérez and López-Bigas 2011.)

orthologs) from other species and computes how often the amino acid
changes to another amino acid with similar or different biochemical properties, and on this basis generates a score representing whether the observed
change is likely to affect the protein function. PolyPhen2 incorporates limited structural information as well as multiple sequence alignments that
provide information on conservation, and it trains a classifier on a data set
of amino acid changes that are suspected to be damaging. The HumDiv
training set consists of 3155 damaging alleles documented in the UniProt
database of Mendelian mutations that affect protein stability or function,
while the more inclusive HumVar set includes all 13,032 Mendelian mutations in UniProt. When asked to predict whether mutations from both sets
of damaging mutations or a matched set of presumptive benign mutations
were damaging, PolyPhen2 predicted that 92% of the HumDiv and 73%
of the HumVar mutations would be damaging, while including only 20%
of the benign ones.

The overlap in prediction using these two tools is, however, far from
complete. A comparison of five methods found that only 73% of the Hum-
Div variants, and only 58% of the HumVar ones, were predicted to be
damaging by all five algorithms. However, the observation that in each case

98% were predicted by at least three algorithms led the developers of Condel (consensus deleteriousness score; bg.upf.edu/condel/home) to generate a weighted average score that improves the prediction accuracy to close to 90%, with a 10% false positive rate (see Figure 7.6). In this idealized situation, then, fairly robust bioinformatic prediction of deleteriousness can be obtained, but application to novel data sets is not expected to be as good. Nevertheless, there is clear enrichment of variants with high Condel scores in the very rare fraction of human exome variants, suggesting that this approach is a relatively efficient method for focusing attention on the most damaging substitutions.

Less damaging polymorphisms may nevertheless have important influences on disease, particularly if they act in combination with other rare variants (Sunyaev 2012). For this reason, ongoing efforts are directed at generating deleteriousness scores that are specific for individual families of proteins. These scores will account for specific structural features of protein folding topology as well as the biophysical attributes of individual amino acids and their propensity to affect attributes such as protein stability, flexibility, and contribution to protein-protein interactions (Liberles et al. 2012). Similarly, structural information can be used to enhance prediction of the potential influence of regulatory sequence changes on transcription factor binding. Inference of the functional impact of regulatory variants is less advanced than for protein variants, which is one reason why research tends to focus on exomes, but advances from the ENCODE project (discussed in Chapter 9) provide hope that computational assessment of this important class of polymorphisms is within reach.

Variants are also typically evaluated relative to their frequency in healthy control individuals, such as the set of 6503 whole-exome sequences in the Exome Variant Server (EVS: evs.gs.washington.edu/EVS/). Analysis of this data set has led to the realization that different genes—and indeed, regions within genes—have different intrinsic loads of mutations that reflect their tolerance for disruption. This realization led Petrovski et al. (2013) to develop an intolerance score that ranks genes with respect to the likelihood that they carry pathogenic mutations. Genes that harbor Mendelian mutations are less tolerant of damaging mutations in general, and this should be taken into account in interpreting personal genomes. Over the next 10 years, as millions of exome sequences accumulate, we will likely have a comprehensive database of the domains that are empirically enriched for rare variants in people who have rare diseases.

Candidate Gene Mapping

In some cases, there may already exist strong candidate genes that are a priori more likely to harbor mutations than the rest of the exome. For example, human genes may cause mutant phenotypes that are related to those caused by their orthologs in model organisms. What constitutes a good candidate gene will vary from study to study. Sometimes the hypothesis of deleteriousness derives from model organism studies of a well-characterized mutation that mimics a human phenotype. Other times it derives from integrative genomic strategies, such as gene expression experiments or metabolomic and biochemical profiling, that suggest a lesion in a gene that regulates an abnormal process. Once one gene in a family has been implicated, it is often not a stretch to suppose that related genes may be responsible. Again, interaction screens in model organisms may suggest a candidate; these screens may identify genes from different gene families that nevertheless produce similar phenotypes or modify the phenotype that is due to a known gene. Similarly, protein-protein interactions can suggest that mutations in a gene may contribute to the phenotype. Of course, candidate gene studies can be very misleading and cause years of anguish for students or postdocs, so it is safest to proceed where there are also linkage data that place the candidate in a LOD interval.

Furthermore, candidate gene mapping is not always possible, for example, where a condition is rare and insufficient family data are available to support linkage studies, or where the phenotype is only partially penetrant. In this case, genes already identified with some certainty in particular families may become candidates for causation in other families for which there are few data. An example is provided by scoliosis, or curvature of the spine (Maisenbacher et al. 2005). Working backward from rodent studies, researchers detected rare homozygous mutations in four genes that all have roles in development of the somites from which vertebrae derive: *DLL*, *MESP2*, *LFNG*, and *HES7*. Heterozygous mutations in two of these genes have been identified in other families with congenital scoliosis, but in which the disease does not follow Mendelian segregation patterns. Haploinsufficiency for the mouse homologs produces incompletely penetrant vertebral defects as well, but the condition is severely exacerbated by hypoxia during the critical phase of fetal development (Sparrow et al. 2012). Thus, a gene-by-environment interaction, in which the environmental factor is insufficient oxygen provided through the placenta, is suggested from candidate gene studies as a plausible mechanism for some of the phenotypic variation of a relatively common disease.

Consideration of this example raises the issue of what constitutes evidence that a mutation is contributing, if not causal. As we have seen, all humans carry dozens of potentially deleterious mutations in heterozygous form, some of which may be interacting with other polymorphisms and/or uncharacterized environmental factors. It has been suggested that at least three lines of evidence are required to establish causality. That evidence may be some combination of additional linkage or association evidence from distinct pedigrees, possibly involving different mutations; experimental evidence either from animal models or in vitro assays establishing that the variant influences gene activity; in silico evidence that the mutation disrupts an evolutionarily conserved site and/or alters the structure of the protein and its stability or flexibility; and gene expression data showing that aberrant transcription is associated with pathology. Software is emerging that allows researchers to incorporate knowledge of gene function and of patient symptomatology into their assessments of likely causality; examples include Phen-Gen (www.phen-gen.com) and eXtasy (homes.esat.kuleuven.be/~bioiuser/eXtasy/).

Emerging Technologies

Although next-generation sequencing is currently dominated by the three technologies discussed at the start of this chapter, there are a handful of emerging "single-molecule" methods that may result in another order-of-magnitude jump in sequencing volume and cost effectiveness.

One of these methods is Pacific Bioscience's SMRT (single-molecule real-time) sequencing (Carneiro et al. 2012; **Figure 7.7A**). The sequencing cell has a large number of wells, each of which contains a single DNA polymerase molecule. At each step of strand synthesis, a special phospholinked nucleotide with one of four fluorescent dyes is held momentarily at the bottom of the well, where special optics allow it to be read. Furthermore, modified bases, such as methylated cytosines, take slightly longer for the incorporation to occur, and this longer interval can be used to infer not just the nucleotide, but also whether or not it is modified in the genomic DNA. Very long reads of up to 20 kb are possible, and even the average read of 3–5 kb provides significant advantages for original assembly of microbial genomes, for phasing of extended haplotypes, and for sequencing intransigent loci such as the fragile X region (Loomis et al. 2012). Concerns over the relatively low accuracy of this method have prevented its widespread adoption, but the technology is being used alongside next-generation methods and for targeted resequencing.

Nanopore single-molecule sequencing (Stoddart et al. 2009) has been touted as the next wave for almost a decade, but at the time of this writing remains stalled by technical issues, despite the release of Oxford Nanopore's GridION system (Ayub and Bayley 2012; **Figure 7.7B**). The simple idea underlying this method is that a 1 nm pore anchored in a lipid membrane is just wide enough to allow a single strand of DNA to migrate through it, tens of kilobases at a time. As each base passes through the pore, it allows current that is characteristic of that base or modified base to flow. The pore is made of a protein such as α-hemolysin (or in the future, solid-state synthetics) and can be modified to support sequencing of RNA or proteins. A miniaturized MinION device is being developed that is designed to slot into a USB drive on a laptop, potentially facilitating genome-scale sequencing outside the lab environment.

All this sequencing is set to generate terabytes of data for individual projects, presenting massive challenges for data storage, access, and analysis. Large bioinformatics labs backed by high-performance computing networks are well equipped to deal with these data, but clinicians and groups focused on molecular and cell biology tend not to be. For this reason, genomic service providers are beginning to offer cloud-based solutions: as it is generated, the data will be directly uploaded to the cloud, where dedicated analytic software will perform the initial short-read alignment and

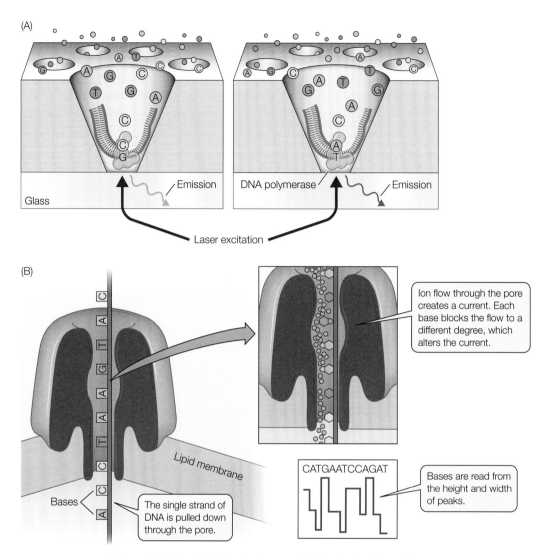

Figure 7.7 Single-Molecule Sequencing Technologies. (A) Pacific Bioscience's SMRT sequencing employs an array of tiny wells, each containing a single molecule of DNA polymerase and a window that allows the fluorescence associated with incorporation of one new base to the strand to be detected. (B) Oxford Nanopore uses an array of tiny pores that allow protons to cross the membrane in proportion to the width of the channel that remains as bases of different sizes occupy the pore.

variant annotation, delivering processed data to the customer (Thakur et al. 2012). A large community of users has already adopted the open-source Galaxy Web-based platform (galaxyproject.org), which promotes "accessible, reproducible, and transparent" computational genomics (Giardine et al. 2005; Goecks et al. 2010).

Summary

1. Next-generation sequencing methods have brought the costs of genome-scale sequencing within the budgets of standard researchers and are set to revolutionize clinical genetics.

2. Three motivations for genome-scale sequencing are identification of very rare protein-coding variants, direct measurement of Goldilocks variants in support of GWAS, and capture of the full range of non-coding regulatory variation.

3. Whole-genome sequencing (WGS) refers to deep sequencing of total genomic DNA, whereas whole-exome sequencing (WES) refers to deep sequencing of only the 1% of the genome that consists of exons.

4. Three technologies, produced by Illumina, Life Technologies, and Complete Genomics, dominate the next-generation sequencing landscape in the first half of the 2010s.

5. Read depths of at least 30X coverage of each base ensure that the majority of heterozygous polymorphisms are detected. This requires 100 Gb of whole-genomic DNA or 1 Gb of exome sequence.

6. Mendelian mutations can be identified by sequencing of as few as a half dozen proband exomes.

7. Family-based whole-exome sequencing is a promising approach for precise mapping of mutations that were previously known only as linkage peaks in pedigree analysis.

8. Sequencing of unrelated individuals from the extremes of the phenotype distribution, or with very rare diseases, may also identify rare variants that contribute genetic variation for disease, but significant statistical challenges must be overcome.

9. De novo variants are novel mutations not present in either parental genome. Each person is thought to carry an average of one potentially deleterious new mutation.

10. Whole-genome sequencing is revealing new secrets about human evolution.

11. Computational algorithms can help investigators assess whether a newly discovered variant is likely to be deleterious. They work best when several different approaches are combined.

12. The next wave of single-molecule sequencing may be driven by single-molecule real-time sequencing from Pacific Biosciences and nanopore sequencing from Oxford Nanopore.

References

Adzhubei, I. A., Schmidt, S., Peshkin, L., Ramensky, V. E., Gerasimova, A., et al. 2010. A method and server for predicting damaging missense mutations. *Nat. Methods* 7: 248–249.

Ayub, M. and Bayley, H. 2012. Single molecule RNA base identification with a biological nanopore. *Biophys. J.* 102: 429.

Bamshad, M. J., Ng, S. B., Bigham, A. W., Tabor, H. K., Emond, M. J., et al. 2011. Exome sequencing as a tool for Mendelian disease gene discovery. *Nat. Rev. Genet.* 12: 745–755.

Bell, C. J., Dinwiddie, D. L., Miller, N. A., Hateley, S. L., Ganusova, E. E., et al. 2011. Carrier testing for severe childhood recessive diseases by next-generation sequencing. *Sci. Transl. Med.* 3: 65ra4.

Botstein, D., White, R. L., Skolnick, M., and Davis, R. W. 1980. Construction of a genetic linkage map in man using restriction fragment length polymorphisms. *Am. J. Hum. Genet.* 32: 314–331.

Carneiro, M. O., Russ, C., Ross, M. G., Gabriel, S., Nusbaum, C., and DePristo, M. 2012. Pacific Biosciences sequencing technology for genotyping and variation discovery in human data. *BMC Genomics* 13: 375.

Choi, M., Scholl, U. I., Ji, W., Liu, T., Thikonova, I. R., et al. 2009. Genetic diagnosis by whole exome capture and massively parallel DNA sequencing. *Proc. Natl. Acad. Sci. U.S.A.* 106: 19096–19101.

Cirulli, E. T. and Goldstein, D. B. 2010. Uncovering the roles of rare variants in common disease through whole-genome sequencing. *Nat. Rev. Genet.* 11: 415–425.

Curran, M. E., Splawski, I., Timothy, K. W., Vincent, G. M., Green, E. D., and Keating, M. T. 1995. A molecular basis for cardiac arrhythmia: *HERG* mutations cause LongQT syndrome. *Cell* 80: 795–803.

Dewey, F. E., Grove, M. E., Pan, C., Goldstein, B. A., Bernstein, J. A., et al. 2014. Clinical interpretation and implications of whole-genome sequencing. *JAMA* 311: 1035–1044.

Freimer, N. and Sabatti, C. 2004. The use of pedigree, sib-pair and association studies of common diseases for genetic mapping and epidemiology. *Nat. Genet.* 36: 1045–1051.

Fromer, M., Moran, J. L., Chambert, K., Banks, E., Bergen, S. E., et al. 2012. Discovery and statistical genotyping of copy-number variation from whole-exome sequencing depth. *Am. J. Hum. Genet.* 91: 597–607.

Fu, W., O'Connor, T. D., Jun, G., Kang, H. M., Abecasis, G., et al. NHLBI Exome Sequencing Project, Akey, J. M. 2013. Analysis of 6,515 exomes reveals the recent origin of most human protein-coding variants. *Nature* 493: 216–220.

Giardine, B., Riemer, C., Hardison, R. C., Burhans, R., Elnitski, L., et al. 2005. Galaxy: A platform for interactive large-scale genome analysis. *Genome Res.* 15: 1451–1455.

Goecks, J., Nekrutenko, A., Taylor, J., and The Galaxy Team. 2010. Galaxy: A comprehensive approach for supporting accessible, reproducible, and transparent computational research in the life sciences. *Genome Biol.* 11: R86.

González-Pérez, A. and López-Bigas, N. 2011. Improving the assessment of the outcome of nonsynonymous SNVs with a consensus deleteriousness score, Condel. *Am. J. Hum. Genet.* 88: 440–449.

Itsara, A., Wu, H., Smith, J. D., Nickerson, D. A., Romieu, I., et al. 2010. De novo rates and selection of large copy number variation. *Genome Res.* 20: 1469–1481.

Kong, A., Frigge, M. L., Masson, G., Besenbacher, S., Sulem, P., et al. 2012. Rate of de novo mutations and the importance of father's age to disease risk. *Nature* 488: 471–475.

Krumm, N., Sudmant, P. H., Ko, A., O'Roak, B. J., et al., and NHLBI Exome Sequencing Project. 2012. Copy number variation detection and genotyping from exome sequence data. *Genome Res.* 22: 1525–1532.

Kumar, S., Dudley, J. T., Filipski, A., and Liu, L. 2011. Phylomedicine: an evolutionary telescope to explore and diagnose the universe of disease mutations. *Trends Genet*. 27: 377–386.

Lachance, J., Vernot, B., Elbers, C. C., Ferwerda, B., Froment, A., et al. 2012. Evolutionary history and adaptation from high-coverage whole-genome sequences of diverse African hunter-gatherers. *Cell* 150: 457–469.

Ladouceur, M., Dastani, Z., Aulchenko, Y. S., Greenwood, C. M. T., and Richards, J. B. 2012. The empirical power of rare variant association methods: Results from Sanger sequencing in 1,998 individuals. *PLoS Genet*. 8: 1002496.

Levy S., Sutton, G., Ng, P. C., Feuk, L., Halpern, A., et al. 2007. The diploid genome sequence of an individual human. *PLoS Biology* 5: e254.

Li, H. and Durbin, R. 2009. Fast and accurate short read alignment with Burrows-Wheeler transform. *Bioinformatics* 25: 1754–1760.

Li, H., Handsaker, B., Wysoker, A., Fennell, T., Ruan, J., et al., and 1000 Genome Project Data Processing Subgroup. 2000. The Sequence Alignment/Map format and SAMtools. *Bioinformatics* 25: 2078–2079.

Liberles, D. A., Teichmann, S. A., Bahar, I., Bastolla, U., Bloom, J., et al. 2012. The interface of protein structure, protein biophysics, and molecular evolution. *Protein Sci*. 21: 769–785.

Liu, X., Jian, X., and Boerwinkle, E. 2011. dbNSFP: A lightweight database of human nonsynonymous SNPs and their functional predictions. *Hum. Mutat*. 32: 894–899.

Loomis, E. W., Eid, J. S., Peluso, Yin, J., Hickey, L., et al. 2012. Sequencing the unsequenceable: Expanded CGG-repeat alleles of the fragile X gene. *Genome. Res*. 23: 121–128.

MacArthur, D. G., Manolio, T. A., Dimmock, D. P., Rehm, H. L., Shendure, J., et al. 2014. Guidelines for investigating causality of sequence variants in human disease. *Nature* 508: 469–476.

Maisenbacher, M. K., Han, J-S., O'Brien, M. L., Tracy, M. R., Erol, B., et al. 2005. Molecular analysis of congenital scoliosis: A candidate gene approach. *Hum. Genet*. 116: 416–419.

McKenna, A., Hanna, M., Banks, E., Sivachenko, A., Cibulskis, K., et al. 2010. The Genome Analysis Toolkit: A MapReduce framework for analyzing next-generation DNA sequencing data. *Genome Res*. 20: 1297–1303.

Merriman, B., Ion Torrent R&D Team, and Rothberg, J. M 2012. Progress in Ion Torrent semiconductor chip based sequencing. *Electrophoresis* 33: 3397–3417.

Metzker, M. L. 2010. Sequencing technologies—The next generation. *Nat. Rev. Genet*. 11: 31–46.

Meyerson, M. and Gabriel, S., Getz, G. 2010. Advances in understanding cancer genomes through second-generation sequencing. *Nat. Rev. Genet*. 11: 685–696.

Need, A. C., Shashi, V., Hitomi, Y., Schoch, K., Shianna, K. V., McDonald, M. T., et al. 2012. Clinical application of exome sequencing in undiagnosed genetic conditions. *J. Med. Genet*. 2012: 100819

Ng, P. C. and Henikoff, S. 2003. SIFT: Predicting amino acid changes that affect protein function. *Nucl. Acids Res*. 31: 3812–3814.

Ng, P. C., Levy, S., Huang, J., Stockwell, T., Walenz, B., et al. 2008. Genetic variation in an individual human exome. *PLoS Genet*. 4: 1000160.

Ng, S. B., Buckingham, K. J., Lee, C., Bigham, A. W., Tobor, H., et al. 2010. Exome sequencing identifies the cause of a Mendelian disorder. *Nat. Genet*. 42: 30–35.

Niederriter, A. R., David, E. E., Golzio, C., Oh, E. C., Tsai, I. C., Katsanis, N. 2013. In vivo modeling of the morbid human genome using *Danio rerio*. *J. Vis. Exp*. (78): e50338.

Pelak, K., Shianna, K. V., Ge, D., Maia, J. M., Zhu, M., et al. 2010. The characterization of twenty sequenced human genomes. *PLoS Genet.* 6: 1001111.

Peters, B. A., Kermani, B. G., Sparks, A. B., Alferov, O., Hong, P., et al. 2012. Accurate whole-genome sequencing and haplotyping from 10 to 20 human cells. *Nature* 487: 190–195.

Petrovski, S., Wang, Q., Heinzen, E. L., Allen, A. S., and Goldstein, D. B. 2013. Genic intolerance to functional variation and the interpretation of personal genomes. *PLoS Genet.* 9: e1003709.

Rasmussen, M., Guo, X., Wang, Y., Lohmueller, K. E., Rasmussen, S., et al. 2011. An Aboriginal Australian genome reveals separate Human dispersals into Asia. *Science* 334: 94–98.

Risch, N. 1990. Linkage strategies for genetically complex traits. II. The power of affected relative pairs. *Am. J. Hum. Genet.* 46: 229–241.

Rybicki, B. A. and Elston, R. C. 2000. The relationship between the sibling recurrence-risk ratio and genotype relative risk. *Am. J. Hum. Genet.* 66: 593–604.

Schuster, S. C., Miller, W., Ratan, A., Tomsho, L. P., Giardine, B., et al. 2010. Complete Khoisan and Bantu genomes from southern Africa. *Nature* 463: 943–947.

Shendure, J. and Ji, H. 2008. Next-generation DNA sequencing. *Nat. Biotech.* 26: 1135–1145.

Sobreira, N. L. M., Cirulli, E. T., Avramopoulos, D., Wohler, E., Oswald, G., et al. 2010. Whole genome sequencing of a single proband together with linkage analysis identifies a Mendelian disease gene. *PLoS Genet.* 6: 1000991.

Sparrow, D. B., Chapman, G., Smith, A. J., Mattar, M. Z., Major, J. A., et al. 2012. A mechanism for gene-environment interaction in the etiology of congenital scoliosis. *Cell* 149: 295–306.

Stoddart, D., Heron, A. J., Mikhailova, E., Maglia, G., and Bayley, H. 2009. Single-nucleotide discrimination in immobilized DNA oligonucleotides with a biological nanopore. *Proc. Natl. Acad. Sci. U.S.A.* 106: 7702–7707.

Sunyaev, S. R. 2012. Inferring causality and functional significance of human coding DNA variants. *Hum. Mo.l Genet.* 21(R1): R10–R17.

Thakur, R. S., Bandopadhyay, R., Chaudhary, B., and Chatterjee, S. 2012. Now and next-generation sequencing techniques: Future of sequence analysis using cloud computing. *Front. Genet.* 3: 280.

Veltman, J. A. and Brunner, H. G. 2012. De novo mutations in human genetic disease. *Nat. Rev. Genet.* 13: 565–575.

Wheeler, D. A., Srinivasan, M., Egholm, M., Shen, Y., Chen, L., et al. 2008. The complete genome of an individual by massively parallel DNA sequencing. *Nature* 452: 872–876.

Zuk, O., Schaffner, S. F., Samocha, K., Do, R., Hechter, E., et al. 2014. Searching for missing heritability: Designing rare variant association studies. *Proc. Natl. Acad. Sci. U.S.A.* 111: E455–E464.

8

Gene Expression Profiling

We turn now to the first of the functional genomic technologies that are used to profile dynamic aspects of genome activity: namely, gene expression profiling, or "transcriptomics." Whereas genotyping and DNA sequencing are ideal for discovering which mutations and polymorphisms contribute to genetic variation, high-throughput transcriptomics, epigenomics, proteomics, and metabolomics are used to begin to measure how the variants might influence traits. **Gene expression profiling**—the process of systematically documenting the abundance of all transcripts expressed in a tissue sample—was the first of these technologies to be developed, in the mid-1990s (Schena et al. 1995; Lockhart et al. 1996), and it is the most widely used as well as the best developed in theory and practice. Its many applications include the following:

- Cataloging which genes are expressed in which tissues at which times under which conditions

- Comparing healthy and diseased tissues to elucidate the biochemical pathways involved in pathogenesis

- Finding biomarkers of different subtypes of disease or states of disease progression

- Monitoring alternative splicing and transcript isoform usage

- Understanding how the environment influences gene activity

- Annotating disease SNPs by establishing which transcripts they influence the expression of

- Describing the covariance of transcripts and microRNAs in networks

- Supporting systems biology approaches and hypothesis testing in regard to complex trait genetics

In this chapter, we first consider the two main technologies for gene expression profiling, microarrays and RNA sequencing (RNA-Seq), then discuss aspects of data analysis including normalization, statistical inference, and pathway identification. Several representative examples of the applications listed above are then described.

Microarray Analysis

Microarray analysis refers to any approach to gene expression profiling that relies on hybridization of labeled cDNA (RNA that has been reverse-transcribed into complementary DNA) to an array of oligonucleotide probes. In the early days of microarray analysis, before around 2005, the arrays were typically made by robotic spotting of thousands of PCR-amplified cDNA clones onto the surface of a glass microscope slide. Today, at least for human studies, commercial platforms vastly exceed the performance of such home-made arrays in terms of comprehensive coverage, cost-effectiveness, data quality, and reproducibility. There are three main platforms available, supplied by Agilent, Illumina, and Affymetrix (Wolber et al. 2006; Fan et al. 2006; Dalma-Weiszhaus et al. 2006), all of which use content based on gene structures annotated by GenBank and Ensembl in the HuRef19 reference human genome.

The principle behind each of these technologies is that transcript abundance can be estimated by contrasting the intensity of fluorescence produced by dye-labeled cDNA that hybridizes to gene- or exon-specific oligonucleotide probes (**Figure 8.1**). There are various strategies for labeling single- or double-stranded cDNA during or after the reverse transcription, using either polyT or random hexamers to prime the synthesis. Each technology uses a different approach to generating tens of thousands of unique probes, ranging in length from 25 nt to 60 nt. Total cDNA from a sample is hybridized to an array containing all of these probes, and each molecule finds its complementary sequence, binding in proportion to the amount of the transcript in the cDNA. Consequently, if sample 1 has twice the amount of transcript as sample 2, then twice as many cDNA molecules will hybridize to the corresponding probe. When the arrays are scanned with a laser to induce fluorescence, sample 1 will generate twice the signal intensity as sample 2. Comparisons between different probes are only qualitative because the probes differ in GC content as well as in other factors that influence hybridization. Consequently, analysis generally focuses on comparison of the same probe across different samples. For most purposes, data are acquired as 16-bit TIFF files, which means that the signal intensity values range from 0 to 65,536 (that is, 2^{16}). Since the background fluorescence is adjusted to be around 50 units, this means that microarray analysis provides resolution over three orders of magnitude, which is also approximately the range over which most transcript abundance varies in cells.

Agilent Technologies offers three levels of coverage on its SurePrint G3 gene expression microarrays: gene-oriented arrays with 44,000 or 60,000 probes and an exon-oriented 400,000-probe array that includes 233,184 exons in 27,696 genes as well as long noncoding RNAs and other content. Its arrays are printed on glass slides using an ink-jet-like printer designed

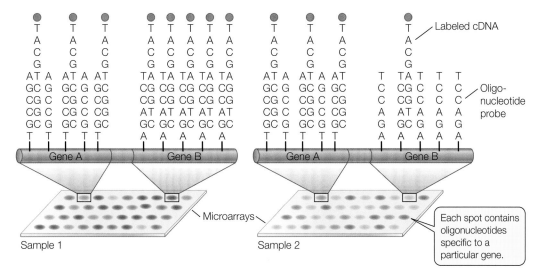

Figure 8.1 Principle of Microarray-Based Gene Expression Profiling.   Thousands of oligonucleotide probes are arrayed in such a way that dye-labeled cDNA generated from a sample will find its complement under optimal hybridization conditions. In this diagram, half the genes expressed in the two samples differ in transcript abundance. In sample 1, gene B is inferred to be more highly expressed than gene A, since more cDNA molecules hybridize to the spot of probes matching that gene. In sample 2, by contrast, gene B is downregulated, whereas gene A is expressed at the same level as in sample 1.

to specifically deposit just picoliters of A, G, C, or T nucleotides on each of the hundreds of thousands of spots from which probes are grown. In each of 240 (= 4 × 60) cycles, a single nucleotide is added, and a chemical deblocking step is used after each cycle to allow the addition of the next nucleotide in the chain. This very flexible manufacturing approach also allows for custom design of arrays targeting genes and exons of specific interest. Each spot is 30 μm in diameter, and the Agilent scanner supports 2 μm resolution, providing approximately 100 pixels of data for each probe, the averaging of which generates the transcript abundance measure. By scanning at two different intensities, it is possible to increase the dynamic range another order of magnitude. Agilent arrays can be used in either one-color or two-color modes, the latter allowing two samples (e.g., an experimental and a reference sample) to be profiled simultaneously (**Figure 8.2A**).

Illumina (**Figure 8.2B**) now exclusively provides the HumanHT-12 Expression Bead Chip, which includes 47,000 probes representing 28,688 well-annotated genes. Its 50 nt probes are synthesized in solution and attached to a unique 29 nt address "barcode" that is cross-linked to a microbead. A pool of these microbeads is spread across the etched surface of an array such that each well contains a single bead that can be interrogated by laser scanning. Every array has a random assortment of 1.5 million beads, with an average of 30 beads per probe, the identity of which is decoded by Illumina using the barcodes before the array is sent to the customer. Thus, the positions of the

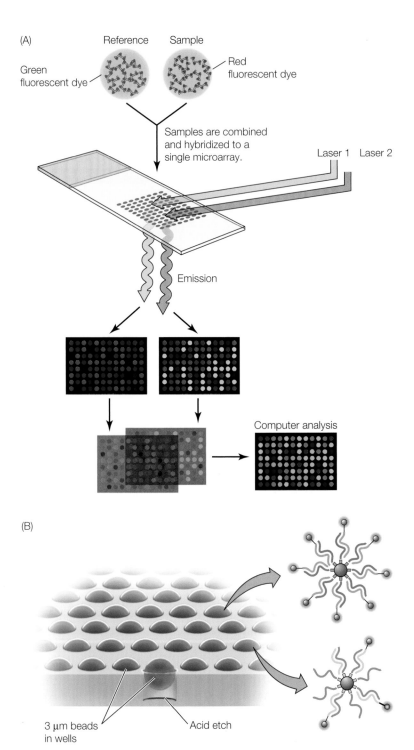

(A)

Reference   Sample

Green
fluorescent dye

Red
fluorescent dye

Samples are combined
and hybridized to a
single microarray.

Laser 1   Laser 2

Emission

Computer analysis

(B)

3 μm beads
in wells

Acid etch

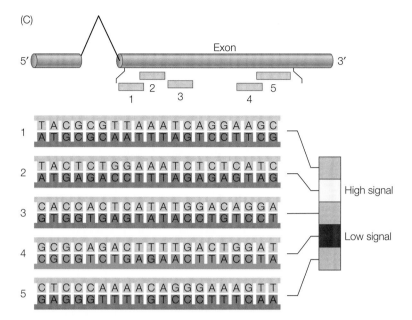

Figure 8.2 Commercial Microarray Platforms. (A) Agilent employs spots of oligonucleotide probes printed on a glass slide by a modified ink-jet-like technology. In two-color mode, the experimental sample is labeled with one dye and a reference sample with another. Fluorescence is induced by lasers of two different wavelengths, and the signals are merged to generate ratios of experimental sample to reference sample, which provide a measure of quality control for differences in spot size and quality. (B) Illumina deposits oligonucleotide-coated beads into an array of etched wells and reports the simple fluorescence intensity of each bead. Signals from several dozen identical beads spread across the array are averaged to give the estimated abundance of each transcript. (C) Affymetrix uses sets of up to a dozen shorter oligonucleotides that provide coverage of a gene or exon and may overlap. A weighted average of the fluorescence intensity from each probe provides the signal for that gene or exon. If there is a mismatch between the probe and the cDNA (as shown for probe 4), the signal intensity will drop, which can reduce the average signal for the gene by as much as 10%. This is in the range of differential expression. This problem, which is also an issue with the other array technologies, can be adjusted for statistically.

probes are unique for each individual array, which minimizes a source of technical error that arises when probes are always at the same place. Signal intensity is computed as a trimmed average of the signals from the 30 beads by Illumina GenomeStudio software. Some genes are represented by up to five different probes, some of which are exon-specific and some common. Illumina also offers a modified protocol and array, the DASL assay, which supports transcriptome analysis of degraded RNA samples such as those obtained from formalin-fixed, paraffin-embedded clinical specimens.

Affymetrix (**Figure 8.2C**) provides both the PrimeView and U133 families of short (25 nt) oligonucleotide arrays on which each transcript is represented by a probe set consisting of 11 partially overlapping probes.

These arrays come in cartridge format for single samples with a core set of 18,400 probe sets (U133A) or an extended set of 47,400 (U133A Plus) or 49,395 probe sets (PrimeView), and in plate format for 24 or 96 samples representing 39,000 probe sets on dual-array plates. The oligonucleotides are synthesized on the surface of a silicon chip using a series of masks to permit light to be focused on spots where the growing oligonucleotide needs to be chemically deblocked to allow incorporation of the next base. One hundred cycles, each with a different mask, are utilized in succession, cycling through A, C, G, and T nucleotides 25 times. Earlier versions of Affymetrix technology utilized "mismatch" probes that incorporated an incorrect nucleotide in the middle position, preventing strong hybridization, and these probes were used as negative controls for each perfect-match probe. However, statistical advances have obviated the need for such mismatch probes, though it should be noted that several different algorithms can be used to extract robust signals from the probe sets, and these can yield quite different results (Gautier et al. 2004). One concern with all hybridization technologies, but especially with short oligonucleotide technologies, is that polymorphisms can strongly influence hybridization (a fact that is used by Affymetrix as the core of its original Human SNP Array genotyping platform). Solutions to this dilemma are available through the NetAffx analysis center or with open-access software obtained through the Bioconductor project (www.bioconductor.org). Affymetrix also offers small RNA array solutions, kits for dealing with degraded samples, and a new GeneChip Exon 1.0 array with 1.4 million probe sets that provides unprecedented coverage of the vast majority of all exons (Xu et al. 2011).

RNA Sequencing

RNA sequencing, or RNA-Seq for short (Cloonan et al. 2008; Mortazavi et al. 2008; Wang et al. 2009), is gradually replacing microarray analysis for many applications. The costs of the two technologies are now comparable, running between $130 and $1000 per sample, depending on read depth, economies of scale, and access to next-generation sequencers. The perceived advantages of RNA-Seq include the following:

- Digital estimation of transcript abundance based on actual counts of molecules

- Comprehensive coverage of all RNAs independent of the content included on arrays

- Ability to distinguish alternative exon usage and to estimate transcript isoform abundance

- Resolution of duplicate genes that would be likely to hybridize to the same probes

- Ability to measure allele-specific expression by comparison of the relative abundances of both alleles

However, there are also drawbacks to RNA-Seq that need to be considered:

- Relatively low sensitivity to low-abundance transcripts and/or transcript isoforms

- The as yet poorly understood statistical properties of RNA-Seq-based transcript profiles

- Intense computational demands, including many opportunities for analytic mistakes

- Data storage burdens and associated issues with data sharing

- Concerns that sequence information may compromise the privacy of individual identity

A basic RNA-Seq workflow diagram is shown in **Figure 8.3**. The first few steps after mRNA has been converted to cDNA closely resemble those in the genomic NGS workflow diagram in Figure 7.2. Rather than synthesizing full-length cDNA, which results in 3′ end biases, the RNA is fragmented and asymmetrical adapters, which conveniently provide information on the orientation of the strand, are incorporated in the reverse transcription

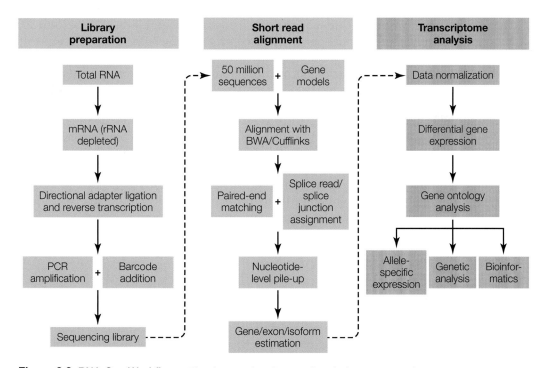

Figure 8.3 RNA-Seq Workflow.   The three major phases of analysis are preparation of the cDNA library, alignment of short paired-end reads to an annotated genome, and downstream inference of differential expression and bioinformatics.

step. The cDNA is thus amplified "directionally." Libraries are then pre-pared for sequencing, usually on an Illumina HiSeq or ProtonTorrent platform to ensure that a sufficient number of reads, usually 50 million per sample, is obtained. Typically, short hexamer barcodes are added to each fragment in a single sample, so that four or more samples can be run in parallel on one lane, a process known as **multiplexing**. After the sequencing run, each read is assigned to the appropriate sample on the basis of the barcode. The reads are aligned to the reference human genome with the Burrows-Wheeler Aligner (BWA) or a similar short-read aligner (Langmead et al. 2009) and piled up so that counts of the number of reads per exon are obtained (**Figure 8.4**). Paired-end information improves alignment accuracy by ensuring that both ends are matched to the same gene. One popular open-source analytic procedure is the Tuxedo protocol, consisting of Bowtie and TopHat for alignment, Cufflinks for inference of transcript abundance and differential expression, and CummeRbund for data visualization (cufflinks.cbcb.umd.edu; Trapnell et al. 2009, 2012). A peculiar issue with RNA-Seq data is that a large proportion of reads cross splice junctions and so do not align contiguously to genomic DNA. TopHat and related algorithms include routines that rescue these short reads and use paired-end information to span introns and create alignments that are most likely to match the true gene structure.

The raw output of an RNA-Seq experiment is a file containing the number of reads that align to each base (see Figure 8.4C). These data are converted to reads per exon based on the known or inferred location of exon-intron boundaries. Because genes and exons have different lengths, and because experiments have different read depths, normalization of the raw reads is required before statistical inference can be initiated. Cufflinks converts the raw counts to RPKM values, or their mate-pair equivalent FPKM (*reads/fragments per kilobase of exon per million mapped reads*). The adjustment for each kilobase of exon allows direct comparison of genes with one another, while the adjustment for mapped reads ensures that sam-ples can be compared without the influence of variation in library quality and quantity. As many as a third to a half of all reads in some experiments do not align unambiguously to the reference genome (due to repeats, low sequence quality, micro-exons, and other factors), so the normalization is to the number of mapped reads rather than the total reads. FPKM values are, however, just one way to represent transcript abundance. Another popular procedure is to use SAMtools, Perl scripts, and open-source software such as edgeR (Robinson et al. 2010) or DESeq (Anders and Huber 2010) to per-form initial normalization without adjusting for exon length. Refinements that adjust for nucleotide content and the negative binomial distribution of counts have been reported to improve accuracy (Anders et al. 2013).

Furthermore, read counts can be analyzed at three different levels, cor-responding to genes, exons, or transcripts. Gene counts may be the most robust of these measures, since summation over all exons is least suscep-tible to stochastic noise with the smaller numbers of reads for each exon. For many purposes, it is more than adequate for a first-pass analysis. Exon

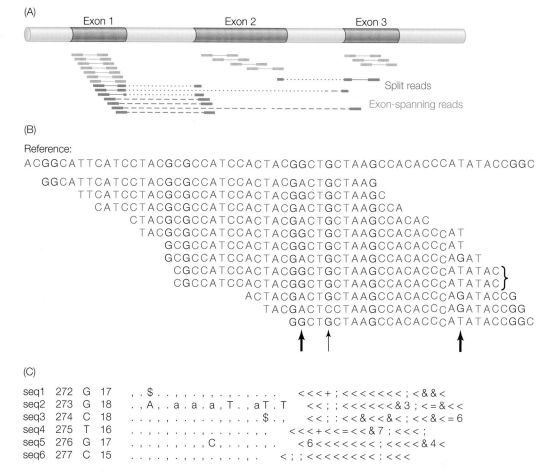

Figure 8.4 Representations of Read Alignment. (A) In this example, 22 pairs of short reads align to a gene with three exons. Four of the reads (blue) have one end split between two exons. Four of the reads (red) completely span two exons, and the remainder (green) have both ends within one exon. Each read is typically 100 bases long, and the two ends are separated by 100 to 200 bases. (B) An alignment of one end of each of 12 reads (half of which would be expected to have been read from the opposite strand). The bracket indicates a potential duplicate that should be removed. The heavy arrows show a potential polymorphism with complete linkage disequilibrium between the two sites. Equal expression of both alleles can be inferred from the first (G/A) site, but at the second site, the G is slightly underrepresented (3/8) possibly due to sampling variance. The thin arrow shows a probable sequencing error, since just one read has a C instead of a G. (C) A pileup format representation of an extended set of reads starting at base 272, the first base of the bottom read in (B). Commas indicate that the read was from the opposite strand but matches the reference, and the polymorphism is also indicated as the alternative bases that do not match the reference (CAPS for forward strand, lowercase for reverse). $ implies that the base is at the end of a sequence read. All the symbols on the right correspond to base quality measures in the individual reads.

counts take full advantage of the RNA-Seq platform and provide insight into alternative exon usage, which may be critical for many biological processes. Accurate transcript isoform counts are the most desirable measure, but since short reads a few hundred nucleotides in length cover only a portion of each transcript, isoform abundance is estimated only indirectly. As many transcripts will be relatively rare, and different graph theoretical algorithms will give variable results, transcript-level analyses need to be treated with caution. At any site at which individuals are heterozygous, it is also possible to compare the counts of the transcripts from the two chromosomes to obtain allele-specific expression measures. Allele-specific expression can arise due to regulatory variation that acts in *cis* (only on the same chromosome) or because of imprinting or somatic epigenetic modifications. It is an important parameter because it provides insight into how regulatory variants act, generates a sensitive measure of differential expression, and informs us about the additivity of genetic effects.

An important parameter of RNA-Seq is the read depth that is required to support robust inference. Some useful measures in this regard are shown in **Figure 8.5**. The correlation between technical replicates is generally very high, approximately the same as that seen in comparisons of microarray samples, but only when FPKM values are above 5 in a typical experiment involving 50 million reads. Below that level, repeatability is poor, and the sampling error will tend to overwhelm any biological differences. Figure 8.5A shows examples of high, moderate, and low correlation between samples as the biological differences between them increase. Generally, as read depths increase, total counts for each exon increase and FPKM values stabilize, so lower FPKM values can be included. In a typical whole-blood

Figure 8.5 Measures of RNA-Seq Performance.   (A) Each graph shows the relationship between read counts per gene ($\log_2$) for two samples in an experiment examining gene expression in macaques exposed to the antimalarial drug pyrimethamine. The top graph contrasts two successive samples of bone marrow taken from the same individual a week apart and is similar to a technical replicate. Note the greater variance at low levels of expression (bottom left of the graph). The center graph contrasts bone marrow samples taken from two different macaques at the same time; note the slightly reduced correlation overall. The bottom graph contrasts a blood sample and a bone marrow sample taken from the same individual at the same time, showing large-scale differential expression, largely reflecting higher expression of many genes in the bone marrow. (B) Two measures of the effect of read depth on inference of transcript abundance in a study of gene expression in four samples from rat nucleus accumbens. The total number of exons detected increases by 10% moving from 20 million to 40 million reads (upper plot). The correlation of inferred abundance from subsets of the data with that from the full data (78 million reads) shows a clear dependence on read depth that becomes increasingly stronger at lower abundance levels (lower plot). (C) Similarly, for a given read depth, the fraction of reads within 5% of the high-depth (final) value is a function of the RPKM. Plots like these establish 40 million reads as the minimum number required for accurate inference of gene expression. (A, from author's unpublished data; B, after Chen et al. 2011; C, after Mortazavi et al. 2008.)

transcriptome, approximately 10,000 genes can be included with high confidence at FPKM > 5 (see Figure 8.5B). Repeatability can also be assessed by asking how low-read-depth measures correlate with high-read-depth ones, or by asking what percentage of transcripts are within 5% of the high-depth abundance estimate; that percentage is a function of the FPKM range (see Figure 8.5C). For the most highly expressed genes, over 80% are already close to the high confidence value with 20 million reads, but for low-abundance transcripts, fewer than 50% give accurate estimates at that depth. It seems that for humans, at least 40 million reads are required to give accurate estimates for the 10,000 most abundant transcripts. Note

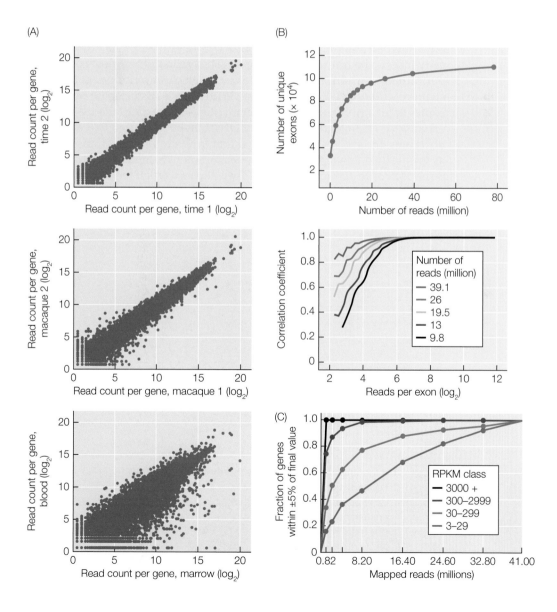

also that transcript abundance is almost always skewed, with most genes represented by just a few transcripts per cell and a small minority of genes represented by more than 100 transcripts.

With the recognition that small RNAs perform important regulatory functions, much attention has turned to profiling of miRNAs in a high-throughput manner. Microarrays and RNA-Seq can both be used for this purpose, but they require specific protocols that enrich for RNAs less than 100 bp in length. These small RNAs need only be sequenced unidirectionally, and with fewer than a thousand miRNA species of most interest, lower read depths are required, and multiplexing can include up to 96 samples on a single lane. As with mRNA, a few species of miRNA tend to account for the majority of the reads.

For some gene expression profiling applications, it is important to validate microarray or RNA-Seq findings with a different assay that specifically targets the transcripts of interest. This validation is most commonly performed by means of the quantitative reverse transcription–polymerase chain reaction (qRT-PCR) technique. With each PCR cycle, the number of copies of an amplified cDNA fragment may double, until the reagents are exhausted. The resulting amount of DNA can be measured in real time by observing the fluorescence of a dye that binds to double-stranded DNA as the product is synthesized. This measurement results in a characteristic S-shaped curve of fluorescence intensity (**Figure 8.6A**). If there is more RNA, and hence cDNA, at the beginning of the reaction, the exponential phase of amplification commences earlier, and the so-called Ct threshold value is

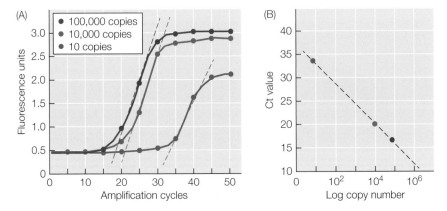

Figure 8.6 Quantitative Reverse-Transcription PCR (qRT-PCR). (A) Fluorescence in each PCR cycle is measured, and as the PCR amplification products accumulate, signal intensity moves from undetectable through an exponential phase to a plateau. High-abundance transcripts appear earlier, since there is more starting material, whereas low-abundance transcripts may take 30 or more cycles to begin to appear. The inflection points of each curve are projected to the x axis to yield Ct values (cycle threshold, dotted lines) representing the amount of each RNA. (B) Abundance is inferred by interpolation of the Ct values onto a control regression line. (After Rasmussen et al. 1998.)

attained earlier than it would be with a smaller amount of starting RNA. Smaller Ct values correspond to greater transcript abundance (**Figure 8.6B**). A qRT-PCR analysis can be performed with nanoscale volumes and massive parallelization with, for example, Fluidigm, Nanostring, or Raindance devices, allowing multiples of 96 transcripts to be validated in parallel for just cents per reaction (Sanchez-Freire et al. 2012). These protocols can also be extended to high-volume profiling of single-cell gene expression.

An emerging application for RNA-Seq is targeted gene expression profiling. Just as exome sequencing targets a subset of exons and can be customized for genes of interest, oligonucleotide pools can be synthesized that will pull down cDNAs of interest that can then be sequenced. This method facilitates higher read depths, and hence greater accuracy, as well as higher levels of multiplexing and reduced costs (Mercer et al. 2012). Along with qRT-PCR arrays, methods for targeted transcript profiling allow extensive experimentation focusing on the genes of interest identified in the unbiased discovery phase of a project. This process leads to more accurate description of alternative splice forms and their abundance, and it can help in the characterization of poorly annotated genes, including noncoding RNAs.

Basic Statistical Issues in Transcriptome Analysis

Whichever technology is used, there are multiple analytic steps to be considered in gene expression profiling. These start with extraction of signal intensity data from the array files or read alignments, including background signal subtraction and adjustment for known biases. Possibly the most important step in any analysis is **normalization** to remove the systematic biases, both technical and biological, that are inherent in all transcriptome data sets. Subsequently, statistical inference is used to select genes that are deemed to be differentially expressed among the various conditions contrasted in an experiment or as compared with data sets available online. Finally, downstream analyses are used to annotate the function of genes and pathways and to attempt to understand the regulatory basis of the differential expression.

Normalization is a tricky and poorly understood business, but it can have a major impact on transcriptome analyses (Qin et al. 2013). The most common issue is that biological factors can be confounded with technical ones: for example, in a study comparing smokers and nonsmokers, if even a subset of one group is profiled at a different time of year, then batch effects could be responsible for what appears to be the biological effect of interest. The other major concern is that different normalization strategies make hidden assumptions about whether it is a change in absolute or relative abundance that is important or interesting. Suppose that gene expression in one group—say, people taking a drug—is more spread out than in the controls, but that the rank order of transcript abundance does not change. In this case, a normalization strategy that equilibrates the variance across the total sample will not detect any differences, but one performed on the absolute abundance measures will find that genes expressed at either extreme differ between the samples.

Figure 8.7 illustrates three approaches to normalization. Figure 8.7A shows the raw distribution of $\log_2$-transformed gene expression intensity measures for six samples (which could be from a microarray or RNA-Seq experiment), three from females and three from males. There is a large peak in the frequency distribution, and there is a long tail extending to the right, corresponding to the relatively small fraction of the transcriptome with high abundance. The expression level of the same gene in each sample is shown by the arrows. The six samples have slightly different means and variances, and at least the means need to be equilibrated, otherwise the female profiles would be inferred to be much more variable than the male profiles, and this inference would affect any hypothesis testing. Figure 8.7B

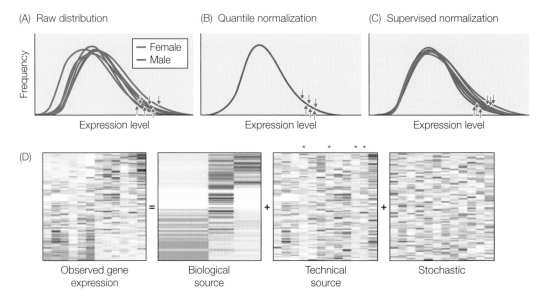

Figure 8.7 Normalization of Gene Expression Profiles. (A) Raw data from different samples tend to have different means due to factors such as the amount of input RNA and laser settings on the scanner (or total number of sequence reads). Statistical power to see differences in expression is reduced if these factors are not accounted for. (B) Variance-based transforms, such as quantile normalization, assume that it is the rank order of expression that is important. Here, this method shows that the male and female samples do not differ in expression of the indicated gene. (C) Supervised normalization procedures, such as SNM, assume that more similar samples have more similar overall profiles, and that it is the absolute level of expression that matters. Here, supervised normalization shows that gene expression does differ between the sexes. (D) The objective of supervised normalization is to partition the observed gene expression into biological and technical sources, while accounting for random variation. A major difficulty is that technical and biological sources can be confounded: for example, three of the four "yellow" lanes with a technical factor (marked with asterisks) are in the right-hand biological group. (D, courtesy of J. D. Storey, Princeton University.)

shows the effect of a procedure known as **quantile normalization** (Hansen et al. 2012), in which the rank of every gene is adjusted to the average value of that gene's rank, as a consequence of which all six profiles appear to be identical. However, the ranks of the individual transcripts will differ among samples, so there may be differential expression. This normalization procedure looks attractive, but actually does not deal with systematic biases and can cause spurious results if a large fraction of the transcriptome varies among samples. It is most appropriate and powerful where only a small number of genes are expected to change expression. Figure 8.7C shows the effect of performing a **supervised normalization** procedure—in this case, supervised normalization of microarrays (SNM), available as open-source code from Bioconductor (Mecham et al. 2010). This analysis deliberately removes technical effects such as the hybridization date and RNA quality differences (see Figure 8.7D), while also adjusting the effect of sex on the assumption that male and female profiles probably do resemble one another more than do opposite-sex profiles. The final distributions retain the overall distribution of abundance, but are less variable than the raw distributions. The extent to which normalization influences the results will vary from experiment to experiment, but it can be quite significant, and it is good practice to compare several approaches.

Statistical hypothesis testing involves formulation of gene-, transcript-, or probe-specific models. The effects of the variable of interest are assessed one by one for each intensity measure across all the samples in the analysis. There are two parameters of interest: the magnitude of difference in expression, and the significance of that difference. These parameters can be plotted against one another, resulting in a **volcano plot** (**Figure 8.8**), in which significance is represented on the y axis as the negative logarithm of the p-value (NLP) associated with the difference of expression, and the difference is represented in $\log_2$ units on the x axis (Wolfinger et al. 2001). Working on the $\log_2$ scale makes it easy to interpret differences: double the intensity has a value of 1, half the intensity a value of −1, four times the intensity a value of 2, and one-fourth the intensity a value of −2. Figure 8.8 shows an example in which approximately 15% of the transcriptome is upregulated and 15% downregulated in one condition relative to the other. In the absence of differential expression, all the points would fall in the light tan sector D.

Volcano plots highlight several reasons why it is important to consider both parameters. You will notice that some genes that show an average twofold increase in expression in one group—say, smokers—appear to be highly significant (NLP > 5), whereas others are not at all significant (NLP < 1.3; that is, $p > 0.05$). The reason is that the variance is greater in the second situation. If all the smokers have twice as much expression of the gene as the nonsmokers, the average log difference is 1, just as it is if one smoker has 6 times higher expression while the others have the same expression as the nonsmokers. Clearly you would trust the more consistent result more, and this is reflected in the more significant test statistic. There are also quite a few genes that appear to be highly significantly differentially expressed even though the magnitude of difference is quite

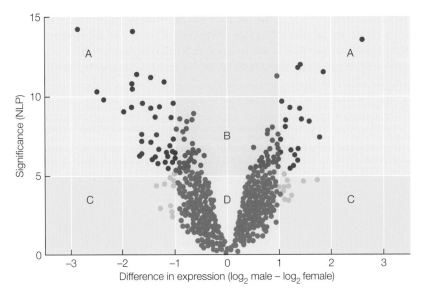

Figure 8.8 A Volcano Plot. Inference of differential expression involves measuring both the magnitude of differential expression (plotted on the x axis) and its significance (plotted on the y axis as the negative $\log_{10}$ of the p-value, a function of sample size, within-sample variance, and between-sample difference). Each transcript is indicated as one point. Those in the A sectors are both significant and show a large difference; those in the B sectors are significant even though the difference is small; those in the C sectors appear to differ a lot, but the effect is not significant; and those in the D sector are not differentially expressed. (After Jin et al. 2001.)

small, only 1.5. This can happen if all the samples have approximately the same small difference. Just slightly elevated expression of a kinase might be biologically very meaningful, or strong upregulation of a gene in only a minority of cells in a sample may result in only a small observed difference but still be functional. On the other hand, it should be recognized that with thousands of measures, the variance of some genes will be underestimated and that of others overestimated, just by chance. Consequently, if a gene of particular interest is only marginally significant, that does not mean it should be ignored. Transcriptome analysis is powerful and very often yields hundreds of significant genes, but, as with all statistical analyses, there will also be false negatives. For this reason, biology is also an important component of interpretation of genome-wide results.

What is the appropriate threshold of significance? Whereas in GWAS, the multiple comparison problem is usually addressed by adopting a strict Bonferroni adjustment for the number of independent loci in the genome, the standard in transcriptome analysis is more liberal. Clearly $p < 0.05$ is an inappropriate standard because, of 10,000 genes, we would expect 500 to be significant by chance. On the other hand, there is little point in focusing on only a handful of transcripts if there is good evidence that several hundred really are differentially expressed. For this reason, the standard procedure is to adopt a false discovery rate (FDR) threshold. With 10,000 genes or transcripts, 10 are expected to be significant at $p < 0.001$ by chance. If we observe that 100 are significant at $p < 0.001$, then we can conclude that 90 of them are true positives and 10 are false positives, and hence the FDR is 10%. We do not know which 10 are false, but it is accepted that this usually will not matter for downstream analyses because the 90% that are true will dominate the signal. A further adjustment can be performed that starts by estimating the true negative rate (π_0) rather than assuming that no genes are differentially expressed. In many biological circumstances, it is not uncommon for 10% or more of the genes to be differentially expressed, and the q-value procedure (Storey and Tibshirani 2003) has been shown to provide a conservative threshold for detection of the genes of interest.

In practice, a wide variety of statistical procedures can be used to formulate p-values (Allison et al. 2006). Significance analysis for microarrays (SAM), which incorporates a modified t-test that is useful for pairwise comparisons, is a popular procedure. For experimental designs that include multiple levels of one factor (e.g., three or more genotypes or environments) and/or multiple factors including interaction effects (e.g., asking whether there is a difference in how males and females respond to smoke inhalation), the analysis of variance (ANOVA) framework is more appropriate. Bayesian procedures are also available, and methods that incorporate information from other transcripts, recognizing that genes tend to covary extensively, are used. Longitudinal analyses provide some particularly interesting challenges.

Gene Ontology and Downstream Analysis

Given a set of differentially expressed genes, the next task is to evaluate what those genes have in common and which physiological, developmental, or biochemical pathways they influence. This process starts with gene ontology (GO), or **gene set enrichment** (GSE), **analysis**. The GO Consortium (www.geneontology.org), introduced in Chapter 2, has developed a classification scheme for all proteins that uses three hierarchies, reflecting the molecular function, the biological role, and the cellular compartment in which the protein is found (Ashburner et al. 2000). Within each hierarchy, genes are assigned to increasingly restrictive groups (see Figure 2.2). Numerous open-source tools allow tests of enrichment in GO groups whose results allow us to evaluate whether the proportion of genes in the group that are in a differentially expressed set is greater than the proportion of genes in the whole genome that are that group. Thus, if 30 out of the 150 genes in a differentially expressed set are annotated to T cell signaling, compared with 250 out of the 25,000 genes in the genome, there is a 20-fold excess of this category (20% versus 1%), and it may be concluded that the process under study relates to T cell activity.

A commonly used tool maintained by the NIH is the Database for Annotation, Visualization and Integrated Discovery (DAVID; david.abcc.ncifcrf.gov; Huang et al. 2008). Rather than simply reporting enrichment p-values, it can map the enriched genes onto pathway maps that show how genes

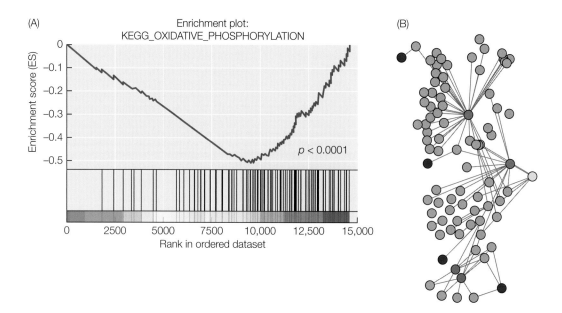

(A) Enrichment plot:
KEGG_OXIDATIVE_PHOSPHORYLATION

$p < 0.0001$

Rank in ordered dataset

Enrichment score (ES)

(B)

work together. There are several popular, highly curated maps of known pathways of metabolism, cellular regulation, and disease processes. KEGG (Kyoto Encyclopedia of Genes and Genomes; www.kegg.org) was one of the first databases to annotate genes in biochemical pathways and has expanded to include other aspects of intracellular regulation. Ingenuity is a pay-for-use site, but offers limited public access to particularly useful tools for examining enrichment in disease networks. A new public platform, Wikipathways, that taps into the wealth of information from the open scientific community is rapidly gaining in use as well. By coloring the genes on a pathway, immediate visualization of where differential expression occurs can be achieved. ToppFun (toppgene.cchmc.org; Chen et al. 2008) is another open-source tool that performs enrichment analyses on several other gene categories, including mutations affecting specific biological attributes in mice or humans, genes known to respond to drug treatment, and potential targets of transcription factors or miRNAs. Gene set enrichment analysis (GSEA; **Figure 8.9A**) is conceptually slightly different: it ranks the test statistics for all genes in an analysis and tests for enrichment of the ranks in the genes of interest (www.broadinstitute.org/gsea; Subramanian et al. 2005).

An alternative way to visualize gene networks that does not require prior knowledge of pathways is to assemble networks based on patterns of interconnectivity (**Figure 8.9B**). Cytoscape (www.cytoscape.org; Saito et al. 2012) is the tool most often used for this purpose. Each gene or protein is represented as a box or circle, according to user-specified criteria, and these shapes are joined by lines that indicate that some property links the two items together. In graph theory, the genes or proteins are nodes and the linkages are edges, and edges may be directed with arrows indicating that one node influences

◀ **Figure 8.9 Downstream Analysis Approaches.**   (A) GSEA generates a ranked list of test statistics for all genes in an analysis comparing two conditions and plots each of the genes in a specified pathway (here, oxidative phosphorylation, as determined by KEGG) with respect to that list (black bars in middle). This gives rise to an enrichment score (top) that can be associated with a *p*-value using permutation procedures. In this case, it is clear that oxidative phosphorylation is downregulated since a preponderance of genes are to the right. (B) Cytoscape generates a network of interactions among all the genes that are found to be co-regulated. In this image, three hubs can be recognized: a microRNA (purple) found to regulate all the red genes at the top; a different miRNA and a transcription factor (yellow) that jointly co-regulate the middle set of genes; and two components of the histone methyltransferase complex (green) that regulate the genes at the bottom. Other genes with putative roles in epigenetic regulation are shown in maroon. (A, after author's unpublished data.)

the other. The connections may have a variety of different sources: known genetic interactions, protein-protein interactions, linkages in pathways, or even the fact that the nodes are mentioned together in the abstract of a paper. Stronger evidence can be represented by thicker lines. An interactive clustering function allows the user to click on a node and drag it around the screen, observing how the edges drag other nodes into perspective. Hubs are genes that have a large number of linkages and tend to be essential genes. Very often, nodes fall into small sub-networks of highly interconnected proteins that are linked to other sub-networks through a small number of key nodes. Exploration of gene expression networks is at the heart of moving from transcript profiling to hypothesis generation and experimental design.

The next four sections of this chapter discuss examples of gene expression profiling and its application to illustrate some generalities. Specific findings are then incorporated into the disease-specific chapters that follow in Part III.

Atlases of Gene Expression

Most journals in the field of genetics require, as a condition of publication of transcriptome studies, that the data, both raw and processed, be uploaded into a public database where they can be accessed by anyone, without restriction. For most studies, the NCBI's Gene Expression Omnibus (GEO; www.ncbi.nlm.nih.gov/geo) and the EBI's ArrayExpress (www.ebi.ac.uk/arrayexpress) serve this purpose, accepting submissions from anywhere in the world as long as they adhere to minimal information for annotation of microarray experiment (MIAME) guidelines (Brazma et al. 2001). These guidelines require that each data submission be accompanied by a link to a description of the gene content on the platform used to perform the profiling, as well as by descriptions of the experimental design and protocols that were used. There are several thousand series of experiments in each of the databases, and hundreds of thousands of transcript profiles are represented in searchable format.

Such an enormous volume of data supports powerful interrogation of all the experiments that have included specific tissues under a variety of conditions. ArrayExpress has automated this process (Parkinson et al. 2005), providing several "Expression Atlas" search engines. One supports query of RNA-Seq data sets for transcript abundance in heart, liver, kidney, testis, and four regions of the brain from ten mammalian species (facilitating studies of the evolution of these organs), and another profiles RNA-Seq data from the Illumina Body Map of 16 human tissues (Kapushesky et al. 2012). A compilation of 5372 Affymetrix arrays from 369 different cells and tissues is accessible at www.ebi.ac.uk/gxa/experiments/E-MTAB-513 (Lukk et al. 2010), and a popular compilation of microarray data from the Genome Institute of the Novartis Research Foundation (GNRF) has migrated to BioGPS.org (Su et al. 2002). Gene-Cards (www.genecards.org) graphically displays information from both BioGPS and the Illumina Body Map by default. The former database includes 32 "major tissues" and 50 "anatomical compartments" from the immune and nervous systems as well as muscle, secretory, reproductive, and other internal organs. The Genotype-Tissue Expression Project (GTEx Consortium 2013) is currently assembling a set of gene expression profiles from 30 different tissues isolated from about a thousand people immediately following their deaths. These profiles will be associated with whole-genome genotypes, facilitating genetic analyses of the regulation of gene expression across tissues.

Other atlases of gene expression are dedicated to specific organs. The Allen Brain Atlas (human.brain-map.org; Hawrylycz et al. 2012) is notable for the inclusion of both microarray and in situ hybridization data (providing visualization of where in the brain genes are expressed) and for a search engine that allows users to query by category of gene or disease. Its Brain Explorer software is a desktop application that provides three-dimensional images of expression localization in hundreds of regions of the cortex, cerebellum, and brain stem from two donors side by side, and is even linked to MRI data on neuronal activity. The Immunological Genome Consortium (www.ImmGen.org; Shay et al. 2013) provides a similar resource for the immune system, supporting, for example, comparisons of 38 human and 244 mouse myeloid and lymphoid cell types.

Much can also be learned simply by deep transcript profiling of specific cell types. As one example, RNA-Seq of pancreatic islets containing insulin-producing β cells provided unique insight into the pathogenesis of type 1 diabetes (Eizirik et al. 2012). Five islet cell samples were cultured with or without exposure to the inflammatory cytokines interleukin-1β and interferon-γ. Almost 30,000 transcripts were detected, expression of one-fifth of which were affected by the cytokines, suggesting alterations in apoptosis and inflammatory responses. Of the 41 genes identified by GWAS as contributing to type 1 diabetes, 24 were found to be expressed in the islets. Some of the genes showed altered splicing or expression in the inflammatory condition, providing biological evidence that those genes are likely to have a direct role in insulin production through their function in pancreatic β cells.

Biomarkers and Disease Progression

Whereas the most obvious application of gene expression profiling in disease is the contrasting of cases and controls, at least where appropriate tissues can be obtained, one of the first applications of microarray analysis was actually the identification of subclasses of disease among cases. Gene expression profiling may, for example, distinguish individuals with a disease with respect to etiology (different causes of disease may produce different transcriptomes), severity, stage, predicted rate of progression, and likely responsiveness to therapy. In some cases, the transcript profile may simply reflect altered abundances of key cell types that contribute to the pathology (e.g., airway neutrophils or granulocytes that contribute disproportionately to airway sputum gene expression according to type of asthma; Baines et al. 2011); in others, it will reveal bona fide differences in gene regulation within cells (as in the pancreatic islet example above).

Altered gene expression in disease can be used either as a biomarker in diagnosis and development of a treatment strategy or to understand more about the molecular mechanisms of the pathology. Both applications have been used extensively in the study of cancer, starting with microarray analysis of blood samples from patients with leukemia. Alizedeh et al. (2000) were the first to show that patients with two different subtypes of diffuse large B cell lymphoma, germinal and activated, that were not detectable histologically could be distinguished on the basis of the expression of hundreds of genes. Furthermore, plotting survival curves for the germinal and activated subtypes revealed a clear difference in likelihood of patient survival: 14 of 19 patients with the germinal subtype were still alive 5 years after diagnosis, but only 5 of the 21 patients with the activated subtype. The nature of the differentially expressed genes also gave hints to the aberrant gene activity underlying cell proliferation and apoptosis as well as lymphocyte differentiation. This line of work has since been expanded to many different cancer types and is now supplemented by sequence-based analyses of the mutations and gene fusions that are likely to drive individual cancers.

Peripheral blood is a particularly attractive tissue for gene expression profiling because it is readily obtained at low cost with minimal invasion or risk to the patient. It also seems likely to provide biomarkers for a wide range of diseases. Starting with 239 blood samples from individuals in eight classes (healthy people and people with immunosuppression, influenza, bacterial infection, systemic lupus erythematosis [SLE], type 1 diabetes, melanoma, and juvenile arthritis), Chaussabel et al. (2008) identified 28 modules of up to several hundred transcripts each that are consistently co-regulated, but differ in abundance among the eight classes (**Figure 8.10**). At least 15% of the genes in 11 of the modules in children with SLE were significantly divergent in expression from those in age-matched healthy children, representing

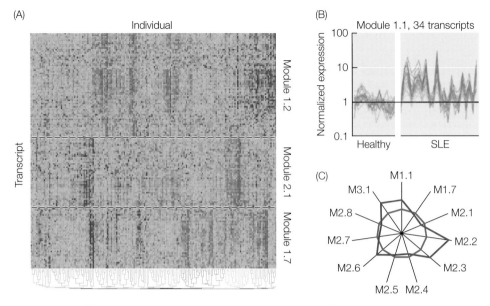

Figure 8.10 Modularity of Gene Expression. (A) A heat map showing the co-regulation of modules of genes in peripheral blood samples of 180 healthy people. Transcript abundance for each gene was standardized by conversion to a z-score (a normal distribution centered on zero with a standard deviation of 1), then coded from low (blue) to high (red). Each individual's expression is represented by a column, and each row represents a gene. Several dozen genes in each of three modules can be seen to covary, and with just three modules, nine classes of individuals (shown by different colors on the dendrogram at the bottom) can be identified. (B) Another module, labeled M1.1, is upregulated in children with systemic lupus erythematosis (SLE): the plot shows the fold difference in the abundance of each of its 34 transcripts (red lines) in healthy children and children with SLE plotted along the x axis, again showing how the genes covary within individuals. (C) Each line on this radar plot indicates the average normalized expression of the genes in one module in healthy children (blue) and in children with SLE (green). Five of the eleven modules (M1.1, M2.2, M2.3, M2.6, and M3.1) are upregulated in the children with SLE, whereas the others show relatively little difference. Different immune-system diseases show perturbation of different modules (A, from author's unpublished data; B and C, after Chaussabel et al. 2008.)

gene activity related to inflammatory, antiviral, and lymphocyte activity. A composite score representing the average abundance of the representative transcripts was then shown to correlate with disease severity.

Longitudinal profiling for the purposes of monitoring disease progression and predicting the course of disease presents particular analytic challenges as well as opportunities for personalizing functional genomic medicine. Desai et al. (2011) provided a comprehensive strategy for matching trajectories of change in expression over 28 days with outcomes in 168 patients who had undergone blunt-force trauma. They identified five modules, each with 37–577 probe sets, that showed consistent dynamic profiles that were to some extent correlated with five clinically assessed multiple organ failure scores. There appeared to be a critical window 2–4 days after an accident in which downregulation of MHC class II and upregulation of p38 MAPK signaling, among other changes, was associated with severe complications or failure to recover. A critical aspect of this study is that careful normalization procedures were used to ensure repeatability across four different study sites, addressing some of the technical issues that have delayed approval of transcript profiling for a wide range of clinical applications.

Genome-Environment Interactions

Given that the environment is just as important as the genome in the origin of many complex diseases, there is a reasonable expectation that gene expression profiling may provide a global view of how the environment influences gene activity. The strategy here is to correlate variation in transcript profiles with environmental variables, which may be as specific as a particular diet or exposure to pollution, as general as urban versus rural lifestyles, or as global as gradients of humidity or altitude. This type of analysis is necessarily simply correlative, as there may be any number of unknown factors that associate with the studied environmental variable and cause the gene expression differences, but it has revealed widespread nongenetic influences and has provided important perspectives on the range of normal variation in healthy people.

One of the most important environmental influences is social stress, as psychosomatic interactions are well known to influence patient recovery as well as susceptibility to disease. **Social genomics** research aims to dissect the mechanisms by which glucocorticoid and other stress hormones mediate signaling from the brain to relevant organs, including the immune system. A remarkable finding is that low socioeconomic status in childhood may have a lifetime influence on lymphocyte gene expression. Miller et al. (2009) compared the transcriptomes of 60 adults aged 20–40 with similar occupations. They found significant differences between two groups distinguished by a measure of their socioeconomic status (SES) in the first 5 years of life, reflecting increased systemic inflammation and response to stress in the lower-SES group. Low early-life SES is thought to more than double the risk of many complex diseases, including diabetes and cardiovascular disease, so important goals now are to understand the mechanisms by which these environmental effects are transduced over several decades and to determine to what extent they are reversible.

Humans occupy very diverse environments, and we have begun to adapt to those environments over thousands of generations of dispersal across the globe. What are the relative contributions of genetic, geographic, and cultural factors to divergence among human populations? As background, consider the observation that 10% of the differences observed between mouse and human gene expression in the liver turn out to mirror the changes observed in mice fed a fast-food humanized diet instead of traditional chow (Somel et al. 2008)! Idaghdour et al. (2010) addressed this question by microarray comparisons of Arabs and Berbers living in urban and in remote rural locations in southern Morocco. They found almost no significant differences between the sexes or between ethnicities in the city of Agadir, but found a complex pattern of differentiation among the rural villages, involving up to a third of the peripheral blood transcriptome, that was best explained by cultural and lifestyle differences. How these effects translate to large and diverse American or European cities, and whether they help explain very large regional differences in disease incidence, including urban-rural contrasts, remains to be determined.

For the most part, individual transcript profiles in healthy people remain constant over time, but the deep profiling of one person in particular shows that changes in health do have a functional genomic signature. Stanford University professor Mike Snyder sequenced his own genome and measured his peripheral blood gene expression at 18 time-points over a 14-month period (along with his proteome, metabolome, and many other features; Chen et al. 2012). His integrated Personal Omics Profile (iPOP) not only showed short-term spikes of immune-system activity accompanying two episodes of respiratory viral infection, but also revealed a longer-term shift involving glucose metabolism and insulin signaling as he entered a prediabetic state prior to initiating a lifestyle change. Interestingly, the RNA-Seq analysis of his transcriptome facilitated monitoring of alternative splicing and dosage of the two alleles of each gene, demonstrating that as many as a thousand transcripts may be directly affected in an allele-specific manner by viral infection.

Genetics of Gene Expression Variation

GWAS can be combined with gene expression profiling to study the genetics of gene regulation. This approach, sometimes called **genetical genomics** (Jansen and Nap 2001), is not only a very powerful method for evaluating the mechanism of action of putative disease-causing SNPs, but also provides an important window on the genetic basis of transcriptional variation (Skelly et al. 2009). The simple idea is to jointly measure genotypes genome-wide, as well as transcript abundance, and to perform association studies treating each of the more than 10,000 transcripts as 10,000 traits. This approach has been applied to autopsy samples of brain and liver as well as biopsies of skin, muscle, and adipose tissue, and more extensively to peripheral blood samples and lymphoblast cell lines derived from blood. To date, individual studies have rarely exceeded a thousand people, but since individual genotypes have a much larger effect on transcript abundance than on disease risk, often explaining 20% or as much as 50% of transcript abundance, it turns out that samples of even a hundred people are powerful enough to detect hundreds of "**expression SNPs**," or **eSNPs**. The term **eQTL** is also used, where QTL stands for quantitative trait locus, and helps to emphasize that eSNPs may tag either the true causal variant or variants at another locus that influences gene expression (just as disease-associated variants most often tag causal disease polymorphisms).

There are two types of results that emerge from eSNP profiling, depending on the proximity of the SNP to the transcript that it affects (**Figure 8.11A,B**). If the polymorphism and transcript are located on different chromosomes or are some distance apart on the same chromosome, they are called **distal eSNPs**. If they are close enough together to support the inference that the SNP is likely to act on the same gene, they are called **local eSNPs**. Operationally, a local eSNP is often defined as lying within 1 Mb of the transcription start site, but it turns out that the vast majority of eSNPs are within 10 kb, and many are as close as 1 kb upstream or downstream, of the transcription start site. The terms *trans*-**eSNP** and *cis*-**eSNP** are used interchangeably with "distal eSNP" and "local eSNP," respectively, and they occur in the literature more commonly, but they technically refer to whether or not the SNP acts on the same physical chromosome (Rockman and Kruglyak 2006). That is, if a SNP on the allele inherited from your mother influences the allele on the other chromosome inherited from your father, then it is acting in *trans*, even if it is a local SNP. Since it is thought that homologous chromosomes are for the most part transcribed independently (but see the next chapter), most local eSNPs are likely to be *cis*-acting. They probably affect the capacity of transcription factors to bind to the regulatory DNA and hence directly influence the rate of transcript initiation or elongation. By contrast, distal eSNPs are assumed to affect the activity of an intermediate protein or RNA that then mediates the regulation of the associated transcript.

The additive nature of SNP effects may be explained by the summation of the influences of regulatory polymorphisms on the two chromosomes of a homologous pair. If an eSNP "G" results in an average of 10 transcripts per chromosome per cell, while the "A" variant produces 20 transcripts per chromosome per cell, then heterozygotes should have intermediate abundance—30 transcripts—relative to the two homozygote classes, which will have 20 and 40 transcripts per cell after summing the contributions of the two chromosomes. Allele-specific gene expression profiling provides some support for this inference. RNA-Seq can demonstrate differential expression of the two alleles because the SNP in the transcript will result in more counts of one type of transcript than the other. The ratio of the two alleles is assumed to associate with the genotype of a regulatory SNP that is in linkage disequilibrium with the expressed SNP (Pickrell et al. 2010). Regulatory SNPs also influence the efficiency of splicing; in fact, the largest study directly matching RNA-Seq in lymphoblast cell lines to whole-genome sequences of hundreds of individuals (Lappalainen et al. 2013) argued that transcript structure is at least as large a source of transcriptome variance as is abundance.

Regulatory eSNPs are pervasive. Despite relatively small sample sizes compared with GWAS of disease, strong effects have been documented for at least 10% of all genes in most tissues that have been examined. Secondary associations indicative of multiple independently acting variants influencing the same gene are also observed. A meta-analysis of blood eQTL studies in 8000 people identified highly significant local regulatory

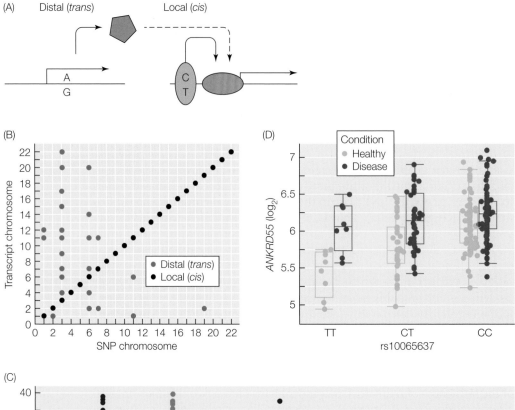

(A) Distal (*trans*)   Local (*cis*)

(B)

Transcript chromosome vs SNP chromosome

- Distal (*trans*)
- Local (*cis*)

(D)

Condition
- Healthy
- Disease

ANKRD55 (log$_2$)

TT   CT   CC

rs10065637

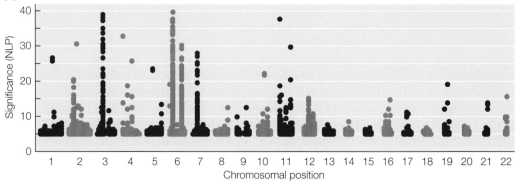

(C)

Significance (NLP) vs Chromosomal position

associations at 44% of all genes and 643 *trans*-eQTL, all of which are documented on the Blood eQTL Browser (genenetwork.nl/bloodeqtlbrowser; Westra et al. 2013; **Figure 8.11C**). Although most studies have concentrated on steady-state transcript abundance, emerging work is showing that eSNPs also affect alternative splicing, transcript stability, interactions with microRNAs, and possibly even RNA editing. Distal or *trans* effects are much less well characterized, in part because there are only a handful that have been robustly demonstrated. This is because the burden of statistical evidence is very high (at least 500,000 SNPs multiplied by 10,000

◀ **Figure 8.11 Expression QTL.**   (A) Local (or *cis*) eSNPs are located adjacent to the gene encoding the target transcript and are thought to act primarily by affecting binding of transcription factors that interact with the basal transcriptional machinery. Distal (or *trans*) eSNPs are located on a different chromosome, or at a different location on the same chromosome, from the target transcript and act through an intermediary regulatory molecule. (B) A plot of transcript location against eSNP location shows *cis*-eSNPs along the diagonal and *trans*-eSNPs off the diagonal. This plot shows all *trans*-eSNPs at NLP > 40 in the Blood eQTL Browser: one locus on chromosome 3 regulates genes on 10 other chromosomes, whereas one locus on chromosome 2 regulates only a target on chromosome 1. (C) A Manhattan plot showing the locations of SNPs along the genome that are involved in each of 4732 *trans* associations with peripheral blood gene expression as documented in the Blood eQTL Browser. Significance on the *y* axis is the negative logarithm of the *p*-value (NLP; 67 associations stronger than 10^{-40} are excluded from the figure). Whereas in disease GWAS, the peaks represent haplotype blocks associated with one trait, many of the peaks on this plot are associated with multiple different target transcripts on different chromosomes. (D) A typical eQTL effect. In this case, the major allele (CC) is associated with higher gene expression. The mean difference between homozygote classes in healthy people (yellow) is 0.5 $\log_2$ units, implying a 1.2-fold increase per C allele. Heterozygotes are intermediate, implying additivity. Under a hypothetical disease condition (green), expression of the gene is higher in all individuals, but the allelic substitution effect is smaller, implying a genotype-by-disease interaction for gene expression.

transcripts demands a significance threshold of $p < 10^{-11}$) and because the effects are indirect and hence usually smaller. Nevertheless, collectively, the vast majority of genetic influences on transcript abundance must be distal, since local eSNPs explain only a small minority of the variance of most transcripts, and in this sense, the regulation of transcription is likely to follow the infinitesimal model, with hundreds of loci contributing to the regulation of each gene. **Figure 8.11D** shows a typical eSNP effect, in which the substitution of each C allele for T tends to increase expression 1.2-fold and in which this SNP explains approximately 20% of the transcript variance. Theoretically, the environment also contributes to variation, and under some circumstances genetic effects vary across environments or under disease conditions.

An important application of eSNP analysis is the annotation of disease SNPs. Estimates of the fraction of genome-wide associations that are attributable to regulatory, as opposed to structural, polymorphism range up to 80%, with compelling evidence for at least 10%. For example, of 1598 disease associations documented in 2010, Nicolae et al. (2010) found that 625 are eSNPs for the adjacent gene at $p < 10^{-4}$ in a sample of European lymphoblast cell lines. Similarly, Westra et al. (2013) found that 103 complex trait GWAS variants are also *trans*-eSNPs. Some of these are likely to be false positives, and there is no guarantee that the same SNP is contributing to both the expression and the disease associations, but the enrichment of eQTL for concordant regulatory trait associations is highly significant. Furthermore, a fraction of regulatory genetic effects are known to be tissue-specific, so failure to detect an eSNP in blood or cell lines does not preclude the possibility that a regulatory effect would be observed in the most appropriate tissue for a disease. For example, an early eSNP study of human liver biopsies (Schadt et al. 2008) identified three strong candidate genes for coronary artery disease after cross-matching of the expression results with the Wellcome Trust Case Control Consortium coronary artery disease (CAD) study as well as a GWAS for LDL cholesterol. This finding implies that genetic regulation contributing to disease can occur in a variety of tissues with direct or indirect roles in the etiology (and often we do not know which tissue is most relevant).

In the next two chapters, we will explore other functional genomic profiling approaches, starting with chromatin profiling in Chapter 9, and then moving to the metabolome, proteome, and microbial metagenome in Chapter 10. SNP associations are also being described with elements of each of these classes of profiles, and the integration of these diverse approaches constitutes the emerging field of systems biology.

Summary

1. Gene expression profiling is the process of systematically documenting the abundance of all transcripts expressed in a tissue sample, giving rise to the transcriptome of an individual.

2. There are three major commercial platforms for microarray analysis, provided by Agilent, Affymetrix, and Illumina, all of which use oligonucleotide probes that hybridize to labeled cDNA.

3. Increasingly, gene expression profiling is performed by RNA sequencing. This technique, which can generate 50 million paired-end reads per sample, allows the abundance of exons, genes, or transcript isoforms to be estimated. Analysis of the data obtained by this method requires advanced computational and statistical expertise.

4. RNA-Seq also facilitates analysis of alternative splicing and allele-specific expression.

5. A critical step in the analysis of all types of gene expression data is normalization to account for technical and biological sources of error that can be confounded with the biological differences of interest.

6. A commonly adopted statistical threshold in gene expression analysis is the false discovery rate (FDR), which specifies an acceptable level of false positives to be included in downstream analyses.

7. Genes tend to be co-regulated, with the result that large numbers of transcripts known as modules are observed to covary across samples

8. Gene ontology, or gene set enrichment, analysis is used to make inferences about the common biological functions of sets of co-expressed transcripts that distinguish different classes of samples.

9. Gene expression atlases document which genes are expressed in which cell types under which conditions. They facilitate annotation of gene function as well as interpretation of differential expression.

10. Comparative profiling is used to identify subclasses of patients who may have distinct types of gene activity that indicate likely disease progression or can be targeted for individualized treatment.

11. The environment has a pervasive influence on gene activity, so transcript profiling can be used to study interactions between genetics, culture, and environment.

12. eSNP or eQTL analysis is the study of the genetic basis for variation in gene expression. It has been employed to detect a highly significant concordance between disease SNPs and regulatory polymorphisms.

References

Alizadeh, A. A., Eisen, M. B., Davis, R. E., Ma, C., Lossos, I. S., et al. 2000. Distinct types of diffuse large B-cell lymphoma identified by gene expression profiling. *Nature* 403: 503–511.

Allison, D. B., Cui, X., Page, G. P., and Sabripour, M. 2006. Microarray data analysis: From disarray to consolidation and consensus. *Nat. Rev. Genet.* 7: 55–65.

Anders, S. and Huber, W. 2010. Differential expression analysis for sequence count data. *Genome Biol.* 11: R106.

Anders, S., McCarthy, D. J., Chen, Y., Okoniewski, M., Smyth, G. K., et al. 2013. Count-based differential expression analysis of RNA sequencing data using R and Bioconductor. *Nat. Protoc.* 8: 1765–1786.

Ashburner, M., Ball, C. A., Blake, J. A., Botstein, D., Butler, H., et al. 2000. Gene Ontology: Tool for the unification of biology. *Nat. Genet.* 25: 25–29.

Baines, K. J., Simpson, J. L., Wood, L. G., Scott, R. J., and Gibson, P. G. 2011. Transcriptional phenotypes of asthma defined by gene expression profiling of induced sputum samples. *J. Allergy Clin. Immunol.* 127: 153–160.

Brazma, A., Hingamp, P., Quackenbush, J., Sherlock, G., Spellman, P., et al. 2001. Minimum information about a microarray experiment (MIAME)—Toward standards for microarray data. *Nat. Genet.* 29: 365 – 371.

Chaussabel, D., Quinn, C., Shen, J., Patel, P., Glaser, C., et al. 2008. A modular analysis framework for blood genomic studies: Application to systemic lupus erythematosus. *Immunity* 29: 150–164.

Chen, H., Liu, Z., Gong, S., Wu, X., Taylor, W. L., et al. 2011. Genome-wide gene expression profiling of nucleus accumbens neurons projecting to ventral pallidum using both microarray and transcriptome sequencing. *Front. Neurosci.* 5: 98.

Chen, J., Bardes, E. E., Aronow, B. J., and Jegga, A. G. 2008. ToppGene Suite for gene list enrichment analysis and candidate gene prioritization. *Nucl. Acids Res.* 37(Suppl 2): W305–W311.

Chen, R., Mias, G. I., Li-Pook-Than, J., Jiang, L., Lam, H. Y. K., et al. 2012. Personal omics profiling reveals dynamic molecular and medical phenotypes. *Cell* 148: 1293–1307.

Cloonan, N., Forrest, A., Kolle, G., Gardiner, B., Faulkner, G., et al. 2008. Stem cell transcriptome profiling via massive-scale mRNA sequencing. *Nat. Methods* 5: 613–619.

Dalma-Weiszhaus, D. D., Warrington, J., Tanimoto, E. Y., and Garrett Miyada, C. 2006. The Affymetrix GeneChip platform: An overview. *Meth. Enzymol.* 410: 3–28.

Desai, K. H., Tan, C. S., Leek, J. T., Maier, R. V., Tompkins, R. G., and Storey, J. D. 2011. Dissecting inflammatory complications in critically injured patients by within-patient gene expression changes: A longitudinal clinical genomics study. *PLoS Med.* 8: e1001093.

Eizirik, D. L., Sammeth, M., Bouckenooghe, T., Bottu, G., Sisino, G., et al. 2012. The human pancreatic islet transcriptome: Expression of candidate genes for type 1 diabetes and the impact of pro-inflammatory cytokines. *PLoS Genet.* 8:e1002552.

Fan, J-B., Gunderson, K. L., Bibikova, M., Yeakley, J. M., Chen, J., et al. 2006. Illumina Universal Bead Arrays. *Meth. Enzymol.* 410: 57–73.

Gautier, L., Cope, L., Bolstad, B. M., and Irizarry, R. A. 2004. affy—analysis of Affymetrix GeneChip data at the probe level. *Bioinformatics* 20: 307–315.

GTEx Consortium. 2013. The genotype-tissue expression (GTEx) project. *Nat. Genet.* 45: 580–585.

Hansen, K. D., Irizarry, R. A., and Wu, Z. 2012. Removing technical variability in RNA-seq data using conditional quantile normalization. *Biostatistics* 13: 204–216.

Hawrylycz, M. J., Lein, E. S., Guillozet-Bongaarts, A. L., Shen, E. H., Ng, L., et al. 2012. An anatomically comprehensive atlas of the adult human brain transcriptome. *Nature* 489: 391–399.

Huang, D. W., Sherman, B. T., and Lempicki, R. A. 2008. Systematic and integrative analysis of large gene lists using DAVID Bioinformatics Resources. *Nat. Protoc.* 4: 44–57.

Idaghdour, Y., Czika, W., Shianna, K. V., Lee, S. H., Visscher, P. M., et al. 2010. Geographical genomics of human leukocyte gene expression in southern Morocco. *Nat. Genet.* 42: 62–67.

Jansen, R. C. and Nap, J. P. 2001. Genetical genomics: The added value from segregation. *Trends Genet.* 17: 388–391.

Jin, W., Riley, R., Wolfinger, R., White, K., Passador-Gurgel, G., and Gibson, G. 2001. Contributions of sex, genotype, and age to transcriptional variance in *Drosophila*. *Nat. Genet.* 29: 389–395.

Kapushesky, M., Adamusiak, T., Burdett, T., Culhane, A., Farne, A., et al. 2012. Gene Expression Atlas update—A value-added database of microarray and sequencing-based functional genomics experiments. *Nucl. Acids Res.* 40(D1): D1077–D1081.

Langmead, B., Trapnell, C., Pop, M., and Salzberg, S. L. 2009. Ultrafast and memory-efficient alignment of short DNA sequences to the human genome. *Genome Biol.* 10: R25.

Lappalainen, T., Sammeth, M., Friedländer, M. R., 't Hoen, P. A. C., Monlong, J., et al. 2013. Transcriptome and genome sequencing uncovers functional variation in humans. *Nature* 501: 506–511.

Lockhart, D. J., Dong, H., Byrne, M. C., Follettie, M. T., Gallo, M. V., et al. 1996. Expression monitoring by hybridization to high-density oligonucleotide arrays. *Nat. Biotechnol.* 14: 1675–1680.

Lukk, M., Kapushesky, M., Nikkilä, J., Parkinson, H., Goncalves, A., et al. 2010. A global map of human gene expression. *Nat. Biotechnol.* 28: 322–324.

Mecham, B. H., Nelson, P. S., and Storey, J. D. 2010. Supervised normalization of microarrays. *Bioinformatics* 26: 1308–1315.

Mercer, T. R., Gerhardt, D. J., Dinger, M. E., Crawford, J., Trapnell, C., et al. 2012. Targeted RNA sequencing reveals the deep complexity of the human transcriptome. *Nat. Biotech.* 30: 99–104.

Miller, G. E., Chen, E., Fok, A., Walker, H., Lim, A., et al. 2009. Low early-life social class leaves a biological residue manifested by decreased glucocorticoid and increased pro-inflammatory signaling. *Proc. Natl. Acad. Sci. U.S.A.* 106: 14716–14721.

Mortazavi, A., Williams, B. A., McCue, K., Schaeffer, L., and Wold, B. 2008. Mapping and quantifying mammalian transcriptomes by RNA-Seq. *Nat. Methods* 5: 621–628.

Nicolae, D. L., Gamazon, E., Zhang, W., Duan, S., Dolan, E., and Cox, N. J. 2010. Trait-associated SNPs are more likely to be eQTLs: Annotation to enhance discovery from GWAS. *PLoS Genet.* 6: e1000888.

Parkinson, H., Sarkans, U., Shojatalab, M., Abeygunawardena, N., Contrino, S., et al. 2005. ArrayExpress—A public repository for microarray gene expression data at the EBI. *Nucl. Acids Res.* 37(Database): D868–872.

Pickrell, J. K., Marioni, J. C., Pai, A. A., Degner, J. F., Engelhardt, B. E., et al. 2010. Understanding mechanisms underlying human gene expression variation with RNA sequencing. *Nature* 464: 768–772.

Qin, S., Kim, J., Arafat, D., and Gibson, G. 2013. Effect of normalization on statistical and biological interpretation of gene expression profiles. *Front. Genet.* 3: 160.

Rasmussen, R., Morrison, T., Herrmann, M., and Wittwer, C. 1998. Quantitative PCR by continuous fluorescence monitoring of a double strand DNA specific binding dye. *Biochemica* 2: 8–11.

Robinson, M. D., McCarthy, D. J., and Smyth, G. K. 2010. edgeR: A Bioconductor package for differential expression analysis of digital gene expression data. *Bioinformatics* 26: 139–140.

Rockman, M. V. and Kruglyak, L. 2006. Genetics of global gene expression. *Nat. Rev. Genet.* 7: 862–872.

Saito, R1, Smoot, M. E., Ono, K., Ruscheinski, J., Wang, P. L., et al. 2012. A travel guide to Cytoscape plugins. *Nat. Methods* 9: 1069–1076.

Sanchez-Freire, V., Ebert, A. D., Kalisky, T., Quake, S. R., and Wu, J. C. 2012. Microfluidic single-cell real-time PCR for comparative analysis of gene expression patterns. *Nat. Protoc.* 7: 829–838.

Schadt, E. E., Molony, C., Chudin, E., Hoa, K., Yang, X., et al. 2008. Mapping the genetic architecture of gene expression in human liver. *PLoS Biol.* 6: e107.

Schena, M., Shalon, D., Davis, R. W., and Brown, P. O. 1995. Quantitative monitoring of gene expression patterns with a complementary DNA microarray. *Science* 270: 467–470.

Shay, T., Jojic, V., Zuk, O., Rothamel, K., et al., ImmGen Consortium. 2013. Conservation and divergence in the transcriptional programs of the human and mouse immune systems. *Proc. Natl. Acad. Sci. U.S.A.* 110: 2946–2951.

Skelly, D. A., Ronald, J., and Akey, J. M. 2009. Inherited variation in gene expression. *Annu. Rev. Genom. Hum. Genet.* 10: 313–332.

Somel, M., Creely, H., Franz, H., Mueller, U., Lachmann, M., et al. 2008. Human and chimpanzee gene expression differences replicated in mice fed different diets. *Plos ONE* 3: e1504.

Storey, J. D. and Tibshirani, R. 2003. Statistical significance for genomewide studies. *Proc. Natl. Acad. Sci. U.S.A.* 100: 9440–9445.

Su, A. I., Cooke, M. P., Ching, K. A., Hakak, Y., Walker, J. R., et al. 2002. Large-scale analysis of the human and mouse transcriptomes. *Proc. Natl. Acad. Sci. U.S.A.* 99: 4465–4470.

Subramanian, A., Tamayo, P., Mootha, V. K., Mukherjee, S., Ebert, B. L., et al. 2005. Gene set enrichment analysis: A knowledge-based approach for interpreting genome-wide expression profiles. *Proc. Natl. Acad. Sci. U.S.A.* 102: 15545–15550.

Trapnell, C., Pachter, L., and Salzberg, S. L. 2009. TopHat: Discovering splice junctions with RNA-Seq. Bioinformatics 25: 1105–1111.

Trapnell, C., Roberts, A., Goff, L., Pertea, G., Kim, D., et al. 2012. Differential gene and transcript expression analysis of RNA-seq experiments with TopHat and Cufflinks. *Nat. Protoc.* 7: 562–578.

Wang, Z., Gerstein, M., and Snyder, M. 2009. RNA-Seq: A revolutionary tool for transcriptomics. *Nat. Rev. Genet.* 10: 57–63.

Westra, H. P., Peters, M. J., Esko, T., Yaghootkar, H., Schurmann, C., et al. 2013. Systematic identification of *trans* eQTLs as putative drivers of known disease associations. *Nat. Genet.* 45: 1238–1243.

Wolber, P. K., Collins, P. J., Lucas, A. B., De Witte, A., and Shannon, K. W. 2006. The Agilent in-situ synthesized microarray platform. *Meth. Enzymol.* 410: 28–57.

Wolfinger, R. D., Gibson, G., Wolfinger, E. D., Bennett, L., Hamadeh, H., et al. 2001. Assessing gene significance from cDNA microarray expression data via mixed models. *J. Comput. Biol.* 8: 625–637.

Xu, W., Seok, J., Mindrinos, M. N., Schweitzer, A. C., Jiang, H., et al. 2011. Human transcriptome array for high-throughput clinical studies. *Proc. Natl. Acad. Sci. U.S.A.* 108: 3707–3712.

9

The Epigenome

E ver since the draft human genome sequence revealed that a mere
1% of the human DNA sequence encodes proteins, there has been
much debate over whether the remainder of the genome is functional.
Clinical geneticists tend to focus on actionable mutations, the majority of
which affect protein sequences, as far as is known, and thus they regard
the vast expanse of intergenic DNA as a luxury for basic researchers to
explore. Evolutionary geneticists tend to regard evolutionary conservation
as a mark of functional significance, and since only approximately 8% of
the genome is highly conserved, they are skeptical about the importance of
the remainder. It has even been called "junk DNA," which implies that it
is just dispensable filler. However, for some functions, it is the structure of
the folded nucleotide sequence that matters, and so long as this structure is
retained, an element may remain functional even in the absence of sequence
conservation. Many functional genomicists also point out that transcripts
are generated from large expanses of the genome, and although most are
transcribed only rarely, spatial and temporal expression of these transcripts
is often conserved in rodents or other species. There is increasing evidence
that these rarely transcribed sequences do have biological functions via
RNA products or some influence on chromatin structure. For this reason,
the Encyclopedia of DNA Elements (ENCODE) Project was launched in
the mid-2000s to catalog the function of as much of the genome as possible, focusing on the profiling of biochemical properties of the chromatin
(ENCODE Project Consortium 2011, 2012).

Another rationale for studying chromosomes and DNA directly is epigenetics. Classically, **epigenetics** is the study of traits that are inherited
without being attributable to any discernible difference in the sequence
of the DNA (Goldberg et al. 2007). In plants in particular, it is well known
that chemical or environmental treatments can elicit trait changes that
will be transmitted for one or more generations, but which can revert
spontaneously. In humans, the phenomenon of **parental imprinting**, in
which the phenotype seems to be inherited in a parent-specific manner
(that is, either maternally or paternally), is a good example of epigenetic

inheritance (Hall 1990). Recently, the term has also come to encompass somatic inheritance and thus is no longer restricted to trans-generational effects. Early-life influences that persist in their effects on health, even after the individual moves away from or outgrows the influence, may have a basis in epigenetic modulation of gene activity (Jirtle and Skinner 2007). The idea is that some molecular mechanism independent of sequence variation influences transcription or protein function in a manner that is transmitted through cell divisions, or at least maintained for many years. By this definition, an epigenetic phenomenon is any long-term change in gene function that is independent of the DNA sequence. The most likely candidate mechanism is DNA methylation, but other types of chromatin modification might also be transmitted. It is also conceivable, at least somatically, that the logic of regulatory interactions can produce state switches that maintain networks of gene activity independently of the DNA.

In this chapter, we will survey the major technologies used for profiling chromatin structure, which make use of DNase hypersensitivity, cross-linking of adjacent elements, methylation analysis, and histone modification. The focus in this chapter is on techniques, since specific applications are described in Part III. We conclude with a discussion of the use of chromatin immunoprecipitation to monitor protein-DNA interactions. First, however, we start with an overview of the ENCODE Project.

The ENCODE Project

Initiated in 2003 by the U.S. National Human Genome Research Institute (NHGRI), the ENCODE Project Consortium, originally consisting of 27 research groups, aims to provide the scientific community with high-quality, comprehensive annotations of all functional elements of the genome that either have a defined product (RNA or protein) or a reproducible biochemical signature (**Figure 9.1**). By 2007, the first phase of the project had documented elements in 1% of the genome, developed a common set of tools, and produced the informatics and database infrastructure to proceed to analysis of the whole genome. In 2012, a set of 30 papers were published that delivered on this promise. A useful "User's Guide" was published in *PLoS Biology* in 2011. The project has a coordinating portal at encodeproject. org, and all the data are accessible through the UCSC genome browser at genome.ucsc.edu/ENCODE as well as at individual project sites.

Data have been generated in three tiers. Tier 1, at the highest depth, focuses on immortalized B cell (GM12878) and leukemia (K562) cell lines as well as an embryonic stem cell line, H1. Tier 2 adds two more cancer cell lines, the well-known HeLa cervical carcinoma derivative and the HepG2 hepatocarcinoma, as well as the HUVEC umbilical vein endothelial cell line. Tier 3 is so far described at lower resolution, but includes an additional hundred cell types. In all cases, the initial focus has been comprehensive annotation of the structure and coding potential of all transcribed gene

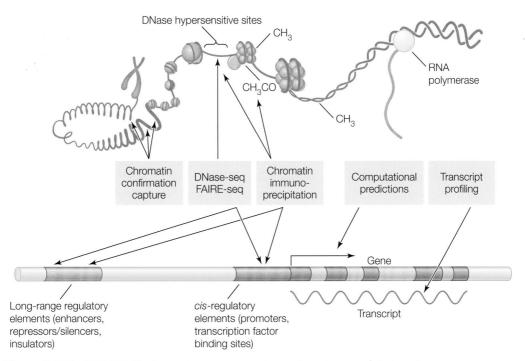

Figure 9.1 The ENCODE Project. This schematic shows the major assays designed to identify long-range regulatory elements, chromatin-binding proteins, transcripts, and structural elements in the DNA. These assays are explained in the text later in this chapter. (After ENCODE Project Consortium 2011.)

regions, including pseudogenes, and documentation of robust evidence for their biochemical features.

Novel methods such as cap analysis of gene expression (CAGE) and paired-end ditag (PET) sequencing are being used to annotate the 5′ and 3′ ends of transcripts. Much has been learned about the rich diversity of antisense and intron-derived transcripts that were hitherto ignored. Subcellular localization of transcripts is also being performed, along with determination of the network of RNAs bound by common ribosomal binding proteins (RIP-Seq). All these data are made freely available to the scientific community essentially as soon as basic quality control checks have been completed. In order to protect the rights of the researchers who have generated the data, a voluntary 9-month embargo known as the Toronto Genomic Data Use Agreement precludes other investigators from publishing before the consortium publishes its findings. Because the data are so rich, and because there are so many technical and statistical issues associated with their interpretation, both genome-wide and locus-specific analyses by researchers throughout the world are encouraged.

The 30 papers that were published in 2012 appeared in three different journals: *Nature, Genome Biology,* and *Genome Research.* In order to facilitate consumption of those papers, *Nature* has developed a novel set of 13 "threads" that link together paragraphs, figures, and tables that deal with a common theme (**Table 9.1**). For example, the first thread covers transcription factor motifs by bringing together 16 figures and tables from 10 of the papers. Some threads are feature-oriented (e.g., thread 7, DNA methylation), others are purely computational (thread 10, Characterization of network topology), and still others are integrative (thread 12, Impact of functional variation on understanding variation). An additional 14 associated papers provide individual perspectives on the types of analyses that will be forthcoming—notably, from a genetics perspective, those that relate the ENCODE features to findings from genome-wide association and gene expression studies.

Chromatin Profiling with DNase Hypersensitive Sites

Most DNA in chromatin is wrapped around nucleosomes in a very ordered manner that prevents its digestion by DNase I or other endonucleases. Prior to the initiation of transcription, DNA-binding complexes begin to assemble at promoter regions and distal enhancers, displacing the nucleosomes and opening up the chromatin, which becomes two orders of magnitude

TABLE 9.1   "Threads" that link related materials from the ENCODE Project papers

Thread	Topic
1	Transcription factor motifs
2	Chromatin patterns at transcription factor binding sites
3	Characterization of intergenic regions and gene definition
4	RNA and chromatin modification patterns around promoters
5	Epigenetic regulation of RNA processing
6	Non-coding RNA characterization
7	DNA methylation
8	Enhancer discovery and characterization
9	Three-dimensional connections across the genome
10	Characterization of network topology
11	Machine learning approaches to genomics
12	Impact of functional information on understanding variation
13	Impact of evolutionary selection on functional regions

Source: www.nature.com/encode/#/threads

more sensitive to digestion. Consequently, the identification of **DNase hypersensitive sites** (**DHS**) is a powerful approach to global definition of regulatory regions. Not all DHS represent actively transcribed genes, and not all transcription factors leave a DHS footprint, but the enrichment of transcribed regions in the vicinity of DHS is sufficiently strong that much effort has been given to defining DHS across a wide variety of cell types.

The DNase-Seq protocol that is used to map DHS is based on the idea that the locations of breaks induced by endonuclease treatment of intact nuclei can be defined by sequencing DNA fragments that terminate at those cleavage sites. DNase I treatment generates nicks that can be ligated to an adapter that is itself linked to biotin (Crawford et al. 2006; **Figure 9.2A**). The biotin facilitates easy capture of the fragments as it binds to the streptavidin-coating on magnetic beads which can then be readily pulled out of solution. The adapter also has a binding site for a restriction enzyme, MmeI, which cuts 20 bp downstream, liberating short tagged fragments that can be easily separated from large undigested DNA fragments and amplified, then sequenced in massively parallel fashion. The short sequence reads are aligned to the complete reference genome sequence, then computationally piled up to identify regions of enrichment—namely, DHS—that can be visualized on a browser (**Figure 9.2B**). There are approximately 200,000 DHS per cell type, covering 1% of the genome at an average of 10 per gene.

A second method for mapping open chromatin, known as FAIRE (*formaldehyde-assisted isolation of regulatory elements*), appears to be more efficient for isolating distal regulatory elements. It relies on the ability of formaldehyde to cross-link proteins to DNA. Histone-bound regions cross-link heavily and are removed by phenol extraction, leaving exposed regions in solution, from which they can be purified for sequencing (Giresi et al. 2007).

The vast majority (98.5%) of known transcription factor binding sites (TFBS) fall within DHS in at least one of the ENCODE cell lines. A minority, about 14%, of DHS were found to be specific to a single one of the twenty cell types surveyed by Natarajan et al. (2012). These tend to be enriched in intergenic regions, whereas DHS associated with transcription start sites (TSS) are more widespread across cell types. Where the transcription factor is well characterized, the overlap of high-resolution DHS with TFBS is accurate almost to the nucleotide level (**Figure 9.2C**). When averaged across multiple binding sites for the same factor, the DHS footprints of transcription factors can match the sequence conservation of the binding site remarkably well. Just a small number of transcription factors appear to bind outside DHS, and these are known to be repressive heterochromatin-binding factors such as TRIM28, SETDB1, and ZNF274.

There appears to be a tight relationship between DHS and eQTL/eSNPs (Maurano et al. 2012). Slightly more than three-fourths of 5134 SNPs catalogued by GWAS are found to be in high linkage disequilibrium with SNPs located in DHS, and one in five are coincident with DHS. This finding is consistent with the idea that eSNPs may affect transcription factor binding, and in turn transcription. Another study of a sample of 70 lymphoblast cell lines from Yoruban Africans in the HapMap collection found that of 8902 DNase

(A)

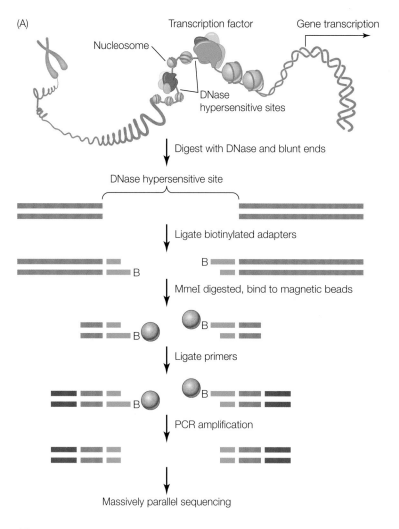

(B)

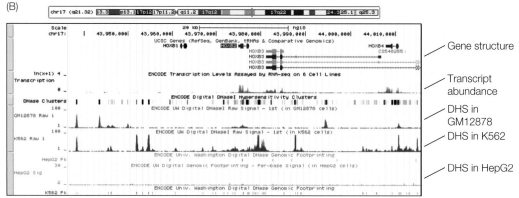

◄ **Figure 9.2 DNase Hypersensitivity.**   (A) The DHS assay aims to discover regions of open chromatin that DNase I can cleave. Active chromatin is unwound to some degree, which makes it accessible to DNase I, but proteins bound to regulatory sites protect segments of up to 20 bp from cleavage. The DNase-Seq protocol is described in the text. (B) Screenshot of typical profile of ENCODE data from the UCSC browser after selection of the appropriate tracks, in this case, showing numerous peaks in K562 cells that are for the most part missing in GM12868 cells, which have their own unique peaks. HepG2 cells do not have DHS at this locus. (A, after Song and Crawford 2010; B, from genome.ucsc.edu)

hypersensitivity–regulating SNPs, or "dsSNPs," that are associated with the read depth of DHS, 16% are also eSNPs (Degner et al. 2012). Conversely, 55% of eSNPs show plausible signs of affecting DNase hypersensitivity. Even more strikingly, according to Maurano et al. (2012), different GWAS SNPs associated with the same disease tend to disrupt DHS that correspond to similar transcription factors. Networks of such DNA-binding factors represent plausible transcriptional regulators of the disease class. Thus, POU2F1 and NFKB1, among others, associate with autoimmune disease and with DHS in relevant genes, while FOXO3 and ETS1 are more likely to associate with DHS in genes for various classes of cancer.

Chromatin Conformation Capture

Since DNA regulatory regions can exert their influences over long distances, up to or sometimes exceeding 1 Mb, techniques have been developed to detect long-range interactions. The idea underlying these techniques is that if DNA-binding proteins mediate the conjoining of a distal enhancer with a promoter, looping out the intervening DNA, then it should be possible to capture and sequence just the fragments that are brought together (Dekker et al. 2013). In the **Hi-C chromosome conformation capture** method (Lieberman-Aiden et al. 2009; **Figure 9.3A**), chromatin complexes are cross-linked, digested with a restriction enzyme, and the free ends are ligated back together. Performing the reactions under dilute conditions ensures that most ligations are within a cross-linked complex, effectively joining together two fragments that are normally a long way apart. The joined fragments are purified via biotin that is attached to nucleotides at the junction, the cross-linking is reversed, and after shearing of the DNA, millions of short fragments are sequenced in massively parallel fashion. This method provides a genome-wide readout of DNA features that are physically proximal to one another in the nuclei. A series of derivative methods known as 3C, 4C, and 5C can be used to target specific loci instead of the entire genome. In the **ChIA-PET method** (*ch*romatin *i*nteraction *a*nalysis by *p*aired-*e*nd *t*ag sequencing; Li et al. 2010; **Figure 9.3B**), adapter sequences are ligated to fragment ends on cross-linked complexes, the adapter sequences ligate to themselves, and then MmeI digestion releases short paired fragments, which are sequenced. In both cases, use of two different restriction

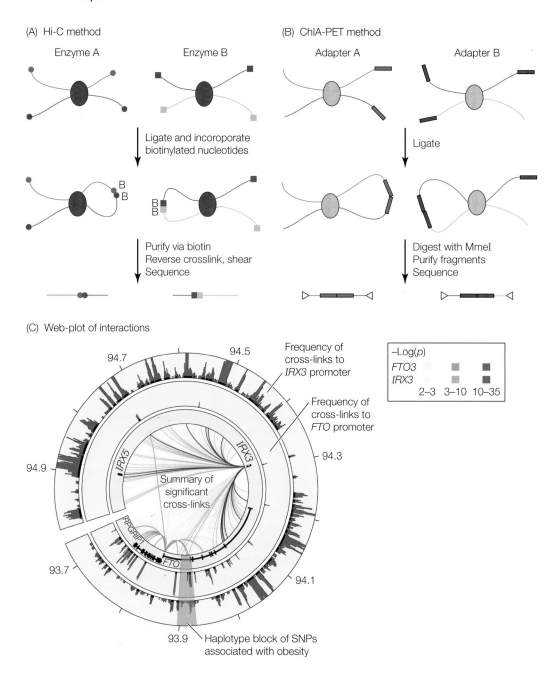

(A) Hi-C method

Enzyme A   Enzyme B

Ligate and incoroporate
biotinylated nucleotides

B
B

B
B

Purify via biotin
Reverse crosslink, shear
Sequence

(B) ChIA-PET method

Adapter A   Adapter B

Ligate

Digest with MmeI
Purify fragments
Sequence

(C) Web-plot of interactions

94.7   94.5

Frequency of
cross-links to
IRX3 promoter

Frequency of
cross-links to
FTO promoter

94.9

94.3

IRX5

IRX3

Summary of
significant
cross-links

RPGRIP1

FTO

93.7

94.1

93.9

Haplotype block of SNPs
associated with obesity

−Log(*p*)

FTO3
IRX3

2–3   3–10   10–35

enzymes or adapters can control for inter-complex ligations, as only a very small proportion of mismatches between the paired ends is expected.

Applied on a massive scale to 1% of the genome by the ENCODE Project (Sanyal et al. 2012), both methods have revealed that most TSS physically engage in an average of 3.9 long-range interactions with distal

◀ **Figure 9.3 Chromatin Conformation Capture.**   (A) The Hi-C method involves digestion of cross-linked chromatin complexes with a restriction enzyme followed by a series
of enzymatic steps that incorporate biotinylated nucleotides ("B" in the figure) at the
cleaved site as well as ligation of the free ends. After reversal of the cross-linking, the
DNA is sheared into fragments of less than 500 bp. These fragments are purified and
then sequenced en masse. Pairs of matched fragments indicate pieces of DNA that are
adjacent in the nucleus. (B) The ChIA-PET method is similar, but in this case, adapters
are added to sheared DNA, and the pairs of "ditags" produced by MmeI cleavage of
ligated ends are sequenced en masse. (C) An example using the targeted 4C method,
showing proximities of fragments across almost 1 Mb of DNA in the *FTO/IRX3* interval.
All fragments joined to the promoters of *FTO* (magenta) and *IRX3* (blue) are shown in
the center, and significant enrichments are highlighted around the outside circles. The
haplotype block of SNPs near the *FTO* gene that associate with obesity and BMI are
highlighted in tan at the bottom of the figure. Magenta and blue curved lines emanating
from this region make contacts with the promoters of both *FTO* and *IRX3*, but only *IRX3*
is highly expressed in the brain. (C, after Smemo et al. 2014.)

regulatory elements. Conversely, distal elements interact with an average
of 2.5 TSS. Tracks representing chromatin conformation data are accessible through the major genome browsers, facilitating in silico studies
without the need for individual labs to perform experiments themselves.
There is cellular heterogeneity to the interactions, and they tend to engage
DNase-sensitive regions or particular histone types. An important implication of these analyses is that they can help identify the genes that are
most likely to be regulated by disease SNPs: it is not always the closest
gene that is influenced by an enhancer, so chromatin structure provides
strong hints concerning which promoters are engaged with regulatory
polymorphisms. An excellent example is provided by the demonstration
that obesity-associated variants at the *FTO* locus significantly cross-link
to the promoter of the *IRX3* gene half a megabase away, and that it is the
IRX3 gene, rather than *FTO* itself, that is expressed in the hypothalamus,
where it is thought to mediate appetite regulation (**Figure 9.3C**; Smemo
et al. 2014).

Most of the interactions detected by chromatin conformation capture
assays are likely to be intrachromosomal—namely, *cis*-acting effects that
bring together parts of the same DNA molecule. However, unambiguous cases of *trans*-acting interactions have also been detected and shown
to influence complex patterns of gene regulation. For example, olfactory
receptor genes are found in large complexes, but only one of the several
thousand actual genes is transcribed per neuron. Studies in mice suggest
that the promoter of each neuron's specific receptor is brought into contact with the H enhancer element, either in *cis* or in *trans*. High-resolution
antibody staining of interphase nuclei also supports the idea that the two
copies of a single gene can assemble into a transcriptional factory (Bickmore 2013). In fact, interactions are also seen between completely different chromosomes, such as that between the interferon-λ gene on chromosome 10 and the T$_H$2 cytokine "locus control region" on chromosome 11.

High-throughput methods such as Hi-C are not designed to detect such interactions systematically, but all the targeted derivative methods are now being used to define just how pervasive interchromosomal pairing is as a mechanism for ensuring the co-regulation of unlinked genes.

Methylation Profiling

The most extensively used method of chromatin profiling is **methylation analysis** (Jaenisch and Bird 2003). Methylation in regulatory regions of genes is generally associated with silencing of transcription, whereas in gene bodies it may accompany upregulation of transcription (**Figure 9.4**). As organismal development proceeds, methylation is thought to stabilize patterns of gene activity, facilitating the robust maintenance of cell types and preventing reversion of differentiating cells to unwanted states. It also plays important roles in the suppression of endogenous retroviruses, in the inactivation of one of the X chromosomes in female cells, and in the genomic imprinting of several hundred genes (where only the allele transmitted from the mother, as in *H19*, or father, as in *IGF-2*, is expressed). In carcinogenesis, aberrant methylation contributes to heritable alteration of gene activity that can promote tumor progression (Weber et al. 2005), but further discussion of this very active field is deferred to Chapter 14. **Box 9.1** conveys an impression of some of the exciting areas of inquiry that methylation analysis has opened up.

The majority of methylation occurs at CpG dinucleotides, although isolated A and C residues can also be modified. **CpG islands** are locations in the genome, usually in upstream regulatory regions, where there is an

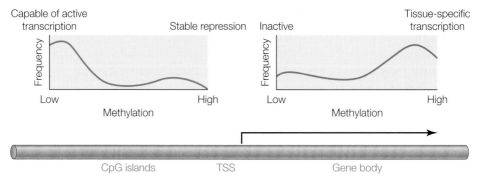

Figure 9.4 Methylation and Gene Expression. Transcription start sites (TSS) and CpG islands tend to be hypomethylated (red), whereas gene bodies show the opposite trend toward hypermethylation (blue). High level of methylation of promoters is associated with stable long-term repression of transcription and in some cases with imprinting. By contrast, methylation of gene bodies tends to be tissue-specific and dynamic, and it usually indicates active transcription. These relationships are trends that do not necessarily apply to specific genes or conditions.

enrichment of these dinucleotides relative to gene bodies and intergenic regions. Because methylated cytosines spontaneously deaminate to form T residues, there has been a genome-wide depletion of cytosines over evolutionary time, such that there is a gross deficiency of CpGs throughout the genome. The islands are what remain, and they seem to be maintained because of their functional importance. The CpG islands have "shores" that have intermediate frequencies of CpGs, and rates of methylation vary widely according to the location and frequency of CpGs. Methylation is established, or transmitted through mitosis, by various members of the DNA methyltransferase (DNMT) family of enzymes. These enzymes are inactive during meiosis, thus allowing demethylation to occur so that gene activity in the zygote is generally reset. Transcriptional suppression through methylation is thought to occur either directly, by disruption of the DNA binding sites for transcriptional activators, or indirectly, by recruitment of protein complexes that restructure the chromatin, for example, by modifying the pattern of histone acetylation (Jones 2012).

There are two major methods for profiling methylation genome-wide. These are essentially modifications of the established techniques for genotyping and DNA sequencing (Laird 2010). Both rely on the process of bisulfite conversion, which is the conversion of unmethylated C residues in CpG dinucleotides into U residues that appear as TpG in subsequent genotypes, whereas methylated sites remain as CpGs. Illumina produces a genotyping chip with 450,000 CpG sites that are known to be commonly methylated, facilitating methylation profiling at essentially all genes for a few hundred dollars a sample (Sandoval et al. 2011). Whole-genome sequencing following bisulfite conversion is possible, but since this is very expensive, methods have been developed for first purifying the methylated fraction of the genome, which is enriched for CpG islands, and then sequencing just this fraction. This method is sometimes called reduced representation BS-Seq, or Me-DIP, for *me*thylated *D*NA *i*mmuno*p*recipitation followed by bisulfite sequencing (Meissner et al. 2005). Whereas the array-based approach interrogates sites that are known to be commonly methylated, the sequencing requires comparison with the individual's untreated DNA in order to distinguish whether observed TpGs are due to bisulfite conversion or a polymorphism.

The ENCODE Project Consortium found that 96% of CpG sites are differentially methylated across 82 cell types, with most of the variation in gene bodies and intergenic regions (Thurman et al. 2012). Within DHS, heavily footprinted regions are less likely to be methylated, which is consistent with the idea that methylation prevents the binding of transcription factors. However, other data suggest that in general, methylation occurs after transcription factors vacate binding sites: for many transcription factors, there is a negative correlation between the abundances of their own transcripts and the methylation of their suspected binding sites. Some genes and transcription factors show the opposite trend, indicating that the relationship between methylation and transcription at the level of individual genes is complex. Methylation profiling and transcript profiling are thus

BOX 9.1
Methylation Analysis in Biology

Much contemporary research in human epigenetics builds on the experience gained through several decades of study of methylation. This research has captured the attention of thousands of investigators, and epigenetics now occupies a prominent place in the contemporary genetics literature. Here, we can only begin to convey some of the excitement in the field from the perspectives of both basic research and clinical practice.

A good place to start is X inactivation. Early in female mammalian development, one of the two X chromosomes must be turned off, or inactivated, lest the ratio of sex chromosomal to autosomal gene expression differ between the two sexes. In flies, evolution hit upon the opposite solution: namely, upregulation of gene expression from the single X in males to match expression from the two copies in females. In humans, however, halving of X chromosome gene expression in females is achieved by random condensation of one of the X chromosomes, which becomes a Barr body that is visible in karyotypes. This process is driven in part by hypermethylation. Since X inactivation occurs sometime after development has begun to pattern the future body, some tissues and organs can end up mosaic for the copy of the X chromosome that is inactive. For example, some women have two different-colored eyes, or even patches of different colors within the iris of one eye. Similarly, the photoreceptors in different portions of a woman's eye might express different alleles that perceive red light at slightly different wavelengths. Famously, the coloration of calico cats is attributed to a similar process, and a human analog is the activity of sweat glands, which can sometimes vary across a woman's body. In Chapter 2 we also saw the importance of careful regulation of gene dosage of the sex chromosomes, since rare X0 and XXY individuals have specific syndromes.

Differential methylation also accompanies the differentiation of cell types. For this reason, exhaustive maps of the methylation profiles of tissues are being generated, as they will help to define which loci and genes are inactivated and provide clues to mechanisms of gene regulation (Rakyan et al. 2008). There is some debate over whether methylation is causal in or responsive to regulatory switches, but most researchers agree that it is a critical component of the maintenance of stable cell states (Jones 2012). Conversely, disrupted methylation is observed in multiple types of cancer, and it accompanies activation of oncogenes and repression of tumor suppressors. Methylation also provides critical biomarkers for subtypes of cancer, such as a form of colon cancer that appears to be driven by altered DNA methyltransferase activity and is hence more likely to be responsive to drugs that target epigenetic effector molecules, including 5-azacytidine (Vidaza) and 5-aza-2′deoxycytidine (Dacogen).

Since hypermethylation is focused at critical CpG islands, it is emerging as an important biomarker for a wide range of diseases, from neurodevelopmental to metabolic and autoimmune disorders. Relatively simple PCR-based assays are routinely employed for the early detection of colon and prostate cancer, now using noninvasive assays of stool, saliva, blood, or urine, which can contain free tumor DNA released into circulation. For example, the *GSPT1* gene, whose product is a mediator of detoxification of carcinogens, appears to be turned off by hypermethylation in over a third of prostate cancers, and it now provides an early screening tool alongside PSA, as shown in the figure (after Heyn and Esteller 2012). As more genes are added to hypermethylation detection platforms, there are high hopes for detection of the majority of cancers with a low false positive rate. In the case of glioblastoma, hypermethylation of the promoter of the *MGMT* gene has pharmacogenetic value, as the gene's protein product repairs DNA adducts introduced by alkylating drugs designed to damage DNA of dividing cells; patients who have an inactive gene thus respond better to the therapy.

A novel area of enquiry arises from concerns that differentiated cells derived from embryonic stem cells, or from induced pluripotent stem cells, are likely to have methylation profiles different from

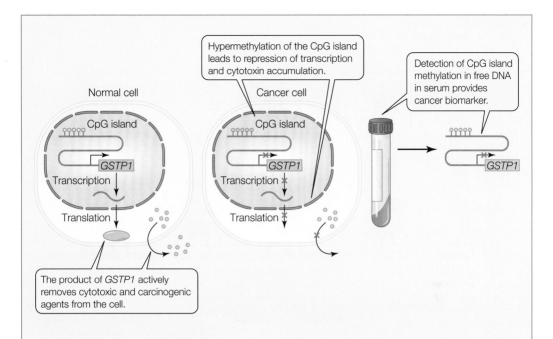

Hypermethylation of the CpG island leads to repression of transcription and cytotoxin accumulation.

Detection of CpG island methylation in free DNA in serum provides cancer biomarker.

Normal cell

Cancer cell

CpG island

CpG island

GSTP1

GSTP1

GSTP1

Transcription

Transcription

Translation

Translation

The product of *GSTP1* actively removes cytotoxic and carcinogenic agents from the cell.

those that occur naturally (Nishino et al. 2011). It is not known whether these differences might contribute to altered cellular physiology, with results such as premature aging or a higher risk of cancer, if and when stem cell treatments become common. Dolly the sheep, the world's first fully cloned animal, had shortened telomeres and early-onset arthritis, conditions that may reflect epigenetic defects. It is well established that a percentage of the genes in all cloned animals have defective expression, probably largely due to aberrant methylation, which is responsible for low rates of completely normal development (Humpherys et al. 2001). Thus, epigenetic research is likely to play a key role in the clinical implementation of stem cell technology. Similarly,

it is being used to investigate limitations on the pluripotency of stem cells and to help devise new strategies that may overcome blocks to driving cells along desired developmental trajectories. It will be particularly interesting to see whether interindividual differences in methylation patterns are important.

At the conclusion of this book, we will encounter the phenomenon of methyl-age, in which methylation status at hundreds of loci provides a combinatorial marker of an individual's chronological age. This phenomenon raises all sorts of possibilities in forensics, prediction of likely rate of aging and disease risk, and no doubt many other areas of clinical relevance.

complementary approaches that illuminate different dynamic and mechanistic aspects of gene regulation.

The Histone Code

No less important for gene regulation than the direct modification of chromatin is the modification of proteins that bind to chromatin, starting with histones (Strahl and Allis 2000). Histones are the basic proteins around which double-stranded DNA winds to form nucleosomes, which in turn

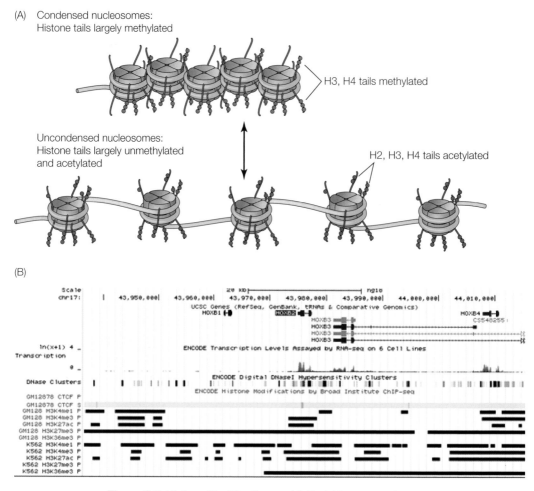

Figure 9.5 Histone Modifications. (A) Each histone core is an octamer of four proteins, each represented twice. The tails of those proteins can be modified at specific sites, four of which on histone H3 are particularly well characterized with respect to their general associations with transcript abundance, as described in the text. (B) Tracks summarizing the prevalence of each histone mark along a chromosomal interval can be displayed at the UCSC browser. Comparison of the two cell types shows differences in the prevalence of marks indicated at the right. (A, after Gilbert 2013; B, from genome. ucsc.edu)

seed the aggregation of higher-order chromatin domains. The core histone octamer consists of two dimers of H3/H4 and H2A/H2B, and the octamers can be linked together by the H1 histone. Chemical modifications of histones regulate their activity; these modifications, often referred to as "marks," include well-characterized patterns of methylation (of arginine or lysine residues), acetylation (of lysine), and phosphorylation (of serine, threonine, or tyrosine) as well as ubiquitination, citrullination, SUMOylation, and ADP-ribosylation.

The relationship between histone modification and gene activity appears to constitute a code, but it is a complex combinatorial code (Jenuwein and Allis 2001; Zhou et al. 2011). Trimethylation of Lys4 of histone H3 (H3K4Me3) bound to the promoter region is strongly associated with active transcription (**Figure 9.5A**), as is trimethylation of Lys36 at the gene body (H3K36Me3), which may help to stabilize the processivity of the enzyme responsible for transcript elongation. By contrast, H3K27Me3 is a marker of transcriptional repression, which probably occurs via interaction with the Polycomb complex of repressors. Similarly, di- or trimethylation of Lys9 is associated with low gene activity in heterochromatin. Acetylation of the same residue promotes transcription. In some cases, histone function is modified by the combination of multiple marks and is dependent on their location with respect to the promoter. There are literally hundreds of billions of possible combinations of histone marks, and direct observation demonstrates that at least 4000 different combinations of 40 basic H3 methylation and acetylation events occur, more than half at a single promoter. Yet surveys over several thousand promoters in different cell types are beginning to untangle statistical correlations between the histone code, transcription factor occupancy, chromatin accessibility, and transcript abundance (Kundaje et al. 2012). Tracks are available at the UCSC browser that allow investigators to visualize the histone profiles defined by the ENCODE Project Consortium (**Figure 9.5B**).

Chromatin Immunoprecipitation

Protein-DNA interactions are most commonly assayed in a high-throughput manner by chromatin immunoprecipitation (ChIP) followed by next-generation sequencing (ChIP-Seq; Valouev et al. 2008; Park 2009). This method can be applied to modified histones; common components of the transcriptional machinery, including RNA polymerase II; and site-specific transcription factors. The strategy (**Figure 9.6**) is to lightly cross-link these proteins to the DNA in active nuclei, isolate the DNA, immunoprecipitate (capture) the associated proteins with a high-specificity antibody, and then reverse the cross-linking allowing the liberated DNA fragments to be sequenced.

The interpretation of ChIP-Seq data is heavily dependent on data quality and analytic algorithms, so the research community has developed strict guidelines that facilitate comparison across experiments (Landt et al. 2012). Some transcription factors have very well-defined binding motifs, while others seem only to have a preference for low-complexity repeats and/ or may require the co-binding of other factors (Vaquerizas et al. 2009). Consequently, data quality and quantity vary for different types of factors. Absence of detected binding does not necessarily imply absence of actual binding, nor can it be assumed that the strongest binding sites are the most important—in fact, biophysical modeling suggests that weak binding sites may often be the ones that provide sharp thresholds of activity in

Figure 9.6 Chromatin Immuno-precipitation Sequencing. The workflow for ChIP-Seq involves cross-linking proteins to DNA in active nuclei, using high-frequency sound waves to isolate protein–DNA complexes, precipitating the protein of interest with antibody-covered beads, reversing the cross-links, and sequencing the liberated DNA fragments, which are then aligned to the reference genome. (After Cooper and Hausman 2013.)

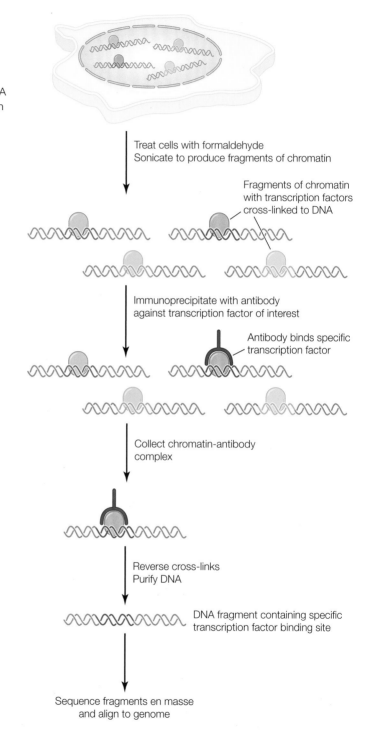

Treat cells with formaldehyde
Sonicate to produce fragments of chromatin

Fragments of chromatin with transcription factors cross-linked to DNA

Immunoprecipitate with antibody against transcription factor of interest

Antibody binds specific transcription factor

Collect chromatin-antibody complex

Reverse cross-links
Purify DNA

DNA fragment containing specific transcription factor binding site

Sequence fragments en masse and align to genome

cooperation with other factors. Nevertheless, the vast quantity of emerging data is extremely useful for a wide variety of applications.

DNA-protein interaction data are publically accessible through portals such as FactorBook (www.factorbook.org; Wang et al. 2012). At this site, users can search the database for specific transcription factors or families of related DNA-binding proteins and download raw data for all available cell lines in which ChIP-Seq has been performed. Transcription factor–specific pages then report results. For example, the low-specificity ZZZ3 protein appears to prefer AGAGAGAGAG repeats, whereas the interferon regulatory factor IRF1 has several different motifs among the top 500 sequences bound by the protein. **Figure 9.7A** illustrates two of these, the relatively highly specific affinity for AANNGAAA(G/C)(T/C) GAAA(G/C)(T/C), found in four-fifths of its targets, and a specific affinity for GTTCTAGG in just one-tenth of them. Also shown in **Figure 9.7B,C**

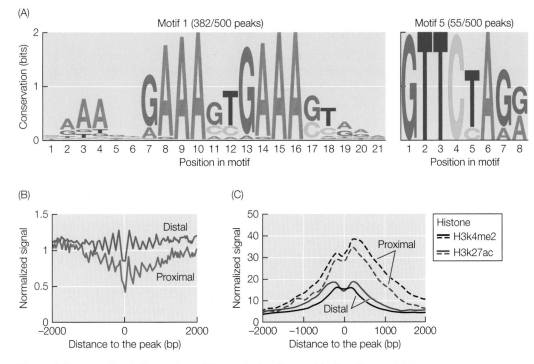

Figure 9.7 FactorBook Illustration of Transcription Factor Binding Sites. (A) Two of the top five DNA sequence motifs found in fragments precipitated by IRF1. Motif 1 is found in 382 of the 500 fragments (and is also bound by PRDM1), and Motif 5 is found in 55 of them. The height of each letter is proportional to the likelihood that the nucleotide is at the respective position. (B) Nucleosomes are positioned approximately every 200 bp on either side of the binding site, but asymmetrically, with a deficit of nucleosomes proximal to the binding site (that is, where the peak is within 1 kb of an annotated TSS). (C) Similarly, there is an asymmetrical distribution of some of the 11 recorded histone modifications proximal, but not distal, to the binding site. (After www.factorbook.org)

are the distributions of nucleosomes and of histone modifications within 2 kb on either side of the peak of transcription factor binding. Comparison with other factors that have similar binding profiles supports generation of hypotheses about how the factors may combine to regulate gene expression. Several other important databases, including JASPAR, TFD, and TRANSFAC, provide users with access to two decades of accumulated information on the approximately 1400 known human transcription factors.

Summary

1. Chromatin profiling is motivated by the finding that as cells adopt different states during development and pathogenesis, they show persistent "epigenetic" modifications that result in the transmission and maintenance of states of gene activity.

2. Heritable epigenetic modifications may also contribute to stable transmission of gene expression differences from parent to child, a special case of which is genomic imprinting, in which only the paternal or maternal allele is transcribed.

3. The ENCODE project, initiated in 2003, aims to document an encyclopedia of functional DNA elements and to define the structure of all genes in the human genome.

4. *Nature* provides access to key results published in 2012 in the novel form of threads of insight from multiple papers. All the ENCODE data are also accessible through the project genome browser.

5. DNase hypersensitive sites are regions of the chromatin that are accessible to digestion, most often because the binding of transcription factors has freed the DNA from tightly wound nucleosomes and compact chromatin domains.

6. High-resolution DHS mapping provides nucleotide-level imprints of the binding footprints for specific transcription factors, allowing investigators not only to localize regulatory regions, but also to infer which regulatory genes are acting on the target.

7. Chromatin cross-linking methods such as Hi-C and ChIA-PET have shown that DNA from distal regulatory regions loops out to contact promoter regions through protein intermediates.

8. Most chromatin contacts are *cis*-acting (within the chromosome), but some occur between homologs and even between different chromosomes.

9. Gene bodies tend to be hypermethylated while promoter regions are hypomethylated, except at CpG islands, which typically become methylated when a gene is stably transcriptionally repressed.

10. Genome-wide methylation profiling can be performed either by genotyping or sequencing of bisulfite-converted samples.

11. The histone code refers to the hypothesis that a combination of chemical modifications of histones, particularly methylation and acetylation, can predict gene expression levels.

12. Interactions between DNA and DNA-binding proteins, including transcription factors and modified histones, can be detected in a high-throughput manner by ChIP-Seq.

References

Bickmore, W. A. 2013. The spatial organization of the human genome. *Annu. Rev. Genomics Hum. Genet.* 14: 67–84.

Cooper, G. M. and Hausman, R. E. 2013. *The Cell: A Molecular Approach*, 6th ed. Sinauer Associates, Sunderland, MA.

Crawford, G. E., Holt, I. E., Whittle, J., Webb, B. D., Tai, D., et al. 2006. Genome-wide mapping of DNase hypersensitive sites using massively parallel signature sequencing (MPSS). *Genome Res.* 16: 123–131.

Degner, J. F., Pai, A. A., Pique-Regi, R., Veyrieras J-B, Gaffney, D. J., et al. 2012. DNase I sensitivity QTLs are a major determinant of human expression variation. *Nature* 482: 390–394.

Dekker, J., Marti-Renom, M. A., and Mirny L. A. 2013. Exploring the three-dimensional organization of genomes: Interpreting chromatin interaction data. *Nat. Rev. Genet.* 14: 390–403.

ENCODE Project Consortium. 2007. Identification and analysis of functional elements in 1% of the human genome by the ENCODE pilot project. *Nature* 447: 799–816.

ENCODE Project Consortium. 2011. A User's Guide to the Encyclopedia of DNA Elements (ENCODE). *PLoS Biol.* 9: e1001046.

ENCODE Project Consortium. 2012. An integrated encyclopedia of DNA elements in the human genome. *Nature* 489: 57–74.

Gilbert, S. F. 2013. *Developmental Biology*, 10th ed. Sinauer Associates, Sunderland, MA.

Giresi, P. G., Kim, J., McDaniell, R. M., Iyer, V. R., and Lieb, J. D. 2007. FAIRE (Formaldehyde-Assisted Isolation of Regulatory Elements) isolates active regulatory elements from human chromatin. *Genome Res.* 17: 877–885.

Goldberg, A. D., Allis, C. D., and Bernstein, E. 2007. Epigenetics: A landscape takes shape. *Cell* 128: 635–638.

Hall, J. G. 1990. Genomic imprinting: Review and relevance to human diseases. *Am. J. Hum. Genet.* 46: 857–873.

Heyn, H. and Esteller, M. 2012. DNA methylation profiling in the clinic: Applications and challenges. *Nat. Rev. Genet.* 13: 679–692.

Humpherys, D., Eggan, K., Akutsu, H., Hochedlinger, K., Rideout, W. M., III, et al. 2001. Epigenetic instability in ES cells and cloned mice. *Science* 293: 95–97.

Jaenisch, R. and Bird, A. 2003. Epigenetic regulation of gene expression: How the genome integrates intrinsic and environmental signals. *Nat. Genet.* 33(Suppl): 245–254.

Jenuwein, T. and Allis, C. D. 2001. Translating the histone code. *Science* 293: 1074–1080.

Jirtle, R. L. and Skinner, M. K. 2007. Environmental epigenomics and disease susceptibility. *Nat. Rev. Genet.* 8: 253–262.

Jones, P. A. 2012. Functions of DNA methylation: Islands, start sites, gene bodies and beyond. *Nat. Rev. Genet*. 13: 484–492.

Kundaje, A., Kyriazopoulou-Panagiotopoulou, S., Libbrecht, M., Smith, C. L., Raha, D., et al. 2012. Ubiquitous heterogeneity and asymmetry of the chromatin environment at regulatory elements. *Genome Res*. 22: 1735–1747.

Laird, P. W. 2010. Principles and challenges of genome-wide DNA methylation analysis. *Nat. Rev. Genet*. 11: 191–203.

Landt, S. G., Marinov, G. K., Kundaje, A., Kheradpour, P., Pauli, F., et al. 2012. ChIP-seq guidelines and practices used by the ENCODE and modENCODE consortia. *Genome Res*. 22: 1813–1831.

Li, G., Fullwood, M. J., Xu, H., Mulawadi, F. H., Velkov, S., et al. 2010. ChIA-PET tool for comprehensive chromatin interaction analysis with paired-end tag sequencing. *Genome Biol*. 11: R22.

Lieberman-Aiden, E., van Berkum, N. L., Williams, L., Imakaev, M., Ragoczy, T., et al. 2009. Comprehensive mapping of long-range interactions reveals folding principles of the human genome. *Science* 326: 289–293.

Maurano, M. T., Humbert, R., Rynes, E., Thurman, R. E., Haugen, E., et al. 2012. Systematic localization of common disease-associated variation in regulatory DNA. *Science* 337: 1190–1195.

Meissner, A., Gnirke, A., Bell, G. W., Ramsahoye, B., Lander, E. S., and Jaenisch, R. 2005. Reduced representation bisulfite sequencing for comparative high-resolution DNA methylation analysis. *Nucleic Acids Res*. 33: 5868–5877.

Natarajan, A., Yardimci, G. G., Sheffield, N. C., Crawford, G. E., and Ohler, U. 2012. Predicting cell-type-specific gene expression from regions of open chromatin. *Genome Res*. 22: 1711–1722.

Nishino, K., Toyoda, M., Yamazaki-Inoue, M., Fukawatase, Y., Chikazawa, E., et al. 2011. DNA methylation dynamics in human induced pluripotent stem cells over time. *PLoS Genet*. 7: e1002085.

Park, P. J. 2009. ChIP–seq: Advantages and challenges of a maturing technology. *Nat. Rev. Genet*. 10: 669–680.

Rakyan, V. K., Down, T. A., Thorne, N. P., Flicek, P., Kulesha, E., et al. 2008. An integrated resource for genome-wide identification and analysis of human tissue-specific differentially methylated regions (tDMRs). *Genom. Res*. 18: 1518–1529.

Sandoval, S., Heyn, H., Moran, S., Serra-Musach, J., Pujana, M. A., et al. 2011. Validation of a DNA methylation microarray for 450,000 CpG sites in the human genome. *Epigenetics* 6: 692–702.

Sanyal, A., Lajoie, B. R., Jain, G., and Dekker, J. 2012. The long-range interaction landscape of gene promoters *Nature* 489: 109–113.

Smemo, S., Tena, J. J., Kim, K. H., Gamazon, E. R., and Sakabe, N. J. 2014. Obesity-associated variants within *FTO* form long-range functional connections with *IRX3*. *Nature* 507: 371–375.

Song, L. and Crawford, G. E. 2010. DNase-seq: A high resolution technique for mapping active gene regulatory elements across the genome from mammalian cells. *Cold Spring Harb. Protoc*. doi: 10.1101/pdb.prot5384

Strahl, B. D. and Allis, C. D. 2000. The language of covalent histone modifications. *Nature* 403: 41–45.

Thurman, R. E., Rynes, E., Humbert, R., Vierstra, J., Maurano, M. T., et al. 2012. The accessible chromatin landscape of the human genome. *Nature* 489: 75–82.

Valouev, A., Johnson, D. S., Sundquist, A., Medina, C., Anton, E., et al. 2008. Genome-wide analysis of transcription factor binding sites based on ChIP-Seq data. *Nat. Methods* 5: 829–834.

Vaquerizas, J. M., Kummerfeld, S. K., Teichmann, S. A., and Luscombe, N. M. 2009. A census of human transcription factors: Function, expression and evolution. *Nat. Rev. Genet.* 10: 252–263.

Wang, J., Zhuang, J., Iyer, S., Lin, X. Y., Whitfield, T. W., et al. 2012. Sequence features and chromatin structure around the genomic regions bound by 119 human transcription factors. *Genome Res.* 22: 1798–1812.

Weber, M., Davies, J. J., Wittig, D., Oakeley, E. J., Haase, M., et al. 2005. Chromosome-wide and promoter-specific analyses identify sites of differential DNA methylation in normal and transformed human cells. *Nat. Genet.* 37: 853–862.

Zhou, V. W., Goren, A., and Bernstein, B. E. 2011. Charting histone modifications and the functional organization of mammalian genomes. *Nat. Rev. Genet.* 12: 7–18.

Integrative Genomics

We conclude this survey of the methods of contemporary human genome analysis with a brief description of three additional types of "omic" analyses and some of the ways in which they can be integrated into systems biology research. **Proteomics** is the study of the distribution and abundance of all the proteins expressed in a cell or tissue; it also includes studies of posttranslational modification and of protein interaction networks. **Metabolomics** is the study of the abundance of small molecules—usually in body fluids such as serum, sputum, or urine—that are synthesized by the body or ingested or absorbed from the environment, including sugars, amino acids, lipids, and other complex organic compounds. Longitudinal studies of changes in the metabolome over time, or in association with pathology, are referred to as metabonomics. **Metagenomics** is the study of the composition of microbial or viral communities, in the human context, from various body cavities or surfaces; it includes surveys of species, gene, and in some cases, transcript content. All these types of analyses can be integrated with genotype and transcriptome studies using the tools of quantitative genetics; a few examples of this approach, variously known as **integrative genomics**, or **systems biology**, are provided in the last section of the chapter. An entire chapter (indeed, an entire book) could be written about each of the diverse bioinformatics approaches, increasingly dominated by machine learning and Bayesian approaches, that are being developed for systems biology. Here, I have chosen to focus instead on the types of data involved, trusting that readers will get a sense of the computational methods when they read the primary literature.

Proteomics

There are several reasons why profiling of the proteome may be useful. The most obvious is that, in general, it is proteins that perform biochemical and cellular functions. If the tools for comprehensive protein profiling were as robust and readily available as those for the transcriptome, more resources would be applied to the proteome. Although there is often a good correlation between protein and mRNA abundance, that is not always the case: proteins can "perdure" (remain in a cell long after the RNA decays), posttranscriptional regulation ensures that mRNAs are translated at different rates, and secreted proteins may not be represented by mRNA in a particular cell type. Furthermore, it is not just protein abundance that needs to be profiled: protein function is widely influenced by posttranslational modifications (phosphorylation, glycosylation, and cleavage, to name just three) that affect protein activity, localization, and protein complex formation. Thus, comprehensive proteomics aims to ascertain not only the abundance of protein isoforms, but also their modification state, their subcellular distribution, and the full range of protein-protein and protein–small molecule interactions.

There are two main methods used for high-throughput proteomic analysis. Two-dimensional differential gel electrophoresis (DIGE; Friedman 2007) separates proteins according to their size and charge, generating several thousand spots corresponding to the most abundant proteins that are amenable to extraction under the experimental conditions employed (**Figure 10.1A**). Spot sizes and locations (which vary subtly according to protein modifications) can be compared semiquantitatively with standards for the tissue of interest, in which the identity of each spot has already been determined by direct peptide sequencing. Options are also available for fluorescent labeling of proteins prior to separation so that two samples can be run on a single gel, which facilitates more quantitative comparisons.

Protein mass spectrometry (MS) is increasingly popular and can be performed proteome-wide or in a targeted fashion (Domon and Aebersold 2006). It provides accurate and quantitative profiling of hundreds to thousands of proteins (**Figure 10.1B**). Typically, a liquid chromatography step is used to fractionate subsets of the proteome by size. Each fraction is then sequentially subjected to mass spectrometry, which is the process of identifying a molecule from the constituent masses of a series of fragments. The mixture of proteins may be digested into peptide fragments by trypsin. These are then ionized from solution by electrospray (ESI) or from a matrix by laser desorption (MALDI). The mass-to-charge ratio (m/z) of the particles is measured by some combination of time of flight (TOF) through a vacuum or release from an ion trap (as in quadrupole or Fourier transform MS). By adding a second step in which individual proteins are

Figure 10.1 Proteomics. (A) Two-dimensional differential gel electrophoresis separates proteins by charge and mass, resulting in a unique spot for each protein isoform. (B) Protein mass spectrometry proceeds by digestion with trypsin of a protein mixture (generally either a fraction from a chromatography column or an affinity-purified complex) into characteristic peptides whose lengths and compositions provide a fingerprint of each protein. Ionization of the fragments (C) allows detection of their mass-to-charge ratios, which are matched to a table of known standards, the theoretical mass spectrum (D). The height of each peak is proportional to the abundance of a peptide with that mass-to-charge (m/z) ratio. In tandem MS, not shown, individual proteins are further fragmented, and the sequence of each peptide can be reconstituted from the small mass differences.

(A)

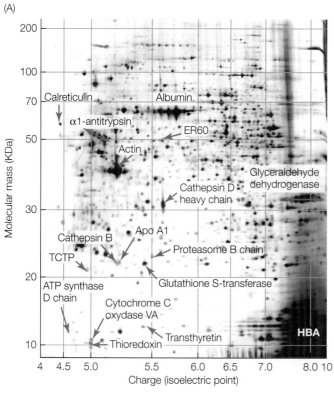

(B)

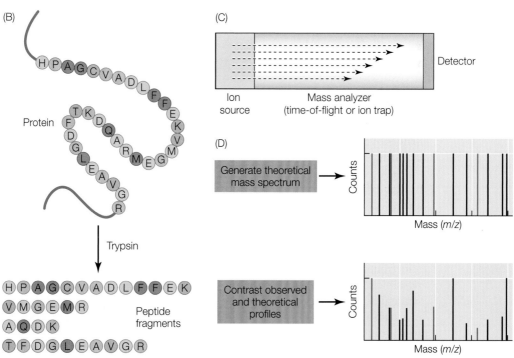

fragmented by high-energy gas collisions, it is possible to generate a series of fragments that provide a diagnostic fingerprint for all proteins that have similar m/z (Hunt et al. 1986). That is to say, the sequences MAGGTYS and MGATGSY have the same mass and charge, but random fragments do not (MAGG and MGAT, or TYS and GSY from the two halves), so the sequence can be reconstituted from a series of small fragment masses. This **tandem MS**, or **MS/MS**, procedure allows highly accurate, sensitive, and comprehensive identification of peptides and their state of modification. For accurate quantification, the proteins can be tagged with stable isotopes (either during cell culture or after initial purification) that result in a very small deviation of the experimental sample peaks relative to a standard. Thus, the m/z ratio of the experimental sample can be compared with that of a reference sample. Bioinformatics has played a critical role in the expansion of the repertoire of MS because a critical bottleneck has been the need to deconvolute myriad small peaks in a high-throughput manner.

Four major applications of protein mass spectrometry are (1) defining patterns of protein co-expression, (2) annotating protein structure, (3) applying comparative biology to the study of disease, and (4) biomarker development. Studies of the malarial parasite *Plasmodium*, for example, have identified posttranslational modifications, potential drug targets embedded in the red blood cell membrane, and novel peptides that have been mapped back to the parasite genome, assisting in annotation (Bautista et al. 2014). Focused studies of molecular machines such as the spliceosome have led to the characterization of high-dimensional protein complexes that can be further studied in combination with the tools of molecular biology. These **tandem affinity purification**, or **TAP**, methods, generally utilize an antibody or an engineered tag to capture a target protein and this facilitates careful characterization of the other members of the complex that are purified along with the targeted protein (Gillette and Carr 2013). Investigators have also worked out how to directly ionize proteins from tissue sections; such **imaging mass spectroscopy** provides detailed spatial profiles of the proteome (Caprioli 2014). Particular challenges in human proteomics include the wide range of protein concentrations (up to ten orders of magnitude in serum), the rich diversity of potentially millions of isoforms and modification states, and inter- and intraindividual variation.

Various other strategies also exist that facilitate generation of complete protein-protein interaction networks. A genetic approach is **yeast two-hybrid analysis** (Fields and Song 1989), in which one protein is used as a "bait" to capture domains ("prey,") that interact with it inside living cells, thereby regenerating a function such as transcription activation that is easily assayed. Two libraries of bait and prey are created and cells containing the recombinant genes are crossed together in massively parallel fashion to screen for possible interaction partners, which then need to be confirmed by other strategies. Bimolecular fluorescence complementation is one such method, in which interacting domains bring fluorescence reporters into close enough proximity to emit a light signal. Subcellular localization is generally pursued by in situ hybridization, in which antibodies cross-linked to a chemical or fluorescence marker, or to a gold particle, bind to proteins in a tissue section or whole mount, and light microscopy is used to image their distribution.

There are many public databases that amalgamate many types of protein information. These databases are too numerous to discuss, but the Expert Protein Analysis System, hosted by the Swiss Institute for Bioinformatics at www.expasy.org, provides links to many of them. These links include access to protein structure and homology modeling tools as well as databases of predicted and verified protein-drug interactions.

Metabolomics

Metabolomics parallels proteomics in many ways, but aims instead to catalog and make sense of the abundance of anywhere from hundreds to tens of thousands of small molecules in a tissue or body compartment under particular conditions (Patti et al. 2012). A convenient source of information about any metabolite of interest is the MetaboCard entry at the HMDB (Human Metabolome Database, www.hmdb.ca). **Figure 10.2** shows just some of the 110 fields for the trimethylamine-N-oxide (TMAO) MetaboCard: it includes the structure and chemical composition (in this case, C_3H_9NO) of the compound and descriptive statistics of its chemical and physical properties. Actual spectra can be downloaded, and the

Metabolite information		External links	
Structure		**DrugBank ID**	Not Available
		DrugBank Metabolite ID	DBMET00513
		Phenol Explorer Compound ID	Not Available
		Phenol Explorer Metabolite ID	Not Available
	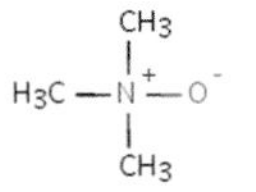 Zoom MOL SDF PDB SMILES InChI View Structure	**FoodDB ID**	FDB010413
		KNApSAcK ID	Not Available
Synonyms	1. N,N-Dimethylmethanamine N-oxide	**Chemspider ID**	1113
	2. TMA-oxide	**KEGG Compound ID**	C01104
	3. TMAO	**BioCyc ID**	TRIMENTHLAMINE-N-O
	4. Trimethylamine oxide	**BiGG ID**	Not Available
	5. Trimethylamine-N-oxide	**Wikipedia Link**	Trimethylamine oxide
	6. Triox	**NuGOwiki Link**	HMDB00925
Chemical Formula	C₃H₉NO	**Metagene Link**	HMDB00925
Average Molecular Weight	75.1097	**METLIN ID**	5876
Monoisotopic Molecular Weight	75.068413915	**PubChem Compound**	1145
IUPAC Name	N,N-dimethylmethanamine oxide	**PDB ID**	TMO
Traditional IUPAC Name	trimethylamine N-oxide	**ChEBI ID**	15724
CAS Registry Number	1184-78-7		
SMILES	C[N+](C)(C)[O-]		
InChI Identifier	InChI=1S/C3H9NO/c1-4(2,3)5/h1-3H3		
InChI Key	UYPYRKYUKCHHIB-UHFFFAOYSA-N		

The structure shown: H₃C — N⁺ — O⁻ with CH₃ groups (Chemical Formula C₃H₉NO)

Figure 10.2 The HMDB MetaboCard Entry for TMAO.   The HMDB website provides detailed data, including structural and chemical data, disease associations, and normal and abnormal abundance measures, for thousands of metabolites. Just a portion of the entry for TMAO is shown. (From www.hmdb.ca/metabolites/HMDB00925)

observed normal concentration range, as well as that in abnormal samples, with links to relevant publications, is provided, followed by a curated summary of clinical phenotypes and disease predispositions associated with the metabolite. Finally, where available, links to biochemical pathway databases are also provided. Three of the most commonly used databases are KEGG, MetaCyc, and Wikipathways. MetaCyc (Caspi et al. 2010; www.metacyc.org) contains over 2000 nonredundant experimentally elucidated pathways and includes a Human-Cyc compendium of known human metabolic pathways.

Most metabolomic analysis is carried out using mass spectrometry. A wide variety of tools are available for targeted analyses, but untargeted metabolome-wide analyses are emerging as well. Many MS studies utilize a well-characterized set of no more than 150 lipids, sugars, and amino acids that are ascertained with confidence. The latest imagers generate 20,000 or more peaks per profile, but unlike in transcriptomics many of these peaks correspond to unknown compounds and/or to multiple possible compounds. Mass-to-charge ratios in the range of a few parts per million of a known compound are suggestive of the likely identity of critical molecules, but confirmation requires comparison with the profile of the pure compound, while a large fraction of peaks remain unknown. The difficulties are compounded by the expectation that many metabolites may be specific either to a subset of individuals or to a specific date and time of sampling, since they represent compounds that have been ingested or taken up from the immediate environment. Two central databases of metabolites, Metlin (Smith et al. 2005; metlin.scripps.edu) and the HMDB (Wishart et al. 2009), maintained at the Scripps Institute and at the University of Alberta, respectively, catalog over 40,000 entries, including rare and common, and water-soluble and lipid-soluble compounds.

Geneticists are just beginning to explore the potential of metabolic profiling in basic, clinical, and translational research. Therapeutic metabolomics refers to the idea that small molecules present in pathological states may be drug targets, or modifiable in some cases by simple dietary supplementation. For example, several types of cancer in which a particular enzyme, ICD1, is mutated show upregulation of the R enantiomer of 2-hydroxyglutarate, and selective inhibition of the production of this compound induces differentiation, stopping the growth of glioma cells (Dang et al. 2010). **Lipidomics** refers to applications that specifically target the tens of thousands of lipids in cell membranes. **Toxicogenomics** and nutritional genomics are particularly relevant from the perspective of environmental health, since they demonstrate the capacity of untargeted metabolic profiling to characterize exposure to poisons that may passively diffuse into our bodies, or to monitor in a highly personalized manner how individuals respond to dietary shifts (Jones et al. 2012). Longitudinal profiling of individuals in hospital environments may be used both for epidemiological purposes, to track sources of variation among patients, and to provide prognostic biomarkers of treatment response (Nicholson et al. 2012). **Therapeutic drug monitoring**, which provides a direct measure of the rate at which patients metabolize drugs, is complementary to pharmacogenetics.

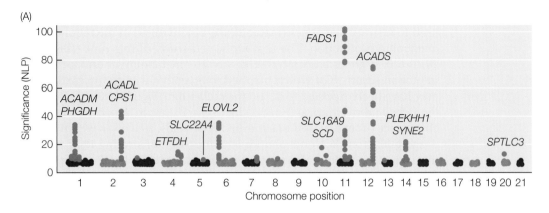

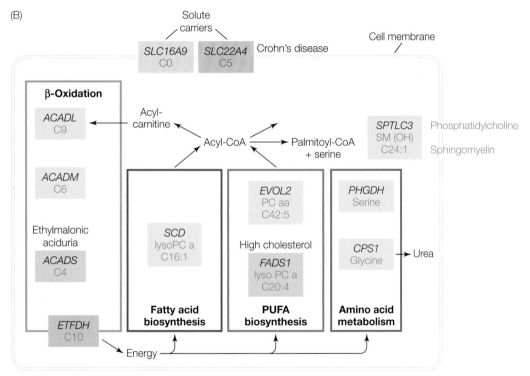

Figure 10.3 Genetics and Metabolomics.   A GWAS for the ratio of the abundance of two metabolites adjacent to each other in a metabolic pathway uncovered 14 independent loci (A) that fall into a half-dozen pathways related to lipid and amino acid metabolism. Red points are significant associations, and chromosomes are distinguished by alternating shades of gray. Each enzyme diagrammed in (B) is associated with the metabolite indicated in red lettering, as well as with the ratio of that metabolite to an adjacent one. The figure shows how genetic analysis of biochemical traits can uncover variation that is potentially clinically relevant, since three of the loci are also associated with diseases. (After Illig et al. 2010.)

Joint genotyping and metabolic profiling leads to GWAS that can be used to identify key genes (metabolite QTL) involved in the regulation of metabolite production. The first replicated study of this type (Illig et al. 2010) found genes associated with 9 of 163 metabolite abundance traits in human serum, 8 of which mapped to an enzyme or solute carrier that may have been expected to act on the metabolite (**Figure 10.3**). Variation in the fatty acid desaturase (*FADS1*) locus, for example, associates with the abundance of several polyunsaturated fatty acids, as well as with the ratio of the C36:3 to C36:4 phosphatidylcholines, demonstrating the principle that some associations also report differences in metabolic rates. Bypassing the genotyping component, another study evaluated environment-wide associations between metabolites and disease status, finding suggestive evidence that families of metabolites, including pesticide-derived epoxides and vitamin E, are associated with type 2 diabetes risk, with effects comparable to strong genotypes (Patel et al. 2010). The limitations on both of these types of analyses are sample size and expense, but clearly the potential for joint consideration of genetic and environmental contributions to disease will promote the field.

Metagenomics

Human metagenomics is the study of the microbial composition of human body cavities and surfaces (Riesenfeld et al. 2004; Cho and Blaser 2012). Since the vast majority of bacterial species cannot be cultured outside the body, the only way to infer their presence is through their biochemical signatures, usually by means of next-generation DNA sequencing. In fact, it is often noted that in terms of the total number of cells in each of our bodies, there is an order of magnitude more bacterial cells than human cells, so we are ten parts microbe to one part human. Hence, no human genome sequence can be considered complete without the metagenome. Another high-throughput method for capturing microbial content is known as **glycomics**, the study of the carbohydrates present in body fluids, many of which are specific to bacterial taxa.

There are three modes of metagenomic sequencing. An overview of the total taxonomic diversity is usually provided by analysis of ribosomal 16S rDNA content in total bacterial extracts. This locus has been so highly conserved throughout prokaryotic evolution that universal primers can be used to amplify the sequence from most species. The precise sequence of each molecule then provides an accurate representation of at least the genus, if not the species, of the bacterium it came from. Comprehensive representation of microbial gene content, by contrast, requires deep sequencing of total microbial DNA, which is aligned against the increasingly representative compilation of whole microbial genome sequences available in public databases (Wooley et al. 2010). Typically, full genomes are obtained for the most prevalent species, partial genomes for others, and some fraction of reads are of undefined origin, yet can still be assigned gene functions. Metagenomic RNA-Seq is used to provide a profile of gene expression in the microbial community, and of course this profile can change within and among individuals under different circumstances, even if the microbial species composition is constant. Mention should also be made of **viral metagenomics**, which is performed by filtering bacteria or human cell–sized matter from a fluid sample and then sequencing the extracted DNA.

The NIH Human Microbiome Project Consortium (www.hmpdacc.org) serves as a data coordination center for research in this domain. A landmark survey of taxonomic diversity in 4788 specimens from 18 body sites in 242 healthy adult volunteers at two medical schools, in St. Louis and Houston, found strong niche specialization (HMPC 2012). To some extent, the oral and nasal cavities, skin, gut, and vagina represent different habitats with quite distinct microbial species composition and complexity (**Figure 10.4A**). Each habitat has a small number of dominant taxa, but no taxa are observed across all body sites, and individuals are distinguished by minor taxa that are sometimes unique to themselves. Gene content is, however, quite constant, and it has been inferred that microbial metabolic capacity—namely, the representation of core biochemical pathways—is conserved across human body habitats. For the most part, barring changes in health, microbial communities remain constant within adults over time, though they take several years to converge on the adult composition, as infant profiles are somewhat different (**Figure 10.4B**). Yet at the higher level of geography, there are certainly differences in the gut microbiome in particular that might be attributed to rural relative to urban lifestyles as well as continental differences (**Figure 10.4C**; Yatsunenko et al. 2012). How this diversity influences disease risk will be a major focus of microbiological research in the coming decades.

Changes in microbial diversity accompany the onset of disease. For example, the gut microbiome, as measured from stool samples, shows a signature drop in diversity associated with obesity, as the composition transitions from a mixture of Bacteroidetes and Firmicutes species to dominance by Firmicutes. This shift changes the metabolic capacity of the microbiome and contributes to increased fatty acid absorption, probably via a number of mechanisms. The shift, which also has possible implications for inflammation and the emergence of insulin resistance and diabetes, is reversible upon weight loss (Ley 2010). Similarly, a perturbed microbiome is implicated in severe malnutrition in Malawian children, as infant twins discordant for the onset of a disease known as kwashiorkor have quite different microbial species composition—and transfer of the kwashiorkor community to mice induces weight loss and metabolic changes (Smith et al. 2013).

The opposite change is observed in women with bacterial vaginosis, in whom disease is associated with a marked increase in bacterial diversity in the body cavity that normally harbors the least complexity (Srinivasan et al. 2012). The microbiome of the respiratory tract is also of interest, particularly in relation to diseases such as cystic fibrosis (CF). The microbiome of airway sputum in patients with CF is clearly differentiated from that in healthy controls, particularly by its greater diversity, although a gradual

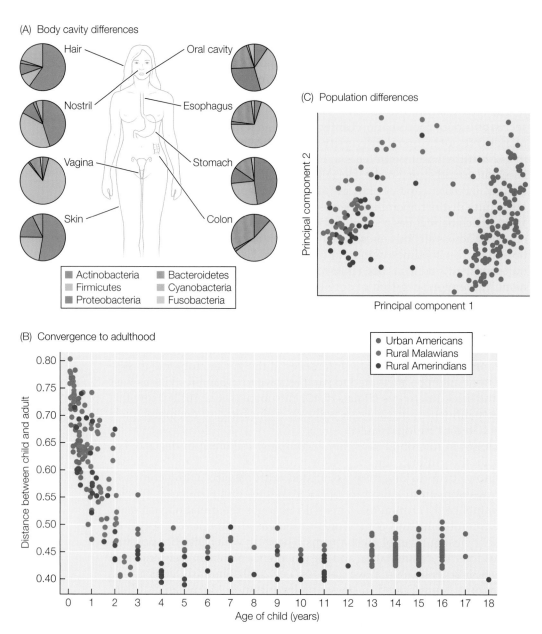

(A) Body cavity differences

Hair

Oral cavity

Nostril

Esophagus

Vagina

Stomach

Skin

Colon

■ Actinobacteria ■ Bacteroidetes
■ Firmicutes ■ Cyanobacteria
■ Proteobacteria ■ Fusobacteria

(C) Population differences

Principal component 2

Principal component 1

(B) Convergence to adulthood

● Urban Americans
● Rural Malawians
● Rural Amerindians

Distance between child and adult

Age of child (years)

Figure 10.4 Metagenomics. (A) Differences in microbial diversity in eight human body cavities as represented by the proportions of six phyla. (B) Fecal bacterial communities converge on a stable adult composition during the first 5 years of life, as indicated by a measure of the distance between the microbiomes of infants and of average adults for each of three population groups. (C) Adults of these three population groups have distinct microbiomes, as shown by principal components analysis of the variation in 16S rDNA species content of fecal samples. (A, after Cho and Blaser 2012; B and C, after Yatsunenko et al. 2012.)

decline in diversity—possibly due to persistent antibiotic usage—is associated with worsening disease (Lynch and Bruce 2013). A recurring theme in all these studies is that it is not just the species and genetic composition of the microbiome that matters for health, but also the microbial ecology, the interactions among species that influence attributes of the community and its interaction with the human host. Studies of bacterial physiology and transcriptional dynamics are also starting to reveal the effects of drugs, antibiotics and other xenobiotics (foreign chemicals not produced by the body) on the activity of the microbiome irrespective of its species composition (Maurice et al. 2013).

Integrative Genomics: Case Studies

We conclude this short chapter with four case studies in which multiple omic profiling strategies have been combined. This integrative genomics approach, also referred to as systems biology, necessarily incorporates advanced statistical methods as well as pathway modeling and computational biology. The core idea, illustrated in **Figure 10.5**, is that joint analysis of genotypes, transcripts, metabolites, proteins, and the microbiome can reveal correlations that help distinguish correlation from causation. Civelek and Lusis (2014) have reviewed the literature on this topic; their review highlights the importance of complementing human studies with animal models—typically nematodes, fruit flies, zebrafish, mice, rats, and nonhuman primates. Work with models offers the particular advantage of mutational genetic analysis—that is, the ability to experimentally confirm the role of a particular gene, or even nucleotide variant, in the production of a phenotype. The advent of high-throughput genome engineering mediated by the CRISPR/Cas9 guide RNA-mediated site-specific mutagenesis system now opens up the possibility of integrating targeted mutagenesis of human cell lines into these analyses as well (Zhou et al. 2014). The succeeding chapters will highlight applications of systems genomics approaches to specific diseases.

Type 2 diabetes risk

Genome-wide association studies have detected several dozen loci that contribute to type 2 diabetes risk, among which one, *KLF14*, has the unusual feature that the effect is maternally restricted: risk is elevated only if the rs4731702 variant 14 kb upstream of the transcription start site is transmitted from the mother. The reason for this appears to be that this eSNP regulates transcription of the gene, which is imprinted such that only the maternal copy is expressed in offspring. In subcutaneous adipose biopsies of 776 female twins as part of the British MuTHER (multiple tissue human expression resource) study, over 400 transcripts were found to be influenced in *trans* by the same variant. This, in turn, is attributed to the fact that the KLF14 protein is a transcription factor, whose DNA binding site is significantly enriched in the regulatory regions of the 121 most strongly affected targets. Furthermore, both rs4731702

Genomics

ACCCACTCTTGGATAGCCGGTTG
TGCGCAGCTGCGTTCAGAGCTTG
TGCGACGCGTCCAGCAGAACTTT
AGCACCGCGGTCATAGCAACACG
ACTGCCTAGGCCAGCAGAAGTCT

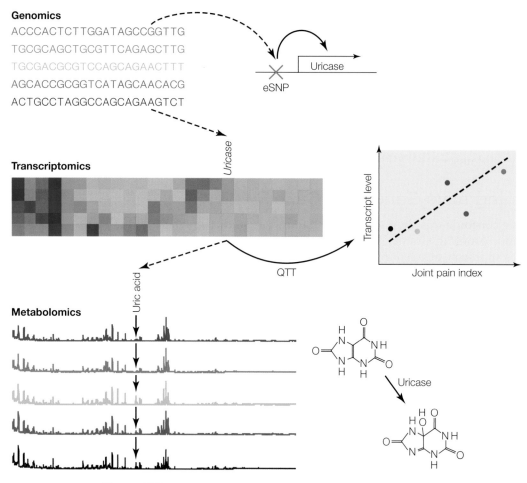

Figure 10.5 Integrative Genomics. This figure illustrates how the correlations among three types of omic data, namely genome sequence variation, transcriptomics, and metabolomics, can be used to enhance inference regarding the source of variation for a phenotype. The hypothetical example shows how DNA sequence variation is associated with the abundance of a transcript (uricase), which in turn correlates with the trait of joint pain. The genotype is called an eSNP (or eQTL), and the transcript is called a quantitative trait transcript (QTT). Furthermore, metabolomics shows that the transcript is also associated with the conversion of uric acid to 5-hydroxyisourate. Note, though, that the human version of the gene is actually nonfunctional, so does not actually contribute to gout—results from systems genomics ultimately need to be validated with targeted experimental manipulation.

and variants in six of the top ten *trans*-regulated loci are associated with a series of metabolic traits, including BMI, HDL cholesterol, triglyceride, and glucose as well as fasting insulin levels. Taken together, these observations (Small et al. 2011) reveal a network in which regulation of the allele of *KLF14* inherited from the mother influences the expression of hundreds of downstream target genes in adipose tissue, contributing to the emergence of peripheral insulin resistance and hence contributing to the risk of diabetes.

Red meat and cardiovascular disease

It is well established that red meat consumption is a risk factor for cardiovascular disease. A combination of metabolomic and metagenomic studies (Koeth et al. 2013) has shown that this risk is mediated in part by the pro-atherogenic properties of a compound known as trimethylamine-*N*-oxide, or TMAO (**Figure 10.6**). L-carnitine, which is an abundant nutrient in red meat, is metabolized to TMAO, primarily by the gut microbiota. TMAO synthesis is not detected in the weeks after experimental depletion of the microbiome by broad-spectrum antibiotic treatment. Furthermore, relative to meat eaters, vegetarians have lower baseline levels of TMAO and a markedly reduced capacity for its synthesis from carnitine, and this difference correlates with the presence of specific bacterial taxa in the intestinal microbiome. Individuals who have undergone major adverse cardiovascular events (heart attacks, stroke, or coronary artery bypass surgery) tend to have higher levels of plasma carnitine, but the risk of such events is much greater among those who also have high levels of TMAO. Additional manipulations in mice confirm all these findings and suggest that TMAO acts by inhibiting the reverse transport of cholesterol from peripheral tissues, including the coronary arteries, to the liver through a mechanism involving regulation of bile acid transporters.

Targeting of prostate cancer

The term "systems biology" is also used to convey the use of genomic data acquisition in the context of traditional hypothetico-deductive science, in which findings are tested iteratively in a cycle of experiments. This strategy, which has been well articulated in cancer studies, is here illustrated by the combination of ribosome profiling, targeted proteomics, and preclinical research (Hsieh et al. 2012; Thoreen et al. 2012). One of the key drivers of prostate cancer progression is constitutive expression of the nutrient biosensor mTOR, which regulates protein translation. Ribosome profiling is a particular type of RNA-Seq in which only RNA fragments protected from digestion by binding to ribosomes in living cells are sequenced. When ribosome profiling was applied to PC3 prostate cancer cells that had been treated with chemical inhibitors of mTOR, 144 transcripts were found to be downregulated specifically at the translational level. Bioinformatic analysis showed strong enrichment of these transcripts for functions in cell proliferation and invasiveness, and it also revealed that almost 90% of the transcripts possessed a modified 5' terminal "TOP" oligopyrimidine tract and/or a novel pyrimidine-rich sequence motif. This finding led to the rational design of a new compound, INK128, that binds to a motif in mTOR and was shown to downregulate a four-transcript signature of metastasis. Targeted knockdown

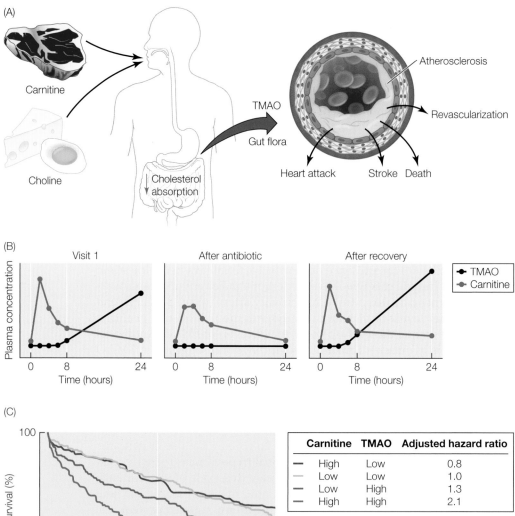

of expression of the genes encoding these four transcripts reduced the invasiveness of PC3 cells. Subsequently, in a mouse model of prostate cancer, the number of metastases and volume of cancer were reduced by treatment with the new drug.

◀ **Figure 10.6 TMAO and Atherosclerosis.** (A) Carnitine ingested with red meat is absorbed in the intestine, where enzymes encoded by the genomes of specific microbial species generate the derivative chemical TMAO. This chemical inhibits cholesterol clearance from the arteries, leading to increased incidence of stroke, heart attack, and death due to cardiovascular disease. (B) The role of gut bacteria in the conversion of carnitine to TMAO is demonstrated by mass spectrometry profiling of plasma. Following a steak meal, plasma carnitine levels rise for up to 8 hours, at which time synthesis of TMAO becomes apparent. However, in individuals treated for a week with broad-spectrum antibiotics to reduce their gut microbiome, TMAO is not produced. After recovery of the microbiome, TMAO synthesis resumes. (C) Survival curves show that individuals with high levels of plasma TMAO (but not carnitine) have a twofold elevated risk of adverse cardiovascular events over a 3-year span. (After Koeth et al. 2013.)

Interindividual immune system variation

Our final example incorporates ex vivo manipulation of the environment of immune system cells to identify common polymorphisms that mediate interindividual variation in pathogen sensing and inflammation. Lee et al. (2014) isolated dendritic cells from peripheral blood samples from 534 people in three population groups and identified eQTL associated with the expression of 415 transcripts under four conditions: resting, stimulation with lipopolysaccharides to mimic bacterial infection, stimulation with an influenza virus antigen, and stimulation with interferon-β. The local genetic influence on 121 genes was observed across all four conditions for only 57 of the genes, indicating how immune activation pervasively alters the regulatory landscape. Several of the effects were mapped to individual eSNPs by a combination of fine mapping across the three population groups in the sample and incorporation of ENCODE data, which showed enrichment for binding sites of the STAT family of immune-responsive transcription factors. Subsequently, luciferase reporter genes were used to confirm the causal role of individual variants, along with CRISPR-mediated modification of one site in its native chromosomal context. Completing the link to disease, the researchers observed that 23 eSNPs that were significant only upon stimulation were identified by GWAS as associated with the inflammatory autoimmune diseases multiple sclerosis, Crohn's and celiac disease, and psoriasis, or with leprosy. This study demonstrates how systems biology can resolve aspects of the innate immune contribution to pathogenesis. A similar analysis of leukoctyes from the same group of researchers (Raj et al. 2014) demonstrated T cell–specific contributions to neurodegenerative as well as autoimmune disease.

Summary

1. Large-scale omics strategies exist for analyzing the proteome, metabolome, and metagenome.

2. Integrative genomics, or systems biology, refers to the combination of multiple large-scale profiling approaches across samples in order to

gain insight into how regulation across molecular domains influences disease.

3. Proteomics is mostly performed by high-resolution mass spectrometry, in which the mass-to-charge ratios (m/z) of thousands of peptide fragments are measured and computationally deconvoluted to infer protein expression levels.

4. Additional proteomic methods are used to infer posttranslational modifications, protein-protein interactions, protein–small molecular interactions, and subcellular localization.

5. Metabolomics is also performed by mass spectrometry, now with modes that allow the parallel measurement of up to 20,000 metabolites from body fluids.

6. Public databases of biochemical pathways, notably KEGG, MetaCyc, and Wikipathways, allow investigators to map metabolites onto chemical interaction networks.

7. As with gene expression and methylation, genotypes that influence the abundance of specific metabolites can be identified, and are called metabolite QTL.

8. Metagenomics is the study of the composition of microbial communities, mostly using deep sequencing strategies. These methods can be used to infer taxonomic diversity, gene content, or gene expression dynamics.

9. The Human Microbiome Project aims to catalog metagenomic diversity within and among individuals in various body habitats.

10. Each body habitat tends to have its own representative microbial species composition, reflecting the environment of the niche in which the bacteria must grow and generate ecological communities.

11. People differ from one another in their metagenomic profiles, which tend to be constant in the absence of a change in diet or health, and hence might be used as biomarkers in personalized medicine

12. Integrative genomic studies that have combined these profiling approaches have enhanced our understanding of biological systems involved in metabolic, cardiovascular, and metastatic disease.

References

Bautista, J. M., Marín-García, P., Diez, A., Azcárate, I. G., and Puyet, A. 2014. Malaria proteomics: Insights into the parasite-host interactions in the pathogenic space. *J. Proteomics* 97: 107–125.

Caprioli, R. M. 2014. Imaging mass spectrometry: Molecular microscopy for enabling a new age of discovery. *Proteomics* 14: 807–809.

Caspi, R., Altman, T., Dale, J. M., Dreher, K., Fulcher, C. A., et al. 2010. The MetaCyc database of metabolic pathways and enzymes and the BioCyc collection of pathway/genome databases. *Nucleic Acids Res.* 38: D473–479.

Cho, I. and Blaser, M. J. 2012. The human microbiome: At the interface of health and disease. *Nat. Rev. Genet.* 13: 260–270.

Civelek, M. and Lusis, A. J. 2014. Systems genetic approaches to understand complex traits. *Nat. Rev. Genet.* 15: 34–48.

Dang, L., White, D. W., Gross, S., Bennett, B. D., Bittinger, M. A., et al. 2010. Cancer-associated IDH1 mutations produce 2-hydroxyglutarate. *Nature* 462: 739–744.

Domon, B. and Aebersold, R. 2006. Mass spectrometry and protein analysis. *Science* 312: 212–217.

Fields, S. and Song, O. 1989. A novel genetic system to detect protein–protein interactions. *Nature* 340: 245–246.

Friedman, D. B. 2007. Quantitative proteomics for two-dimensional gels using difference gel electrophoresis. *Methods Mol. Biol.* 367: 219–239.

Gillette, M. A. and Carr, S. A. 2013. Quantitative analysis of peptides and proteins in biomedicine by targeted mass spectrometry. *Nat. Methods* 10: 28–34.

Hsieh, A. C., Liu, Y., Edlind, M. P., Ingolia, N. T., Janes, M. R., et al. 2012. The translational landscape of mTOR signalling steers cancer initiation and metastasis. *Nature* 485: 55–61.

Human Microbiome Project Consortium (HMPC). 2012. Structure, function and diversity of the healthy human microbiome. *Nature* 486: 207–214.

Hunt, D. F., Yates, J. R., 3rd, Shabanowitz, J., Winston, S., and Hauer, C. R. 1986. Protein sequencing by tandem mass spectrometry. *Proc. Natl. Acad. Sci. U.S.A.* 83: 6233–6237.

Illig, T., Gieger, C., Zhai, G., Römisch-Margl, W., Wang-Sattler, R., et al. 2010. A genome-wide perspective of genetic variation in human metabolism. *Nat. Genet.* 42: 137–141.

Jones, D. P., Park, Y., and Ziegler, T. R. 2012. Nutritional Metabolomics: Progress in addressing complexity in diet and health. *Annu. Rev. Nutrition* 32: 183–202.

Koeth, R. A., Wang, Z., Levison, B. S., Buff, J. A., Org, E., et al. 2013. Intestinal microbiota metabolism of L-carnitine, a nutrient in red meat, promotes atherosclerosis. *Nat. Med.* 19: 576–585.

Lee, M. N., Ye, C., Villani A-C, Raj, T., Li, W., et al. 2014. Common genetic variants modulate pathogen-sensing responses in human dendritic cells. *Science* 343: 1246980.

Ley, R. E. 2010. Obesity and the human microbiome. *Curr. Opin. Gastroenterol.* 26: 5–11.

Lynch, S. V. and Bruce, K. D. 2013. The cystic fibrosis airway microbiome. *Cold Spring Harb. Persp. Med.* 3: a009738.

Maurice, C. F., Haiser, H. J., and Turnbaugh, P. J. 2013. Xenobiotics shape the physiology and gene expression of the active human gut microbiome. *Cell* 152: 39–50.

Nicholson, J. K., Holmes, E., Kinross, J. M., Darzi, A. W., Takats, Z., and Lindon, J. C. 2012. Metabolic phenotyping in clinical and surgical environments. *Nature* 491: 384–392.

Patel, C. J., Bhattacharya, J., and Butte, A. 2010. An environment-wide association study (EWAS) on type 2 diabetes mellitus. *PLoS ONE* 5: e10746.

Patti, G. J., Yanes, O., and Siuzdak, G. 2012. Metabolomics: The apogee of the omics trilogy. *Nat. Rev. Mol. Cell Biol.* 13: 263–269.

Raj, T., Rothamel, K., Mostafavi, S., Ye, C., Lee, M. N., et al. 2014. Polarization of the effects of autoimmune and neurodegenerative risk alleles in leukocytes. *Science* 344: 519–523.

Riesenfeld, C. S., Schloss, P. D., and Handelsman, J. 2004. Metagenomics: Genomic analysis of microbial communities. *Annu. Rev. Genet.* 38: 525–552.

Small, K. S., Hedman, Å. K., Grundberg, E., Nica, A. C., Thorleifsson, G., et al. 2011. Identification of an imprinted master *trans* regulator at the *KLF14* locus related to multiple metabolic phenotypes. *Nat. Genet.* 43: 561–564.

Smith, C. A., O'Maille, G., Want, E. J., Qin, C., Trauger, S. A., et al. 2005. METLIN: A metabolite mass spectral database. *Ther. Drug Monit.* 27: 747–750.

Smith, M. I., Yatsunenko, T., Manary, M. J., Trehan, I., Mkakosya, R., et al. 2013. Gut microbiomes of Malawian twin pairs discordant for kwashiorkor. *Science* 339: 548–554.

Srinivasan, S., Hoffman, N. G., Morgan, M. T., Matsen, F. A., Fiedler, T. L., et al. 2012. Bacterial communities in women with bacterial vaginosis: High resolution phylogenetic analyses reveal relationships of microbiota to clinical criteria. *PLoS ONE* 7: e37818.

Thoreen, C. C., Chantranupong, L., Keys, H. R., Wang, T., Gray, N. S., and Sabatini, D. M. 2012. A unifying model for mTORC1-mediated regulation of mRNA translation. *Nature* 485: 109–113.

Wishart, D. S., Knox, C., Guo, A. C., Eisner, R., Young, N., et al. 2009. HMDB: A knowledgebase for the human metabolome. *Nucleic Acids Res.* 37(Database issue): D603–D610.

Wooley, J. C., Godzik, A., and Friedberg, I. 2010. A primer on metagenomics. *PLoS Comput. Biol.* 6: e10000667.

Yatsunenko, T., Rey, F. E., Manary, M. J., Trehan, I., Dominguez-Bello, M. G., et al. 2012. Human gut microbiome viewed across age and geography. *Nature* 486: 222–227.

Zhou, Y., Zhu, S., Cai, C., Yuan, P., Li, C., et al. 2014. High-throughput screening of a CRISPR/Cas9 library for functional genomics in human cells. *Nature* 509: 487–491.

PART III

Diseases

11

Immune-Related Disease

W̶e will start our tour of the genetics of the major classes of human disease with the immune system. Our discussion will include variation in how people respond to different classes of pathogens, and discussion of the genetic factors promoting inflammatory disorders of the bowel, autoimmune diseases in which the body attacks itself, and primary immune deficiencies. Without doubt, the immune system also plays important roles in many other conditions, including respiratory and musculoskeletal disorders and cancer, which will be touched on in subsequent chapters.

Review of the Immune System

The study of immunology usually starts with the distinction between the innate and adaptive arms of the immune system (**Figure 11.1**), though it is important to recognize that these work closely together (Iwasaki and Medzhitov 2010). The innate arm includes passive defense systems, such as the mucous membranes, that create physical barriers that trap then expel microbes, as well as defensin peptides that target microbes nonspecifically. In the first week or two following infection, it is often immediate cellular responses, mediated by pattern recognition receptors such as the Toll-like receptors (TLRs) and C-type lectins, that ward off the pathogen. These responses lead to cytokine production and inflammation, which eventually also helps to trigger the adaptive immune response. In addition, natural killer (NK) cells destroy human cells that have lost their MHC "self" identification surface markers as a result of viral infection or tumor formation. The major histocompatibility complex (MHC, which in humans is also known as the human leukocyte antigen complex, or HLA) is a highly polymorphic complex of genes with critical functions not only in self identification, but also in foreign antigen recognition and processing.

The adaptive arm of the immune system can be regarded as a complex organ. It is responsible for the ability to recognize foreign entities, which is achieved by way of antibodies or T cell receptors that bind specifically

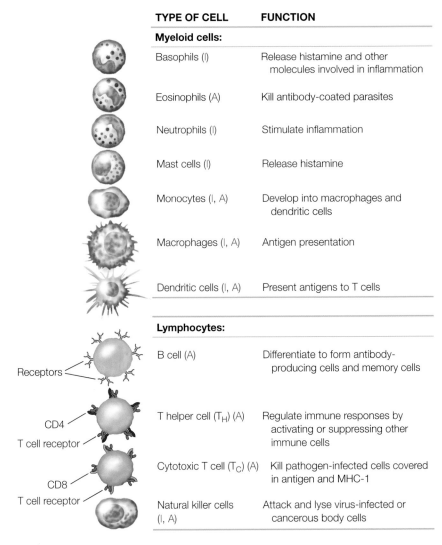

TYPE OF CELL	FUNCTION
Myeloid cells:	
Basophils (I)	Release histamine and other molecules involved in inflammation
Eosinophils (A)	Kill antibody-coated parasites
Neutrophils (I)	Stimulate inflammation
Mast cells (I)	Release histamine
Monocytes (I, A)	Develop into macrophages and dendritic cells
Macrophages (I, A)	Antigen presentation
Dendritic cells (I, A)	Present antigens to T cells
Lymphocytes:	
B cell (A)	Differentiate to form antibody-producing cells and memory cells
T helper cell (T_H) (A)	Regulate immune responses by activating or suppressing other immune cells
Cytotoxic T cell (T_C) (A)	Kill pathogen-infected cells covered in antigen and MHC-1
Natural killer cells (I, A)	Attack and lyse virus-infected or cancerous body cells

Receptors

CD4
T cell receptor

CD8
T cell receptor

Figure 11.1 Overview of the Immune System. The innate arm of the immune system (I) consists of a series of cell types (neutrophils, mast cells, and macrophages from the myeloid lineage; dendritic cells and natural killer cells from the lymphoid lineage) that mount generic inflammatory responses by way of phagocytosis, cytotoxicity, and release of histamines and regulatory peptides that stimulate the adaptive arm. There are three major classes of adaptive immune system cells (A), all of which act through recognition of foreign peptides. B cells secrete immunoglobulin antibodies that circulate in the body and induce the complement cascade after binding foreign peptides. T helper cells recognize foreign antigens presented on host cell surfaces by MHC class II in conjunction with CD4 co-receptors and stimulate other immune system cell types. Cytotoxic T cells recognize viral antigens presented by MHC I in conjunction with CD8 co-receptors and directly kill the infected cells.

to peptides or glycans that are not usually found in the host body. Both of these types of recognition molecules are generated by processes of rearrangement of exons and hypermutation that are specific to the immunoglobulin and T cell receptor genes. B cells produce antibodies that bind to a pathogen directly and initiate the biochemical cascade called the complement system to destroy the invader. T cells recognize molecules that are presented on the surfaces of infected host cells in conjunction with components of the MHC. Cytotoxic T cells are particularly important for antiviral responses; when activated by expression of the cell surface protein CD8, they target virus-infected cells that present antigens with MHC class I. T helper cells, by contrast, bind to antigens in conjunction with MHC class II, using the CD4 co-receptor in a longer-lasting response that requires engagement of hundreds of receptor pairs on each infected cell's surface. This binding causes the release of cytokines and other stimulatory molecules that induce other classes of lymphocytes and macrophages to attack pathogens. Cytokines are small regulatory proteins that promote inflammation and mediate communication among the cells of the immune system; they include the interferon and interleukin families of proteins. Dendritic cells are also key players in antigen presentation and cross talk between the innate and adaptive arms, which also engage hundreds of different cell subtypes too numerous to list here.

One of the more critical concepts in immunology is self/nonself recognition (Burnet 1961; Medzhitov and Janeway 2002). The immune system must be able to identify and eliminate foreign invaders such as viruses, bacteria, fungi, and parasites, which can be unicellular (e.g., *Plasmodium* species in malaria) or multicellular (e.g., nematodes and heartworms). It must also be able to eliminate host cells that have been compromised without harming the self, and it must keep a lifetime memory of invaders so that they can be quickly eliminated if they reappear. Furthermore, people living in different environments and different parts of the world are faced with different immune challenges, and of course, all those potential pathogens are constantly evolving avoidance strategies. All these requirements set up a delicate balance of molecular mechanisms, and in a sense, many immune-related disorders arise when that balance is tipped either toward a hyperactive response or overcautious failure to respond.

Variable Responses to Pathogens

Genetic variation influences whether or not different pathogens are readily cleared by the immune system, or at least tolerated as they persist in the body. In this section we consider response to bacteria, viruses, and bloodborne parasites as three examples.

Bacteria

Billions of people are at risk of infection with the bacteria that are responsible for tuberculosis (*Mycobacterium tuberculosis*), leprosy (*M. leprae*), meningitis (*Neisseria meningitides*), and pneumonia (*Streptococcus pneumoniae* among others) as well as dozens of other life-threatening infections. Candidate gene studies have implicated many genes as sources of variation for infectivity or response, and there is little doubt that there is considerable heritability for susceptibility to bacterial infection. However, the repeatability of these findings has been low, likely for multiple reasons, including genetic heterogeneity across human populations, bacterial genetic diversity, inconsistent diagnosis and assessment of disease, and the low statistical power associated with sample sizes typically in the hundreds rather than tens of thousands.

Genome-wide association studies (GWAS) are beginning to produce more robust evidence for genetic contributions to risk of bacterial infection, as reviewed by Chapman and Hill (2012). Only around 10% of people infected with *M. tuberculosis* develop the disease. Two loci have so far been identified that harbor protective factors (Thye et al. 2012). One is in a 200 kb gene desert (a region without any protein-coding genes) at 18q11.2. The other was first identified in Ghana and then replicated in Gambian, Indonesian, and Russian case-control cohorts. It is located 45 kb downstream of the Wilms tumor 1 (*WT1*) gene on chromosome 11p13, but the affected gene is as yet unknown. Once functional and integrative genomic assays have resolved the identities of affected gene products, these associations will provide insight into the mechanism of resistance.

Sepsis due to meningococcal infection has yielded only a single locus (Davila et al. 2010), but in this case, the identity of that locus is more revealing, as one or two of the genes in the complement factor H (*CFH*) region are affected. *N. meningitides* is thought to evade the complement system by producing a factor that binds to the CFH protein. The frequency of a likely protective variant rises from less than 5% in sub-Saharan Africa to intermediate levels in Europe and almost 50% in Asia, correlating inversely with the incidence of disease on the three continents, but without any evidence for selection having driven the difference in allele frequency.

GWAS of leprosy (Zhang et al. 2009) have been informative in two respects. First, eight loci associated with leprosy have been identified to date, five of which (*CYLD*, *NOD2*, *RIPK*, *TNFS15*, and *TLR1*) fall in the same innate immunity signaling pathway (**Figure 11.2**), while another three (*IL23R*, *RAB32*, and *HLA-DR-DQ*) are also known to have immune functions. The association of these loci with the MHC class II cluster implicates variation in the ability to recognize the mycobacterium via the T helper response. Second, several of the five innate immune loci overlap with the genes identified by GWAS as associated with inflammatory bowel disease (discussed later in this chapter), providing a clear instance of pleiotropy and implicating regulation of the production of specific cytokines in the ability to cope with infections targeting organs as different as the peripheral nervous system and the gastrointestinal tract.

It is further satisfying from a genetic perspective that analyses of very rare but highly penetrant cases of vulnerability to bacterial infection (known as Mendelian susceptibility to mycobacterial disease) also point to the innate immunity signaling pathway through interleukins IL-12 and IL-23, which

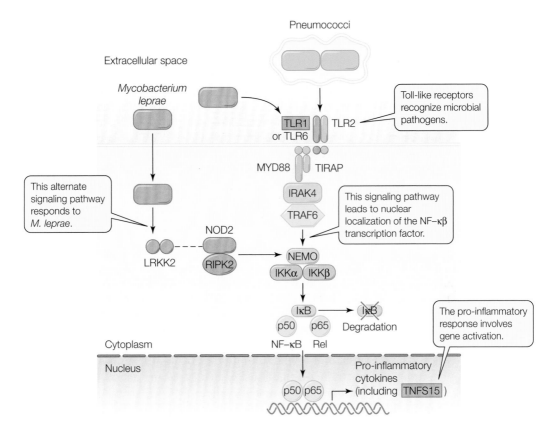

Figure 11.2 Associations with Bacterial Infectious Disease. Five genes each that have been associated with leprosy (green) and pneumococcal infection (orange) fall in immune system signaling pathways that intersect downstream of the Toll-like pattern recognition receptors, implying that genetic variation in response to bacterial recognition mediates resistance to these diseases. (After Chapman and Hill 2012.)

stimulate T helper cell differentiation. The associated loci include mutations in the interferon-γ receptor genes and their downstream activators *STAT1* and *NEMO* as well as in *IL12* itself (Maródi and Notarangelo 2007). Patients with several of these mutations appear to be able to resist many other pathogens. The pattern recognition receptor *TLR1* has also been implicated in risk of gastric ulcers, as seroprevalence of the causative agent, *Helicobacter pylori* (namely, presence of antibodies against this bacterium), is associated with a regulatory polymorphism in the gene (Mayerle et al. 2013).

Viruses

Our knowledge of the genetic regulation of viral resistance is dominated by variants that have a large effect. Hepatitis C infection places over 180 million people worldwide at risk of cirrhosis and liver carcinoma. Resistance to this infection is strongly associated with regulatory variants that increase the expression of the *IL28B* gene and its protein product, interferon-λ3, possibly as a result of loss of sites where methylation is particularly important for hepatocyte expression. These *IL28B* polymorphisms are also associated with spontaneous clearance of the virus in chronically exposed individuals and, importantly, with responsiveness to standard therapy. Only half of all people who receive pegylated interferon prophylaxis are able to clear the virus, and individuals with the responsive *IL28B* haplotype appear to have altered expression of a set of IL28B target genes (interferon-stimulated genes) in hepatocytes that mediate clearance. The variants are now used clinically and may explain differences in response rate between populations. Susceptibility to hepatitis B infection has to date been associated only with variation in MHC class II, implicating viral antigen presentation to T helper cells (Urban et al. 2012).

Human immunodeficiency virus (HIV) enters T lymphocytes through a heterodimeric receptor complex consisting of the CD4 co-receptor and usually either CCR5 or CXCR4. Mutations in both of these latter two co-receptors have been associated with protection against infection and a slow rate of disease progression, whereas upregulation of these receptors may have the opposite effect. The MHC class I appears to be critical in mediating immune system control by way of cytotoxic T cell activation. The diversity of the MHC and linkage disequilibrium patterns in the complex make it difficult to dissect which genes are actually responsible for these associations, but a specific allele, *B*5701*, seems to play a key role in concert with other loci. This polymorphism is associated with the viral load set point, namely with the steady-state level of virus that is maintained after the acute phase of infection, but prior to the onset of acute AIDS. Combined, the *MHC* and *CCR5* loci explain one-fourth of the variation in viral load set point (Fellay et al. 2007), so there is clearly much undiscovered biology not accounted for by these major effects.

Influenza virus must be vaccinated against on a seasonal basis as new strains with different antigens emerge. There is considerable variation among people in their responses to immunization, which can be measured by the production of antibodies. Rather than performing a GWAS, Nakaya et al. (2011) took a systems biology approach in which they characterized differential gene expression in the days and weeks after immunization. Live and attenuated vaccines induced different responses in specific subsets of immune system cells, including, for example, the unfolded protein response in B cells that start generating large numbers of antibodies. Machine learning methods were used to identify novel gene sets that correlated with antibody production, some of which were also observed in a parallel study of vaccination against yellow fever virus. When one of these genes, *CAMK4*, was knocked out in mice, it resulted in an enhanced antibody response, a finding that is consistent with the observation that high levels of transcript in healthy people correlate with a low vaccine response. Further studies will be needed to establish additional genetic and environmental causes for variation among people in their ability to mount robust immune responses to vaccination.

Blood parasites

The third great immunological challenge faced by mammalian species is parasite infections of the blood or intestine. Primary among them in terms of global prevalence and recent impact on the human genome are the malarial parasites *Plasmodium falciparum* and *P. vivax*. Mechanisms of resistance to these parasites, which evolved relatively recently in human history, involve regulation of the expression or activity of hemoglobin, the primary oxygen-carrying protein found in red blood cells. But this protection comes at a cost. Textbooks frequently present globinopathies and thalassemias, both of which arise as an indirect consequence of defenses against malaria, as paradigmatic examples of genetic sources of human disease.

In sickle-cell trait, heterozygosity for a replacement of glutamate with valine at residue 6 of β-globin, encoded by the *HBB* gene, confers some protection from malaria on carriers. However, severely sickle-shaped red blood cells in homozygotes for this recessive trait compromise numerous aspects of red blood cell function, giving rise to sickle-cell disease (**Figure 11.3**). In some parts of West Africa, the abnormal allele attains a frequency as high as 0.15, with the result that as many as 25% of people carry the protective trait, at the cost of 2% of people having the disease. It is almost certainly the best-characterized case of balancing selection known in any organism (Hedrick 2012).

The thalassemias that are found in the Mediterranean regions and Southeast Asia follow a similar principle, but affect the production of globin at the transcriptional level and result in bone and heart disease as well as anemia in homozygotes. Numerous mutations are known that affect production of different globin chains, and over 80 million people carry the major β-thalassemia allele of *HBB* alone. Two other genetic sources of resistance to malaria are absence of the Duffy antigens and deficiency for an enzyme, glucose 6-phosphate dehydrogenase (G6PD), that has a key role in red blood cell metabolism and shows X-linked inheritance.

All these genetic findings were obtained through classical research strategies. GWAS have been less effective in finding additional protective variants, in large part due to difficulties in conducting association mapping in the highly diverse populations of sub-Saharan Africa. The first major study from the Malaria-GEN Consortium, for example (Jallow et al. 2009), scanned 2500 children in The Gambia without initially detecting any loci of genome-wide significance, even including the sickle-cell variant. The reason is that this *HBB* variant is in low linkage disequilibrium (LD) with polymorphisms on the genotyping chip used, and even after correction for a high level of population structure, the strongest signal at the *HBB* locus was only 10^{-6}. This signal was improved to 10^{-14} by imputation of novel variants that the investigators found by sequencing the gene in 62 individuals, and subsequently to 10^{-28} by direct genotyping of rs334, the causal variant in *HBB*. These results imply that weaker associations may

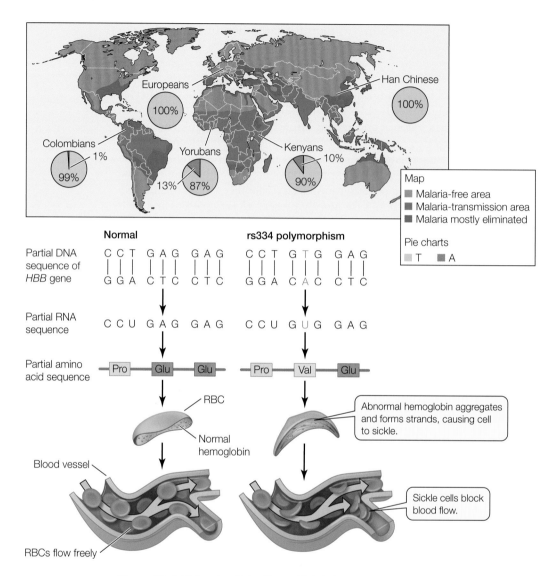

Figure 11.3 Balancing Selection. Malaria has a global distribution throughout the tropics but has been eliminated from large parts of the Americas, Europe, and East Asia. The insets show geographic variation in the frequency of the rs334 polymorphism which is responsible for sickle-cell disease. It leads to the Glu20Val substitution, which causes hemoglobin to aggregate and red blood cells (RBCs) to sickle, in turn causing anemia as a result of reduced oxygen transport. Blockage of blood vessels is responsible for other symptoms of disease. (After Hartl 2004.)

be found only by whole-genome sequencing or by sampling of hundreds of thousands of cases.

Integrative approaches that include genetic analysis of the parasite and its vector dominate research on many other parasites, including tick-borne bacteria such as the *Borrelia* spirochetes that are responsible for Lyme disease. Transmission is accompanied by rapid changes in the surface properties and gene expression of the parasite, which initially establishes a rash underneath the skin, but over the course of months migrates to various depots in the body, including the joints and heart muscle, where it in some cases establishes chronic infection (Rosa et al. 2005). Little is known about resistance factors and the human genes that mediate the severity and persistence of symptoms, but laboratory rodent models facilitate mechanistic hypothesis testing based on detailed knowledge of the spirochete genome.

Inflammatory Bowel Disease

Inflammatory bowel disease (IBD) refers to severe gastrointestinal inflammation leading to diarrhea, often with bloody stools, abdominal pain, vomiting, and other symptoms. Two major types of IBD, referred to as Crohn's disease (CD) and ulcerative colitis (UC), are distinguished by the primary site of pathology as well as by their symptoms (**Figure 11.4A**). Ulcerative colitis usually affects the lower portion of the colon, whereas Crohn's disease tends to affect patchy areas along the small intestine and into the colon. Both diseases have periods of relative quiescence alternating with bouts of inflammation that have their onset most often in young adults, although very severe cases also occur in infants. People with IBD may be treated with corticosteroids to suppress the immune system, with a new range of biologics, such as mesenchymal stem cells, or with monoclonal antibodies, which are administered in refractory cases to suppress the cytokine TNF-α. Estimates of prevalence vary, but both CD and UC are relatively rare, afflicting perhaps 5 in 10,000 people with severe and chronic disease.

Crohn's disease was among the first diseases to yield to GWAS. The initial studies pinpointed a small number of loci with relatively large effects, including those encoding the bacterial cell wall recognition protein NOD2 and the receptor for interleukin 23, IL-23R (Mathew 2008). Environmental contributors have also been suspected, since the disease is more prevalent in industrialized countries. Some have suggested that parasitic hookworms provide some protection against IBD. Others have noted a correlation between the adoption of refrigeration and a rising incidence of disease (suggesting survival of cold-resistant bacteria that become pathogenic upon ingestion). Smoking also appears to be a major risk factor for CD, but may actually be protective against UC (Molodecky and Kaplan 2010).

The most recent report of a GWAS of IBD (Jostins et al. 2012) identified 30 loci associated specifically with CD, 23 associated specifically with UC, and another 110 associated with both diseases, collectively explaining 13.6% and 7.5% of the phenotypic variance for CD and UC, respectively. These figures make IBD among the genetically best characterized of all chronic diseases. Most of the disease-specific loci also show the same direction of effect in the other condition even though it is not statistically significant, implying that there is overwhelming similarity to the architecture of the two forms of IBD (**Figure 11.4B**). Interestingly, though, the

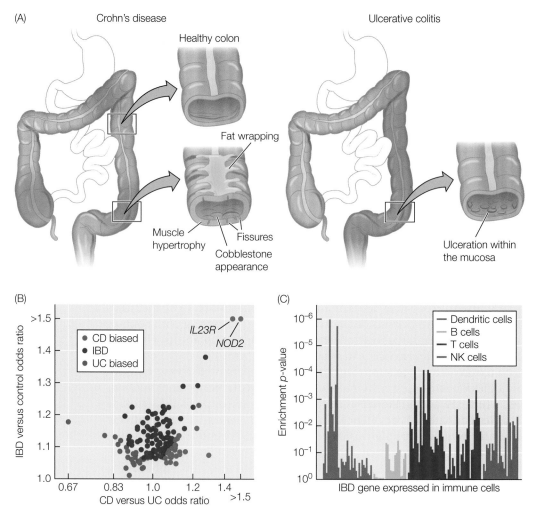

Figure 11.4 Genetics of Inflammatory Bowel Disease (IBD). (A) IBD is divided into two diseases, Crohn's disease (CD), which usually causes characteristic inflammation in the ileum and/or upper colon, and ulcerative colitis (UC), which causes ulceration of the lower colon. (B) Most of the genes associated with these diseases affect both of them, but this plot highlights a subset of loci that have a significant effect on either UC (blue) or CD (red), including the relatively large-effect loci *NOD2* and *IL23R*. (C) Many of the polymorphisms affect gene expression, which is often enriched in subsets of immune system cells, such as dendritic cells, T cells (both CD4+ and CD8+), and natural killer (NK) cells. Each column shows the enrichment *p*-value for one gene in the indicated cell type. (B and C, after Jostins et al. 2012.)

variants at the locus with the largest effect for CD, in *NOD2*, are actually protective against UC. There are 300 candidate genes encoded within 125 of the loci, and only 29 of them are in strong LD with a known common missense mutation, whereas more than twice as many, 64, are associated with regulation of the expression of transcripts, which are thus likely to represent the causal gene. There is also strong enrichment for overlap with GWAS hits for other immune disorders, including psoriasis and type 1 diabetes, as well as for genes involved in primary immunodeficiencies and mycobacterial infection.

Examination of the biological functions of genes within the associated loci, as well as of the cell types within which they are predominantly expressed, clearly establishes host defense against infection as the dominant mechanism behind IBD susceptibility (**Figure 11.4C**). Many of the genes are preferentially expressed in dendritic cells, CD4$^+$ memory cells, or natural killer cells, and/or have roles in interferon signaling and IL-12, IL-10, and TNF-α production. There are signs of both balancing and directional selection, which is consistent with the notion that the dynamics of polymorphism relates to the ever-shifting selection pressures exerted by bacterial infection in particular, where disruption of the delicate balance between pathogenesis and commensalism leads to disease. Stress and the composition of the microbiome surely play an important role in modulating the onset and progression of disease as well.

The prevalence of CD in siblings of people with the disease is approximately 2.5%, which represents a very large sibling relative risk compared with the population prevalence, which is less than 1 in 2000. Use of a machine learning approach on a data set of 20,000 cases and 25,000 controls drawn from 15 European countries produced a polygenic score consisting of 573 SNPs, which proved to be a strong predictor when evaluated in an independent sample (Wei et al. 2013). If 1400 siblings of patients with CD were tested, 35 would be expected to develop the disease, and this genetic predictor would correctly identify 25 of them while failing to call 10. It would also call 215 controls as likely to develop CD.

	Cases	Controls	
Called positive	True positive (TP), 25	False positive (FP), 215	PPV, (TP/[TP + FP]) = 10.4%
Called negative	False negative (FN), 10	True negative (TN), 1150	NPV, (TN/[FN + TN]) = 99.1%
	Sensitivity, (TP/[TP + FN]) = 71%	Specificity, (TN/[FP + TN]) = 84%	

Figure 11.5 shows these calculations projected onto something called a receiver operating characteristic (ROC) curve, in which an area under the curve (AUC) of more than 75% is often regarded as having potential clinical utility.

Sensitivity is the proportion of the cases that are correctly identified, in this case, 25/35, or 71%. **Specificity** is the proportion of controls that are correctly called controls, in this case, 1150/1365, or 84%. In other words, the test correctly identifies 71% of the cases while calling only 16% of the controls cases. This sounds terrific, but the prevalence of the condition also affects the usefulness of the test. Because there are so many more controls than cases, the **positive predictive value** (**PPV**) of the test—that

Figure 11.5 Receiver Operating Curve (ROC) Analysis. ROC curves plot the sensitivity against 1 minus the specificity of a test (in this case utilizing a genetic risk score). The area under the curve (AUC) is 50% for a test that has no value, since each evaluation has a 50–50 chance of being correct. For the example given in the text, the most favorable ratio of sensitivity to specificity occurs where 71% of cases are predicted, while only 16% of controls are called as false positives. Specificity could be increased, say to 84%, but at the expense of doubling the reduction in specificity. For a less accurate test with an AUC of just 70%, the sensitivities at the same two levels of specificity would be just 46% and 63%, respectively.

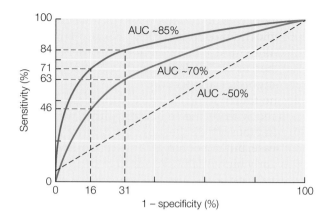

is, the ratio of true positives to all the called positives—is quite low: only 25/(25 + 215), or around 10%. On the other hand, the test has a high **negative predictive value** (**NPV**): over 99% of the called non-cases truly are not likely to develop the disease. The test nevertheless results in a fourfold improvement over simply guessing in this sample, and although its use may generate anxiety for some parents (who would already be aware of the increased familial risk), it potentially alleviates anxiety for a much larger number of people. Importantly, this example shows how genetic risk prediction could help to focus medical resources on those who are at greatest risk. If the proportion of cases were ten times greater, then the PPV would increase to just over 50%, but at the expense of an almost 10% NPV.

Type 1 Diabetes

Type 1 diabetes (T1D) is one of a series of **autoimmune diseases** in which a person's immune system fails to appropriately distinguish self from nonself and consequently attacks the person's own cells, in this case, the β cells of the pancreatic islets of Langerhans that produce insulin. The age of onset is typically in children, hence the former name "juvenile diabetes." Since the capacity to produce insulin is compromised, the disease is also called insulin-dependent diabetes mellitus, or IDDM. The prevalence of T1D ranges from 35 per 100,000 in Scandinavia to just 1 in 100,000 in China and Japan. There is some suggestion that breastfeeding offers protection against T1D, and conversely, factors in infant formula derived from cow's milk have been implicated in promotion of the disease, but given the heterogeneity of epidemiological study outcomes, it is clear that any environmental contributions are as complex as the genetic ones (Patelarou et al. 2012).

The heritability of diabetes is among the highest of any human disease, exceeding 80%. A large study of 22,650 Finnish twin pairs found that more than one-fourth of all monozygotic twin pairs are concordant for T1D, compared with less than 5% of all dizygotic twin pairs (Hyttinen et al. 2003). In other words, genetics has a critical role to play in promotion of the disease, but the penetrance of high-risk genotype combinations is not particularly high (see Chapter 1 for further explanation of these concepts).

Over 40 loci have been associated with T1D risk, many of which are involved in attenuation of T cell receptor signaling (Barrett et al. 2009; Polychronakos and Li 2011). A long repeat polymorphism in the *INS* gene that encodes insulin leads to reduced expression in the thymus, which somehow promotes the insulin peptide as one of the autoantigens recognized by the immune system in T1D. Several downstream components of signaling in response to T cell receptor binding to autoantigens are also implicated, including a tyrosine phosphatase encoded by *PTPN22*, the cytokines IL2 and IL10, and the IL2RA (CD25) receptor. A role for the innate immune system is implicated by variants in the IFIH1 helicase, which elicits the interferon response to viral double-stranded RNA. One of these variants was among the first rare polymorphisms to be discovered by deep sequencing studies, and it actually doubles the odds of protection against IBD (Nejentsev et al. 2009).

However, by far the largest risk factor for T1D is an individual's genotype at the highly polymorphic β-chain of the MHC class II *HLA-DRB* and *HLA-DRQ* loci. Specifically, over 90% of Caucasians with juvenile-onset T1D are carriers for at least one of the haplotypes known as *DRB*301-DQB*201* or *DRB*401-DQA*301-DQB*302*, which account for only 20% of the alleles in the general population. By contrast, strong protection is conferred by *DQB*602*, which has a frequency of 15% in controls but less than 1% in people with T1D. The odds ratios of these *HLA* alleles are approximately 5, which means that their effects are among the largest identified by association studies. Mouse models support the inference that autoimmunity is conferred by altered binding affinity of the high-risk variant proteins for auto-antigen; these alleles have presumably reached high frequency in the population due to their improved recognition of particular pathogen peptides.

Gene expression profiling by sequencing the RNA of isolated pancreatic islet cells has also been used to gain insight into how genes influence T1D (Eizirik et al. 2012). Among almost 20,000 expressed genes, 16% show altered expression after exposure to the inflammatory cytokines IL-1β and interferon-γ, and encode functions related to apoptosis, inflammation and innate immune response, as well as antigen presentation. Splicing of another 548 genes was also affected by cytokine exposure, again implicating cell death and canonical immune system signaling pathways. Furthermore,

25 of 41 candidate genes identified by GWAS are expressed in the islet cells, raising the possibility that their function is required locally, whether in conjunction with or separately from their function in circulating immune system cells.

As with Crohn's disease, a sufficient proportion of the heritability of T1D has been captured by genetic markers to facilitate development of a polygenic risk score that accurately identifies 80% of people with T1D. However, given the very low prevalence of the disease, the positive predictive value is so low that if it were applied to 20% of the population, over 95% of individuals identified as being at high risk would not develop disease. The utility of such a test may lie in targeting early autoantibody detection to children in families, who are most likely to benefit once preventive therapies become available.

Other Autoimmune Diseases

There are several reasons why it is difficult to establish the degree of overlap in genetic etiology among inflammatory autoimmune diseases, yet there is no question that there is considerable sharing of common susceptibility loci (Cotsapas et al. 2011; Parkes et al. 2013). Since power to detect a locus is a function of sample size, which varies greatly across diseases, only the loci of largest effect can be reliably compared across studies. Even in these cases, the fact that the causal SNP is not usually known makes it difficult to know whether the same variant is involved in promoting different diseases. In addition, disease prevalence varies between genders and across populations: for example, systemic lupus erythematosus (SLE) is up to eight times more common in females and five times more common in African than European Americans. A few cases in which the risk allele appears to be the protective allele for another disease have been documented (intriguingly, rs4129267 in *IL6-R* appears to be protective against rheumatoid arthritis while promoting asthma), and there are certainly cases in which variants appear to affect only one disease. On the other hand, variants in *IL2RA*, *IL10*, *PTPN22*, and two other loci have each been reliably associated with four or more diseases, and many other loci are associated with at least two autoimmune conditions.

One recent survey (Ramos et al. 2011) compared 74 GWAS conducted on Caucasian populations for 17 different autoimmune diseases, which collectively have yielded 337 loci (excluding the MHC, different alleles of which appear with most of these diseases). The diseases studied were SLE, rheumatoid arthritis (RA), psoriatic arthritis, juvenile idiopathic arthritis, T1D, multiple sclerosis (MS), ankylosing spondylitis (AS), celiac disease, psoriasis, Kawasaki disease, systemic sclerosis, sarcoidosis, alopecia areata, Behçet's disease, vitiligo, CD, and UC. These diseases affect organs as diverse as the heart, skin, gut, joints, nervous system, pancreas, and hair. Despite this diversity, at least 68 loci spread throughout the genome were reported to show evidence for pleiotropy, some examples of which are shown in **Figure 11.6**.

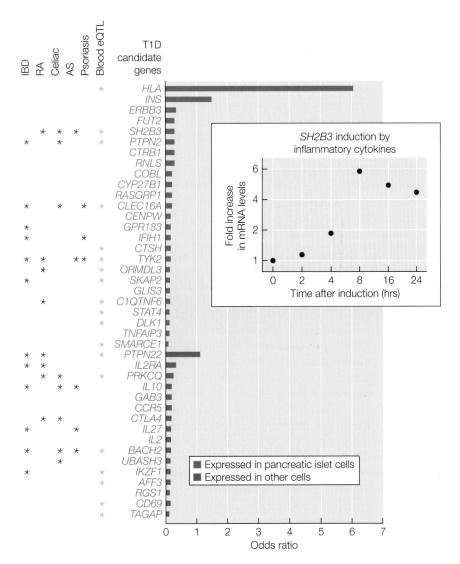

Figure 11.6 Pleiotropy of Loci Involved in Type 1 Diabetes (T1D) Genetic Risk.
Each bar shows the odds ratio for 41 known risk loci for T1D, ordered into those that
are expressed in pancreatic islet cells (red) and those that are not (blue). Yellow asterisks mark genes that also have *cis*-eQTL in peripheral blood cells as reported in the
Blood eQTL Browser. The inset shows how expression of one of the genes, *SH2B3*, is
induced by treatment of an islet cell line with the inflammatory cytokines IL-1β and INF-γ,
measured by the ratio of qRT-PCR product relative to that of a control gene, *GAPDH*.
Green asterisks show that 18 of the loci also show associations with one or more of five
other inflammatory autoimmune diseases (inflammatory bowel disease [IBD], rheumatoid
arthritis [RA], celiac disease, ankylosing spondylitis [AS], and psoriasis), with the proviso
that this does not necessarily implicate the same gene in each case. HLA associations
are excluded, since different variants are associated with each disease. (After Eizirk et al.
2012 with data in green from Parkes et al. 2013.)

The immunogenetics consortia have pooled their members' resources to use a less expensive and higher-resolution genotyping platform, the Immunochip, to assay tens of thousands of cases of each of these 17 autoimmune diseases. The Immunochip reliably measures almost 130,000 variants in 186 loci plus the MHC complex. When applied to 11,475 samples from Europeans with RA, it brought the total number of loci affecting RA to 46 (Eyre et al. 2012). These loci collectively explain just over half the heritability of the disease, or one-fourth of the total risk. At 6 of the genes, there is evidence for at least two different polymorphisms independently contributing variance, including three amino acid substitutions in the *HLA-DRB1* gene. Arthritis also illustrates the concept of disease heterogeneity: clinicians recognize differences between anti-citrullinated peptide antibody (ACPA)–positive and negative types, and it appears that some loci are specific for each of these types. Similarly, juvenile idiopathic arthritis (JIA) shares many loci with RA, but has some unique genetic features.

At least three different models can be invoked to explain how a common set of genes can contribute to such a variety of diseases. One assumes that variation that influences autoimmune signaling results in the recognition of different antigens on the relevant tissues in the context of different environmental stimuli. A second assumes that the activity of common variants is modified by disease-specific polymorphic risk loci that are relevant only to the target tissue. A third assumes that all alleles found in an individual combine to elevate risk for one disease over another, also modifying the contributions of each individual locus through epistasis. SNP-based heritability estimates also provide strong evidence for shared genetic etiology beyond the discovered variants, but it will take comparisons of hundreds of thousands of people with each disease to have robust enough estimates of effect sizes to allow us to distinguish among the different genetic architectures.

In the interim, bioinformatic data mining approaches are being used to formulate hypotheses. After analyzing data from seven autoimmune diseases and introducing a novel test statistic designed to evaluate joint association by comparative analysis of p-value distributions (rather than just cross-listing associations), Cotsapas et al. (2011) argued that of the 107 GWAS-significant SNPs that influence one of the diseases, 44 influence multiple diseases. One of these SNPs, the exonic variant rs3184504 in *SH2B3*, which encodes a negative regulator of cytokine signaling in lymphocytes, probably affects celiac disease, CD, MS, RA, psoriasis, T1D, and SLE. The authors then clustered the 44 genes into four groups and asked whether the genes were transcribed in the same cells, or the proteins formed physical interaction networks, more often than expected by chance, and found this to be the case for three of the groups. Ultimately, this knowledge may be used to target therapies in an individualized manner, assuming that individuals with different risk factors respond to different drugs or other interventions.

Primary Immune Deficiency

Primary immune deficiency (PID) affects at least 1 in 1000 newborn children, and maybe as many as 1 in 500. The absence of components of a functional immune system means that the child cannot eliminate opportunistic pathogens, which lead to symptoms from diarrhea to persistent and recurrent infection, which in turn can lead to failure to thrive, developmental delays, or life-threating infections (Parvaneh et al. 2013). PID is generally attributed to rare or de novo mutations of large effect, many of which remain to be identified. It is diagnosed in the first instance from low lymphocyte counts or immunoreactivity, and subsequently by molecular diagnostics examining immunoglobulin or T cell receptor production. Near absence of lymphocyte function is known as severe combined immune deficiency, or SCID, which is treated by bone marrow transplantation early in life. Milder forms of PID can be treated with immunoglobulin replacement therapy, prophylactic antibiotic usage, and other interventions.

Most cases of SCID are due to mutations in the *IL2R-γ* gene, which encodes a co-receptor for multiple interleukins and is required for B and T cell maturation (Sponzilli and Notarangelo 2011). Inheritance is X-linked recessive, so the disease is more common in boys. Similarly, loss of JAK3 (Janus kinase 3) signaling downstream of this receptor leads to SCID in some cases. Alternatively, in Omenn syndrome, failure to produce immunoglobulins or T cell receptors can arise from mutation of the RAG recombinases that are required for the complex genomic rearrangements that generate the adaptive immune recognition molecules. Bare lymphocyte syndrome, another form of SCID, arises from lack of expression of MHC antigen presentation proteins. Surprisingly, alterations of purine metabolism that reduce dATP or elevate dGTP production have a particularly severe effect on proliferation of lymphocytes that also leads to SCID. Gene therapy has been attempted for several of these cases, but overexpression of IL2R-γ was associated with subsequent leukemia, and these trials have shown limited progress to date (Booth et al. 2011).

Other classes of PID affect single components of the immune system, such as antibody production, regulation of immune system activity in specific tissues, activity of cellular subtypes such as granulocytes, macrophages, or neutrophils, and inflammatory responses. Complement deficiencies arise when the proteins that drill pores in the membranes of detected pathogens are not functioning properly. Yet another cause of PID is defective development of the thymus, which is the organ in the body where T lymphocytes mature and are immunologically selected. Defective development of the thymus is often one of the symptoms of the small 22q11.2 deletion that causes DiGeorge syndrome in 1 in 4000 newborns. The Immune Deficiency Foundation (IDF; primaryimmune.org; **Figure 11.7**) estimates that there are over 200 PIDs; its website, like those of many disease foundations, provides critical resources for patients and families as they deal with the financial and medical consequences of these congenital disorders.

Figure 11.7 Screenshot from the Website of the Immune Deficiency Foundation.
This website provides a valuable resource for families of children living with PIDs. It includes information about the foundation, the genetics and biology of PIDs, resources available to help with coping, community activities, links to healthcare professionals, and social networks. (From primaryimmune.org)

Summary

1. How a person responds to infection with a virus, a bacterium, or a parasite depends on the state of activity in the innate and adaptive arms of the immune system.

2. The major histocompatibility complex (MHC, which in humans is also known as the human leukocyte antigen complex, or HLA) is a highly polymorphic complex of genes with critical functions in foreign antigen recognition and processing.

3. Cytokines are small regulatory proteins that mediate communication among the cells of the immune system, such as macrophages, B and T lymphocytes, and neutrophils, and include the interferon and interleukin families of proteins.

4. Common and rare variants in components of immune signaling regulate susceptibility to pathogenic bacterial infections such as tuberculosis and leprosy.

5. The ability to clear or stabilize levels of the hepatitis and human immunodeficiency viruses has been traced to a small number of common polymorphisms of large effect, but is genetically complex.

6. Resistance to malaria, which has evolved relatively recently in human history, involves regulation of the expression or activity of hemoglobin. A corollary is that a fraction of the population is at risk for thalassemia or sickle-cell disease.

7. The two major classes of inflammatory bowel disease, Crohn's disease and ulcerative colitis, share genetic risk factors at several dozen loci and are among the genetically best characterized of all complex chronic diseases.

8. Three measures of predictive performance of genetic risk scores are sensitivity (the fraction of positives that are detected), specificity (the fraction of controls called negative), and positive predictive value (the proportion of individuals called positive who are truly cases).

9. Type 1 diabetes is an autoimmune disease in which the body destroys its own insulin-producing pancreatic β cells. The major genetic risk factors are located in the HLA complex.

10. There is widespread sharing of genetic risk factors for over a dozen inflammatory autoimmune diseases affecting diverse tissues in the body, including rheumatoid arthritis, multiple sclerosis, systemic lupus erythematosus, and psoriasis.

11. Primary immune deficiencies affect more than 0.1% of newborns and are generally due to rare or de novo mutations of large effect.

References

Barrett, J. C., Clayton, D. G., Concannon, P., Akolkar, B., Cooper, J. D., et al. 2009. Genome-wide association study and meta-analysis find that over 40 loci affect risk of type 1 diabetes. *Nat. Genet.* 41: 703–707.

Booth, C., Gaspar, H. B., and Thrasher, A. J. 2011. Gene therapy for primary immunodeficiency. *Curr. Opin. Pediatr.* 23: 659–666.

Burnet, F. M. 1961. Immunological recognition of self. *Science* 133: 307–311.

Chapman, S. J. and Hill, A. V. S. 2012. Human genetic susceptibility to infectious disease. *Nat. Rev. Genet.* 13: 175–188.

Cotsapas, C., Voight, B. F., Rossin, E., Lage, K., Neale, B. M., et al. 2011. Pervasive sharing of genetic effects in autoimmune disease. *PLoS Genet.* 7: e1002254.

Davila, S., Wright, V. J., Khor, C. C., Sim, K. S., Binder, A., et al. 2010. Genome-wide association study identifies variants in the CFH region associated with host susceptibility to meningococcal disease. *Nat. Genet.* 42: 772–776.

Eizirik, D. L., Sammeth, M., Bouckenooghe, T., Bottu, G., Sisino, G., et al. 2012. The human pancreatic islet transcriptome: Expression of candidate genes for type 1 diabetes and the impact of pro-inflammatory cytokines. *PLoS Genet.* 8: e1002552.

Eyre, S., Bowes, J., Diogo, D., Lee, A., Barton, A., et al. 2012. High-density genetic mapping identifies new susceptibility loci for rheumatoid arthritis. *Nat. Genet.* 44: 1336–1340.

Fellay, J., Shianna, K. V., Ge, D., Colombo, S., Ledergerber, B., et al. 2007. A whole-genome association study of major determinants for host control of HIV-1. *Science* 317: 944–947.

Hartl, D. 2004 The origin of malaria: Mixed messages from genetic diversity. *Nat. Rev. Microbiol.* 2: 16.

Hedrick, P. W. 2012. Resistance to malaria in humans: The impact of strong, recent selection. *Malar. J.* 11: 349.

Hyttinen, V., Kaprio, J., Kinnunen, L., Koskenvuo, M., and Tuomilehto, J. 2003. Genetic liability of type 1 diabetes and the onset age among 22,650 young Finnish twin pairs: A nationwide follow-up study. *Diabetes* 52: 1052–1055.

Iwasaki, A. and Medzhitov, M. 2010. Regulation of adaptive immunity by the innate immune system. *Science* 327: 291–295.

Jallow, M., Teo, Y. Y., Small, K. S., Rockett, K. A., Deloukas, P., et al. 2009. Genome-wide and fine-resolution association analysis of malaria in West Africa. *Nat. Genet.* 41: 657–665.

Jostins, L., Ripke, S., Weersma, R. K., Duerr, R. H., McGovern, D. P., et al. 2012. Host-microbe interactions have shaped the genetic architecture of inflammatory bowel disease. *Nature* 491: 119–124.

Maródi, L. and Notarangelo, L. D. 2007. Immunological and genetic bases of new primary immunodeficiencies. *Nat. Rev. Immunol.* 7: 851–861.

Mathew, C. G. 2008. New links to the pathogenesis of Crohn disease provided by genome-wide association scans. *Nat. Rev. Genet.* 9: 9–14.

Mayerle, J., den Hoed, C. M., Schurmann, C., Stolk, L., Homuth, G., et al. 2013. Identification of genetic loci associated with *Helicobacter pylori* serologic status. *JAMA* 309: 1912–1920.

Medzhitov, R. and Janeway, C. A., Jr. 2002. Decoding the patterns of self and nonself by the innate immune system. *Science* 296: 298–300.

Molodecky, N. A. and Kaplan, G. G. 2010. Environmental risk factors for inflammatory bowel disease. *Gastroenterol. Hepatol. (NY)* 6: 339–346.

Nakaya, H. I., Wrammert, J., Lee, E. K., Racioppi, L., Marie-Kunze, S., et al. 2011. Systems biology of vaccination for seasonal influenza in humans. *Nat. Immunol.* 12: 786–795.

Nejentsev, S., Walker, N., Riches, D., Egholm, M., and Todd, J. A. 2009. Rare variants of *IFIH1*, a gene implicated in antiviral responses, protect against type 1 diabetes. *Science* 324: 387–389.

Parkes, M., Cortes, A., van Heel, D. A., and Brown, M. A. 2013. Genetic insights into common pathways and complex relationships among immune-mediated diseases. *Nat. Rev. Genet.* 14: 661–673.

Parvaneh, N., Casanova, J. L., Notarangelo, L. D., and Conley, M. E. 2013. Primary immunodeficiencies: A rapidly evolving story. *J. Allergy. Clin. Immunol.* 131: 314–323.

Patelarou, E., Girvalaki, C., Brokalaki, H., Patelarou, A., Androulaki, Z., and Vardavas, C. 2012. Current evidence on the associations of breastfeeding, infant formula, and cow's milk introduction with type 1 diabetes mellitus: A systematic review. *Nutr. Rev.* 70: 509–519.

Polychronakos, C. and Li, Q. 2011. Understanding type 1 diabetes through genetics: Advances and prospects. *Nat. Rev. Genet.* 12: 781–792.

Ramos, P. S., Criswell, L. A., Moser, K. L., Comeau, M. E., Williams, A. H., et al. 2011. A comprehensive analysis of shared loci between systemic lupus erythematosus (SLE) and sixteen autoimmune diseases reveals limited genetic overlap. *PLoS Genet.* 7: e1002406.

Rosa, P. A., Tilly, K., and Stewart, P. E. 2005. The burgeoning molecular genetics of the Lyme disease spirochaete. *Nat. Rev. Micriobiol.* 3: 129–143.

Sponzilli, I. and Notarangelo, L. D. 2011. Severe combined immunodeficiency (SCID): From molecular basis to clinical management. *Acta Biomed.* 82: 5–13.

Thye, T., Owusu-Dabo, E., Vannberg, F. O., van Crevel, R., Curtis, J., et al. 2012. Common variants at 11p13 are associated with susceptibility to tuberculosis. *Nat. Genet.* 44: 257–259.

Urban, T., Charlton, M. R., and Goldstein, D. B. 2012. Introduction to the genetics and biology of Interleukin-28B. *Hepatology* 56: 361–366.

Wei, Z., Wang, W., Bradfield, J., Li, J., Cardinale, C., et al. 2013. Large sample size, wide variant spectrum, and advanced machine-learning technique boost prediction for inflammatory bowel disease. *Am. J. Hum. Genet.* 92: 1008–1012.

Zhang, F-R., Huang, W., Chen, S-M., Sun, L-D., Liu, H., et al. 2009. Genomewide association study of leprosy. *N. Engl. J. Med.* 361: 2609–2618.

12

Metabolic Disease

O f all the domains of disease, metabolic syndromes are surely the most avoidable, and yet they are increasing in prevalence globally at an alarming rate. The culprits are well known: a sedentary lifestyle in which days sitting at computers give way to nights in front of the TV, and an imbalanced high-calorie and low-nutrient diet consumed all too rapidly. Nevertheless, there are clearly strong genetic components to metabolic disease risk, and some of these conditions are predominantly genetic. Body mass index (BMI) and obesity, for example, have heritabilities of over 50%, while that of type 2 diabetes is at the lower end of the range for common human diseases, at just around 25%. On the other hand, some relatively rare forms of hyperlipidemia (elevated blood lipids) have a monogenic basis.

Metabolic syndrome is defined loosely as a combination of central obesity and any two of the following: elevated triglycerides, reduced HDL cholesterol, high blood pressure, or elevated fasting blood glucose. The cutoffs for each measure vary among countries and according to which medical society's criteria are adopted, also reflecting differences in whether physicians prescribe medication. As discussed in Chapter 4, obesity in the Western Hemisphere is usually taken as a BMI greater than 30 kg/m², but because body weight also reflects muscle mass, waist-to-hip ratio (WHR) may also be taken into account: the World Health Organization definition of obesity stipulates that WHR must be greater than 0.9 in men or 0.85 in women. Metabolic syndrome is now estimated to be observed in at least one-fourth of Americans and a similar proportion of Europeans and Australasians. It anticipates the likely development of cardiovascular disease or type 2 diabetes.

The genes involved in metabolic disease include genes encoding enzymes and proteins related to blood lipids, genes that regulate central metabolism and insulin production (including the function of pancreatic β cells that produce insulin), and genes involved in the regulation of satiety. In this chapter, we start with a discussion of dyslipidemia, then survey the genetic regulation of cholesterol and triglyceride levels. We then describe the genetics of type 2 diabetes (T2D), formerly known as non-insulin-dependent or adult-onset diabetes, in some detail. This disease has risen in prevalence tenfold in the last 25 years in parallel with the global increase in obesity, such that it now affects some 300 million people. Symptoms include excessive thirst and frequent urination caused by elevated blood glucose levels, which in turn may lead to pathological cell damage, resulting in heart disease, stroke, retinopathy (blindness), and loss of limbs due to vascular damage. Onset and progression can be controlled to some extent by exercise and diet, but, as with most common diseases, people also turn to prescribed drugs for relief. We conclude with a discussion of nutrigenetics, the study of relationships between genotype and diet.

Primary Dyslipidemia and Hypercholesterolemia

Primary dyslipidemias are a series of familial disorders of serum lipid metabolism involving abnormal levels of circulating lipid vesicles. A search of OMIM produces 62 diseases defined by mutations affecting lipoproteins, structural components of lipoprotein vesicles, and fatty acid biosynthesis. The Fredrickson classification of dyslipidemias has been used clinically for almost 50 years (**Table 12.1**). Approximately 1% of individuals have inherited forms of either hypertriglyceridemia (type IV) or combined hyperlipidemia (type IIb), whereas hypercholesterolemia (type IIa) is five times less common. Each of these conditions can also arise through complex mechanisms involving polygenic risk and lifestyle factors. The major symptom is abdominal pain due to pancreatitis, but these conditions are generally also a major risk factor for atherosclerosis (considered in the next chapter). Treatment is with statins, fibrates, and/or vitamin B_3 (niacin), all of which are used to lower cholesterol levels.

There are two major classes of long-chain organic molecules in the blood: triglycerides and cholesterol (**Figure 12.1A**). Triglycerides are esters

TABLE 12.1   Fredrickson classification of familial dyslipidemias

Type	OMIM	Familial name	Deficiency	Excess	Prevalence
Ia	238600	Hyperchylomicronemia	Lipoprotein lipase	Chylomicrons	1/1 million
Ib	207750	Apoprotein C2 deficiency	ApoC2	Chylomicrons	1/1 million
Ic	11830		LPL	Chylomicrons	1/1 million
IIa	143890	Hypercholesterolemia	LDL receptor	LDL	1/500
IIb	144250	Combined hyperlipidemia	LDL receptor	ApoB, LDL	1/100
III	107741	Dysbetalipoproteinemia	ApoE2	IDL	1/10,000
IV	144600	Hypertriglyceridemia	VLDL elimination	VLDL	1/100
V	144650		LPL	VLDL/chylomicrons	?

Source: wikipedia.org/wiki/Hyperlipidemia

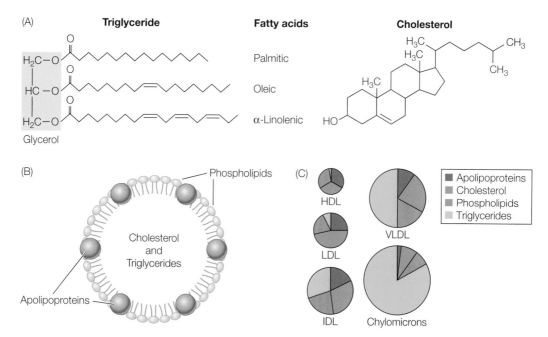

Figure 12.1 Lipoprotein Vesicles. (A) Two predominant classes of lipids, triglycerides and cholesterol, circulate in the blood. (B) These lipids are transported by lipoprotein vesicles that have a variety of different sizes and compositions. (C) The colored segments represent the proportions of each class of lipids in each vesicle type. The size of each pie is proportional to the size of the indicated vesicle type. Vesicle sizes range from 10 μm (HDL) to at least 100 μm in diameter (chylomicrons).

of glycerol with three fatty acids, each of which typically contains from 13 to 17 carbon molecules. If each carbon is bonded to two hydrogen atoms, the fatty acid is saturated, whereas a fatty acid with a double bond between two carbons is unsaturated. Unsaturated fatty acids stack less well, so they tend to be more liquid, as in olive oil, as opposed to the saturated fatty acids in animal-derived lard. Fat is the major energy storage form for the body; it is also required for healthy skin and hair and to buffer some toxic chemicals. Omega-3 fatty acids are a particular type of polyunsaturated triglyceride that is thought to be essential for good health, particularly for cognitive function. Cholesterol is a modified steroid that is synthesized by all cells as a critical constituent of membranes where it regulates their fluidity and function. It is also the precursor for multiple steroid hormones (including estrogen and testosterone), for vitamin D, and for the bile acids that promote processing of dietary fat.

Lipids and cholesterol are carried between tissues in the blood in the form of lipoprotein vesicles, which are classified according to density, size, and content (**Figure 12.1B,C**). High-density lipoprotein (HDL) vesicles, which have approximately equal contents of lipoproteins, cholesterol, and phospholipids, are often called "good cholesterol." They typically cycle cholesterol from fat depots back to the liver. Low-density lipoprotein (LDL) vesicles are twice as large and perform the opposite function, being loaded with cholesterol, most of which is produced in hepatocytes. Very-low-density lipoprotein (VLDL) vesicles are larger still and have a particularly high concentration of newly synthesized triacylglycerides that are being transported from the liver to the fat for storage. VLDL are elevated in hypertriglyceridemia and, in conjunction with LDL, in combined hyperlipidemia. Chylomicrons, which are up to 10 times larger than vesicles of other types, are essentially balls of fat that transport dietary lipids from the intestine to other tissues, where they are unloaded by lipoprotein lipase (LPL).

At least nine genes are known to cause familial hypertriglyceridemia (HTG; Johansen and Hegele 2012). One of the best characterized is in fact *LPL*, loss of function of which results in failure to absorb fatty acids because the secreted enzyme is unavailable, which in turn elevates VLDL and LDL. Though rare, numerous homozygous recessive and compound heterozygous alleles have been documented in familial forms of this disease. Similarly, loss of function of the LPL cofactor encoded by *APOC2* leads to HTG, and several other apolipoproteins have also been implicated both by family studies and by GWAS. By contrast, resequencing of the angiopoietin-like 3 (*AGPTL3*) gene has revealed a probable enrichment of rare variants in individuals with *hypo*triglyceridemia: the corresponding protein is thought to inhibit LPL, so loss of function tends to reduce serum lipids, offering some protection against metabolic disease. All these genes were first implicated by biochemical approaches or mouse studies. In addition, classical linkage analysis in pedigrees found a causal gene for a specific form of familial combined hyperlipidemia to be *USF1*, encoding upstream transcription factor 1, which is an insulin-responsive regulator of many proteins involved in glucose and lipid metabolism. These results emphasize the highly heterogeneous nature of even familial lipid disorders.

Familial hypercholesterolemia (FHC) is similarly caused by dominant and recessive mutations in a handful of genes (Soutar and Naoumova 2007). It is characterized by elevated serum cholesterol that can be deposited in peripheral tissue and skin as yellowish "xanthomas" or in arterial walls, where it contributes to atherosclerosis and eventual clogging of the arteries. The underlying defect is not overproduction of cholesterol, but rather failure to clear LDL, which in turn was shown famously by Brown and Goldstein (1986) to be due to defective function of the LDL receptor pathway in cells that normally take up cholesterol from LDL. While the LDLR receptor itself provided the first characterized mutations, subsequent investigation has uncovered defects in *APOB*, which encodes the ligand on the LDL particles, as well as genes responsible for receptor-ligand internalization (*LDLRAP1*) and for synthesis and recycling of receptors to the cell surface (*PCSK9*). In fact, there are over 800 known mutations that affect cholesterol homeostasis via this pathway.

Confirmation that a newly discovered allele causes FHC requires a combination of strategies. An important question is whether there is co-segregation of the mutation with the phenotype in families. Another is whether the affected amino acid is highly conserved and/or likely to affect protein structure or function. Where possible, functional assays provide the strongest evidence. Even though FHC is highly penetrant, expressivity is clearly modified by environmental (smoking, diet) and genetic factors, as different family members with the same mutation can show quite different phenotypes and respond differently to the same lipid-lowering drugs, primarily statins.

Quantitative Regulation of Lipid Traits

Genome-wide association studies have been used effectively to characterize the genetic basis for normal quantitative variation in lipid traits. Teslovich et al. (2010) performed a meta-analysis of 46 studies engaging over 100,000 Caucasians. They discovered 95 loci that contribute to at least one of four traits—total cholesterol, HDL cholesterol (HDL-C), LDL cholesterol (LDL-C), or triglycerides (TG)—and in most cases, two or more of those traits. Including secondary signals at 26 of the loci, their survey explained, on average, one-fourth of the genetic variance and 12% of the phenotypic variance in serum lipids. Just a handful of the loci showed evidence for sex-specific effects, and none differed significantly in their effects in additional samples of 15,000 East Asians, 9000 South Asians, or 8000 African Americans. Underscoring the overlap with genetic determinants of familial dyslipidemia, 15 of the 19 Mendelian loci known to influence that disease are in the vicinity of one of the 95 loci at which common variants influence cholesterol or fatty acid levels. Furthermore, individuals in the top quartile of a combined genetic risk score for either LDL-C or TG were 13 times more likely to have high LDL-C or 44 times more likely to have high TG than individuals in the bottom quartile, showing that an excess of high-risk variants can also be largely responsible for some individuals having dyslipidemia.

The largest common variant effect detected by GWAS belongs to a cluster of SNPs located in a 6 kb intergenic region at 1p13 on chromosome 1. These SNPs are also associated with a 1.4-fold increase in risk of coronary artery disease in homozygotes for the major, relative to the minor, allele. Their effect on LDL-C is also relatively large, representing a 16 mg/dL difference between homozygous genotypes (relative to a population mean of 125 ± 30 mg/dL). Musunuru et al. (2010) observed that these SNPs have strong eQTL effects on several transcripts in the locus, and they used luciferase reporter gene constructs in hepatic cells to localize the causal variant to rs12740374 (**Figure 12.2**). This T for G substitution creates a binding site for the well-known transcriptional activator C/EBP, which facilitates overexpression of some of the genes in the vicinity. Targeted knockdown of one of these genes, *SORT1*, by RNAi targeted to mouse livers with an engineered virus resulted in substantial increases in LDL-C; by contrast, overexpression of the sortilin gene product reduces LDL-C. The minor allele is derived in humans and has increased in frequency to 0.3 in Europeans, for whom it offers protection against high cholesterol and atherosclerosis, whereas it has decreased to less than 0.05 in East Asians.

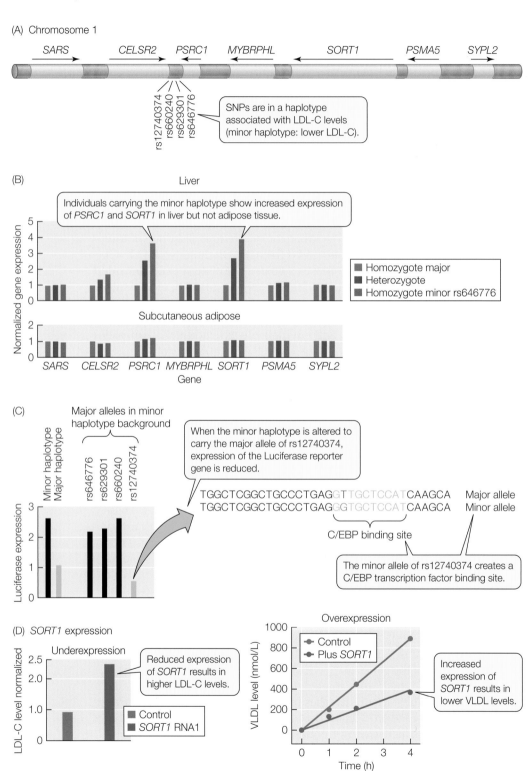

(A) Chromosome 1

SARS   CELSR2   PSRC1   MYBRPHL   SORT1   PSMA5   SYPL2

rs12740374
rs660240
rs629301
rs646776

SNPs are in a haplotype associated with LDL-C levels (minor haplotype: lower LDL-C).

(B) Liver

Individuals carrying the minor haplotype show increased expression of *PSRC1* and *SORT1* in liver but not adipose tissue.

Normalized gene expression

- Homozygote major
- Heterozygote
- Homozygote minor rs646776

Subcutaneous adipose

SARS   CELSR2   PSRC1   MYBRPHL   SORT1   PSMA5   SYPL2
Gene

(C) Major alleles in minor haplotype background

Minor haplotype
Major haplotype
rs646776
rs629301
rs660240
rs12740374

Luciferase expression

When the minor haplotype is altered to carry the major allele of rs12740374, expression of the Luciferase reporter gene is reduced.

TGGCTCGGCTGCCCTGAGGTTGCTCCATCAAGCA   Major allele
TGGCTCGGCTGCCCTGAGGGTGCTCCATCAAGCA   Minor allele

C/EBP binding site

The minor allele of rs12740374 creates a C/EBP transcription factor binding site.

(D) *SORT1* expression

Underexpression

LDL-C level normalized

Reduced expression of *SORT1* results in higher LDL-C levels.

- Control
- *SORT1* RNA1

Overexpression

VLDL level (nmol/L)

- Control
- Plus *SORT1*

Increased expression of *SORT1* results in lower VLDL levels.

Time (h)

◀ **Figure 12.2** *SORT1* rs12740374 Is a Risk Factor for Hypercholesterolemia.
(A) Analyses of eQTL in liver and adipose tissue demonstrated that of the seven genes in
a 200 kb interval around the original association peak from GWAS, *PSRC1* and *SORT1*
are both overexpressed in association with the minor allele. LDL-C, low density lipoprotein cholesterol. (B) Of four SNPs in high linkage disequilibrium with the original SNP
(rs660240), replacement of only rs12740374 in the minor allele haplotype leads to reduction of expression of a reporter gene driving the enzyme luciferase. (C) This observation
implies that rs12740374 is responsible for the increased expression, and that it disrupts
a binding site for the transcriptional repressor C/EBP. (D) Overexpression of *SORT1* in
mouse liver results in decreased cholesterol in the very-low-density lipoprotein fraction,
whereas knockdown by siRNA increases cholesterol. (After Musunuru et al. 2010.)

Lipid traits are also related to the abundance of a number of "adipocytokine" (also called "adipokine") hormones that are secreted by the adipose
tissue, much as cytokines regulate the immune system. These hormones
include the polypeptides adiponectin, leptin, and resistin, all of which
regulate fatty acid breakdown and glucose homeostasis, albeit by different
mechanisms. A multiethnic GWAS on over 45,000 people (Dastani et al.
2012) revealed 12 loci explaining 5% of serum adiponectin levels, only 3 of
which were included in the 95 lipid loci discussed above. Not surprisingly,
the variant of largest effect is in the *ADIPOQ* gene that encodes the hormone itself, while the other loci are suspected to have diverse functions in
adipocytes and the vascular system. A weighted allelic sum score correlates
very significantly with elevated triglycerides and reduced HDL-C, and less
so with many of the other metabolic traits that adiponectin is thought to
influence, including serum glucose, insulin, WHR, and BMI.

Genetics of Blood Glucose and Insulin Levels

An indirect approach to genetic dissection of any disease is to search for
associations with continuous traits that are thought to contribute to disease onset. Diabetes is at its core a disease of impaired glucose tolerance,
which results from impairment of the insulin production in the pancreas,
or emergence of resistance to insulin in peripheral tissues. To the extent that
both production and resistance are modified by obesity, hypothalamic function in regulation of eating behavior is also a major risk factor for diabetes.
However, the most directly relevant quantitative traits for T2D are fasting
glucose, fasting insulin, and 2-hour response to glucose ingestion in a glucose challenge test (which provides a measure of insulin resistance). Scott
et al. (2012) used GWAS meta-analysis to identify a total of 53 glycemic loci,
including 36 associated with fasting glucose, 19 with fasting insulin, and 9
with 2-hour glucose. Several of these loci had pleiotropic effects on two or
more of the traits, and 33 of them were also associated with T2D. Intriguingly,
the observed effect sizes and odds ratios for trait and disease, respectively,
were only mildly correlated, and power considerations imply that at least 15
of the glycemic loci do not contribute to T2D risk. Since less than 5% of the
variance of each trait is currently explained, clearly many more loci remain

to be discovered. However, **text mining** (in which automated algorithms are used to exhaustively search abstracts for co-occurrence of terms related to a condition) suggests that many of the loci detected so far are involved in pancreatic development or insulin secretion.

Since many factors lead to fluctuation in blood glucose levels, some investigators regard glycated hemoglobin (HbA$_{1C}$) in red blood cells as a more reliable indicator of steady-state blood glucose. HbA$_{1C}$ represents the average glucose level over the three-month life span of an erythrocyte, and it has a slightly higher heritability than the plasma concentration of the sugar molecule itself. MAGIC (the *M*eta-*A*nalysis of *G*lucose and *I*nsulin-Related Traits *C*onsortium) performed GWAS on over 45,000 nondiabetics and discovered ten loci influencing HbA$_{1C}$ levels (Soranzo et al. 2010). Three of these loci, *GCK*, *MTNR1B*, and *G6PC2/ABCB11*, are also associated with fasting glucose, which implicates them in glucose homeostasis. By contrast, the other seven loci seem to act independently, perhaps influencing aspects of erythrocyte biology, and in turn are not associated with T2D. Even though these seven SNPs explain less than 2% of HbA$_{1C}$ variability, they may nevertheless have clinical utility: a threshold level of glycated hemoglobin is commonly used to define T2D, but the fraction of the population that is above that threshold was reduced by 2% by incorporating these SNPs into a diagnostic model alongside HbA$_{1C}$.

Conspicuously absent from the list of genes associated with metabolic traits are any enzymatic components of central metabolism. Metabolomic comparisons of individuals with and without diabetes, however, have identified a small number of biomarkers that point to novel features of the emergence of insulin resistance. Five branched-chain or aromatic amino acids (isoleucine, leucine, valine, tyrosine, and phenylalanine) are elevated in individuals who are at highest risk of developing diabetes. By contrast, glycine and a specific lipid, phosphatidylcholine PC(18:2), are both reduced in individuals who are likely to progress to the prediabetic state of impaired glucose tolerance (Wang-Sattler et al. 2012). Further investigation of the effect of genetic variation on the enzymology of these prognostic metabolites is likely to increase our understanding of the path to insulin resistance, which may entail different genetic mechanisms than the pathological onset of diabetes.

An interesting twist on metabolomic analysis is the inclusion of a broad range of environmental chemicals in the profiling and subsequent association with disease status. This environment-wide association study (EWAS) strategy, when applied to diabetes in a pilot experiment (**Figure 12.3**; Patel et al. 2010), suggested that high concentrations of polychlorinated biphenyls (PCBs) and the pesticide derivative heptachlor epoxide may be risk factors for T2D and HDL-C/triglyceride levels. A follow-up study even suggested that there may also be genotype-by-environment interactions in which, for example, polymorphisms in the T2D-associated gene *SLC30A8* modify the influence of β-carotene on diabetes risk (Patel et al. 2013). The odds ratios for these effects are similar to those of strong genetic variants, so sample sizes in the tens of thousands will be needed to validate this

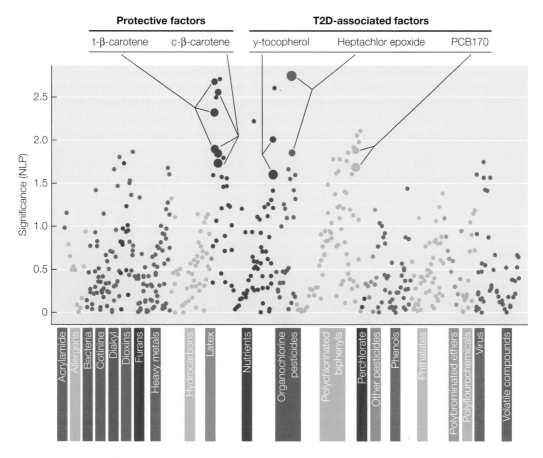

Figure 12.3 Environment-Wide Association Study of Type 2 Diabetes (T2D). This Manhattan plot shows associations between each of 266 environmental agents, grouped into 21 clusters, and T2D in four consecutive cohorts (1999–2006) of the NHANES study. Five examples of metabolites that were associated with T2D in two or three of the cohorts are indicated, where the replication increasing confidence in the effect given that the sample size of fewer than 3320 individuals is too small to obtain significance levels typical of GWAS. (After Patel et al. 2010.)

"exposome" approach (Wild 2005), which combines genetics, epidemiology, and metabolomics to identify novel risk factors for T2D and other diseases.

Genetics of Type 2 Diabetes

The most recent published summary of GWAS for T2D reported 65 autosomal loci plus one X-linked gene, which collectively explain around 6% of the risk, given that 8% of the population will develop the disease (Morris et al. 2012). These loci also account for around 10% of the familial clustering that results in siblings of individuals with T2D having a threefold increased risk relative to unrelated individuals. These results come from meta-analyses of almost 35,000 cases and 115,000 controls performed by the DIAGRAM (*Dia*betes *G*enetics *R*eplication *a*nd *M*eta-analysis) Consortium. Reasonable extrapolations suggest that an additional 500 variants probably influence risk, 200 of which would be discovered through analysis of 100,000 more cases. Odds ratios are generally less than 1.1, implying very small effects. Nevertheless, because the heritability of T2D is modest, at around 25%, common variants do seem to explain the majority of the genetic risk, even if the majority of those variants have effects so small that they are not likely ever to be identified. Four loci contributing to common variant risk—namely, *PPARG*, *KCNJ11*, *WFS1*, and *HNF1B*—are also known to have mutations that cause monogenic forms of diabetes.

Among the 66 discovered variants, two have significant effects in only one of the sexes, while another four trend toward differential effects in men and women. Several of the loci appear to harbor at least two independent variants, and only two show evidence to date that rare variants may be responsible for their association with T2D. Two loci, *FTO* and *MC4R*, probably act through their influence on obesity, while another 14 are unambiguously associated with fasting glucose, and 24 nominally influence glucose metabolism. In general, their effect sizes are similar in obese and in non-obese individuals. Only one of the genes, *BCAR1*, so far shows association with type 1 diabetes, and the influence of this SNP is in the opposite direction for T2D. Many more of the lead SNPs appear to have an effect on the transcription of nearby genes (that is, eSNP effects, typically in adipose tissue) than affect protein structure, establishing that regulation of gene expression is an important component of risk. This observation needs to be considered with the caveat that only in a few instances is the mechanism of action of any of the variants well established. In most cases the documented SNP may not be the causal one—technically it is just the SNP that has led our attention to the locus.

Exome sequencing has led to the discovery of some rare variants responsible for highly penetrant monogenic forms of diabetes, which are important for subclassification of disease (Pal and McCarthy 2013). On the other hand, focused sequencing and genotyping of the zinc transporter gene *SLC30A8* in 150,000 people identified 12 very rare loss-of-function heterozygous mutations in 19 cases and 326 controls that are actually protective against disease at a significance level, 2×10^{-6}, that survives correction for the number of genes in the genome (Flannick et al. 2014). These mutations provide, on average, a 65% risk reduction, but it is unclear whether it results from reduced transporter activity, since Trp325Arg, a common variant of the same protein that also reduces transporter function, is associated with increased risk of T2D at genome-wide significance levels. Soberingly, mouse models have not clarified the molecular basis of these effects on diabetes because genetic backgrounds strongly modify systemic effects, and it is even possible that rodent and human mechanisms are divergent. This study shows how complicated it is going to be to sort out rare variant contributions to common disease risk.

For the 27 T2D loci that clearly do not influence glycemic traits, gene set enrichment analysis was performed and protein-protein interaction networks were interrogated. The results implicated three physiological pathways in the onset of T2D: adipocytokine function, which influences the development of insulin resistance; cell cycle regulation via cyclin-dependent kinases; and transcriptional regulation, particularly in conjunction with the CREBBP chromatin modulator, which probably influences the development of pancreatic β cells. Other studies have suggested that insulin signaling and endoplasmic reticulum stress might be impaired in at-risk individuals. Whether or not this is the case, the GWAS do suggest that specific pathways are affected by variation in multiple genes, and that these pathways influence both the function of the insulin-producing cells and the development of insulin resistance in peripheral tissues of the body. **Figure 12.4A,B** shows two different representations of the connections among the known diabetes loci.

One way to distinguish whether variants affect β cell function or insulin resistance is to plot the magnitude of their association with both traits (Voight et al. 2010), as shown in **Figure 12.4C**. These physiological parameters are commonly modeled by the *ho*meostatic *m*odel *a*ssessment (HOMA) method, which computes two indices from fasting plasma insulin and glucose measurements: HOMA-B, the ratio of insulin to glucose, as a measure of β cell function, and HOMA-IR, the product of the two terms, as a measure of insulin resistance. Such analysis associates loci such as *TCF7L2*, *CENTD2*, and *CDKLA1* with HOMA-B and *PPARG*, *FTO*, and *IRS1* with HOMA-IR. Other measures have been designed that utilize results from glucose challenge tests, including measurement of blood glucose 2 hours after a challenge and of insulin after a hyperglycemic clamp maintains elevated glucose in the blood for a period of time. The disposition index (DI), for example, which measures the insulin response as the ratio of change in insulin to change in glucose, normalized to baseline insulin, has been shown to predict conversion from normal glucose tolerance to impaired glucose tolerance and eventually T2D (Utzschneider et al. 2009).

To date, these genetic discoveries have not been translated into new therapies. Metformin is used to reduce blood glucose levels, particularly in overweight patients, by suppressing gluconeogenesis in the liver. Insulin therapy is also used to overcome insulin resistance, and sulfonylurea drugs are used to stimulate insulin release from the pancreatic β cells. The latter therapy targets the potassium channel KCNJ11, which harbors one of the common variants that increase lifetime risk more than 20%. Several other GWAS-identified loci are also known drug targets, but new drugs have not yet followed from the discovery research. Since the proportion of variance explained by all known variants is less than 10%, genotypic scores have not been used in the prediction of disease onset or as an aide to therapeutic intervention, but this may change in the near future. GWAS are also being utilized to investigate the genetic basis for variation in diabetes-related symptoms such as nephropathy and cataracts as well as the causes of gestational diabetes.

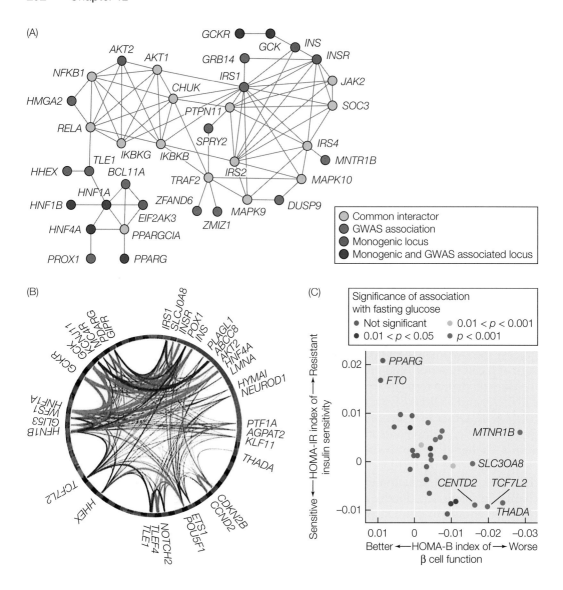

Evolution of Diabetes Risk

The question of why a common disease with a clear genetic basis would be maintained in the population despite its clearly detrimental effect on individual fitness was first posed by Jim Neel in a landmark paper in 1962. According to his "thrifty genotype" hypothesis, alleles that, early in the history of the species, provided a selective advantage in times of famine by promoting the storage of fat would have been rendered detrimental by progress, which has made calorie-rich food amply available to all and hence promotes obesity. Although it is widely accepted, there is in fact little evidence to support the specifics of this hypothesis: known diabetes risk genes do not show signs of strong selection, it is doubtful that famine

◀ **Figure 12.4 Gene Networks for Type 2 Diabetes (T2D) Risk.**   (A) Eighteen of the GWAS loci for T2D and four additional monogenic T2D loci can be assembled into a single network that connects the encoded proteins with fifteen other commonly inter-acting proteins. Six of the GWAS loci are also disrupted by rare alleles that give rise to monogenic forms of disease. This interaction network, which is involved in regulation of adipocytokine activity, was assembled by DAPPLE (disease association protein-protein link evaluator) software from the Broad Institute. (B) Another way to visualize interaction networks is by text mining of abstracts that discuss two or more of the relevant genes, using, for example, GRAIL (gene relationships across implicated loci) software, also from the Broad Institute. Red lines in this plot, which link significant co-occurrences in the literature, connect 35 of the GWAS loci. (C) Plot of HOMA index of insulin sensitivity (proportional to insulin × glucose) against HOMA index of β cell function (proportional to insulin/glucose). Each gene is colored with respect to the strength of its association with fasting glucose, showing that this parameter is correlated with β cell function and that some T2D genes associated with insulin resistance are not associated with fasting glucose. (A and B, after Morris et al. 2012; C, after Voight et al. 2010.)

would have provided the appropriate selection regime, and the genetics of diabetes is somewhat separable from that of obesity and fat mass dis-tribution (Speakman 2008). However, the general principle that there is an imbalance between the demands placed on the human genome in today's cultural environment and the buffering capacity of a genome that evolved under the different conditions that existed throughout much of human evolution is likely true. How this imbalance has shaped the distribution of genetic risk for diabetes and other chronic diseases is an open question.

Although as a rule the incidence of diabetes has increased globally over the past two generations, the disease nevertheless shows marked differ-entiation in prevalence across human populations. Two possible sources of differentiation of genetic risk are divergence of allele frequencies and modulation of allele effect sizes. Most gene discoveries to date have been made in Europeans and later replicated in Asian populations (studies of African Americans are less common due to the methodological challenges raised by their higher polymorphism, lower linkage disequilibrium, and poorly defined population structure). For T2D, there is a characteristic pat-tern of divergence of allele frequencies: a systematic survey of the dozen variants of strongest effect found that all showed a tendency for the risk allele frequency to be greater in Africans than in Europeans, and much lower in East Asians (Chen et al. 2012; **Figure 12.5A**). In several instances, the derived allele is the protective one, implying evolution of genetic resis-tance to diabetes in Asia; interestingly, the same trend is not seen for obesity variants. Offsetting this change in allele frequencies is an opposite trend in allele effect sizes. A study by the PAGE (Population Architecture Using Genomics and Epidemiology) Consortium showed evidence for nominal heterogeneity of effect sizes across human groups for 9 of 20 risk alleles, all of which trended toward smaller effect sizes in African Americans, while 8 alleles had larger effect sizes in Asians (**Figure 12.5B**; Haiman et al. 2012). Generally, though, effect sizes are thought to be very similar across human

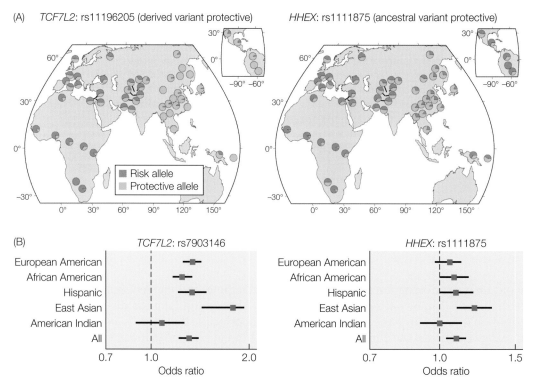

Figure 12.5 Population Differences in Type 2 Diabetes (T2D) Risk. (A) These two maps show the differences in global distribution of two representative T2D risk alleles. In both cases, the protective variant (yellow) is at much higher frequency in Asia than in the rest of the world, but for *TCF7L2* rs11196205, the derived allele is protective, whereas for *HHEX* rs1111875, it is the ancestral allele that is protective. (B) Contrasting with this difference in allele frequency is a possible increase in allele effect size for the two variants (rs7903146 in *TCF7L2* is a proxy for rs11196205) in East Asia relative to Europe and Africa. The forest plots show the 95% confidence intervals for the effect size of each SNP in each cohort studied by the PAGE Consortium. (A, from Chen et al. 2012; B, after Haiman et al. 2012.)

populations. When a polygenic risk score capturing both allele frequencies and effect sizes was generated, it appeared that, on average, Asians have lower susceptibility to T2D than Europeans, who in turn have lower susceptibility than Africans (Chen et al. 2012).

Ethnicities may also differ in the relative contributions of obesity-promoting variants, insulin resistance, and β cell function to their susceptibility. Arab populations, for example, are known to have a relatively high incidence of low HDL-C that may contribute to the prevalence of metabolic syndrome in those populations. A meta-analysis of measures of insulin sensitivity and response by Kodama et al. (2013) indicated that Africans tend to have high insulin resistance and response, whereas East Asians are at the opposite end of the spectrum (**Figure 12.6**). Since glucose tolerance is

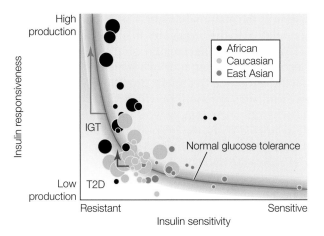

Figure 12.6 Canalization of Blood Glucose Levels.   Glucose homeostasis is
maintained by a balance between insulin production (y axis, in units of an index of acute
insulin response to glucose) and insulin sensitivity (x axis, an insulin sensitivity index). In a
meta-analysis of measures of insulin sensitivity and response, African, European, and East
Asian populations were found to lie at different places along the curve that indicates normal
glucose tolerance, as shown by mean values for each study; each circle is proportional
to the size of the study. Individuals near the midpoint of the curve (green) are best able
to compensate for the increase in insulin resistance that occurs after prolonged periods
on a Western diet by increasing their insulin production. However, as insulin resistance
increases, a larger compensatory response is required; otherwise, the likelihood of fall-
ing into the region of impaired glucose tolerance (IGT) and eventually type 2 diabetes
increases. It is possible that part of the difference in the incidence of T2D among populations arises from decanalization due to perturbation from the healthy curve given background genetic differences. (After Kodama et al. 2013 and Gibson 2009.)

maintained where there is a balance between the ability to produce insulin
when needed and the ability of tissues to respond to insulin, allele frequencies ought to evolve to the equilibrium at which the maintenance of
blood glucose in the optimal zone is most stable. This process, known as
canalization, occurs where perturbations to either resistance or response
are most readily offset by small perturbations in the opposite domain.
The glucose needs of humans have evolved rapidly, the genome is not at
equilibrium, and dietary sources of glucose are very different than they
have been throughout most of human history. These are all conditions that
mitigate against stability, and decanalization theory postulates that more
of the population is at risk of diabetes due to the limited capacity of the
genome to buffer the effects of modern food consumption (Gibson 2009).

The T2D Microbiome

Since most overweight people actually have healthy metabolism, microbiolo-
gists have begun to ask whether the metabolism of food in the gut might be
altered by the content of the microbiome. There are two dominant groups of

probiotic bacteria in the mammalian gut. The Bacteroidetes have decreased representation relative to the Firmicutes in obese people as compared with lean people. The cause-and-effect relationship is complex, since genetically obese mice also have an abundance of Firmicutes, while repopulation of the gut following antibiotic treatment promotes weight gain to a greater degree when Firmicutes are more prevalent (Turnbaugh et al. 2006). Correspondingly, when 12 obese people were placed on either fat-restricted or carbohydrate-restricted diets, the proportion of Bacteroidetes increased as their weight dropped (Ley et al. 2006). The mechanism of this effect likely relates to the effects of the microbes on digestion of polysaccharides, promotion of absorption of polysaccharide derivatives through the intestine, and regulation of host lipid metabolism.

Subsequent quantitative metagenomics analyses of larger samples in which total gut microbial genomic fragments were aligned to over 1000 microbial reference genomes have identified a bimodal distribution of total gene counts (and hence microbial species diversity) in humans. A study of 292 Danes (Le Chatelier et al. 2013) divided participants into those with low gene counts (LGC) and with high gene counts (HGCs). They documented enrichment for LGC profiles in obese participants, whereas lean participants were twice as likely to have a HGC profile (**Figure 12.7A**). Furthermore, the obese LGC people showed, on average, a 3-point increase in BMI over 9 years (over 10 kg of weight increase for a person of average height) compared with half that amount for the HGC people. The LGC profile was characterized by an excess of genes in a small number of taxa that are likely to play roles in coping with oxidative stress while also promoting inflammation and possibly producing toxic compounds. LGC participants had significantly higher insulin and insulin resistance accompanying their diabetes-related lipid and triglyceride metabolome profile.

These results were extended in a parallel dietary intervention study conducted with 49 overweight French individuals, some with HGC and others with LGC profiles, who elected to follow a 6-week protein-based diet that reduced their overall calorie intake by 35%, followed by 6 weeks of a more moderate weight-maintenance diet (Cotillard et al. 2013). In this study, the HGC sample started with lower insulin resistance and a healthier lipid and triglyceride profile, but most notably, showed a greater improvement in these characteristics than the LGC participants, at least after the strict low-calorie dietary intervention (**Figure 12.7B**). Gene richness, a measure of the total number of genes in the microbial metagenome, increased initially in the LGC group, but not to the HGC level, and showed signs of reversion once the intervention was relaxed.

The conclusion is that each person has a particular microbiome profile that tends to be maintained throughout life, but can be modified by diets rich in calories or food sources (vegetables, fish) that modify the intestinal environment. More obese individuals tend to have less microbial diversity as the ecology of the microbial community supports dominance by a

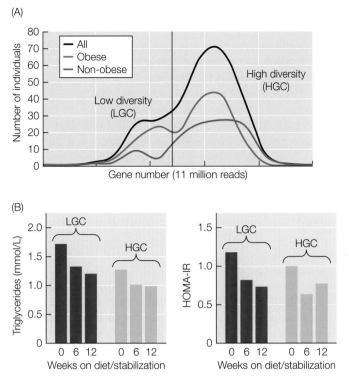

Figure 12.7 Diabetes Metagenomics and Metabolomics.   (A) In a study of obese and non-obese Danish people (*n* = 292), a bimodal distribution of microbial gene diversity in the gut microbiome was observed, albeit with a greater proportion of low gene counts (low species diversity) in obese individuals. (B) Forty-nine obese or overweight French partici-pants were given a 12-week dietary intervention consisting of 6 weeks of a high-protein diet for weight loss followed by another 6 weeks of weight maintenance. The low gene count (LGC) group showed higher triglycerides and insulin resistance in general. Improvements in these and other inflammation-related parameters accompanied shifts in microbial diversity during the intervention. These improvements may be more limited in the LGC group, though the reported significance comparisons were marginal. HGC, high gene count. (A, after Le Chatelier et al. 2013; B, after Cotillard et al. 2013.)

relatively small number of species that promote inflammation. At the other end of the weight spectrum, severely malnourished children in Bangladesh have a grossly altered trajectory of development of the gut microbiome, with long-lasting effects on their ability to absorb food (Subramanian et al. 2014). The obesity results are mostly correlative, so to what extent the low-diversity profile promotes the eventual development of diabetes remains to be determined. It is noteworthy that a genomic profile based on the presence of genes found in just nine microbial species predicts whether an individual is lean or obese with considerably better sensitivity and specificity than a 32-SNP profile from GWAS.

Nutrigenetics

One of the new frontiers for human genetics is making recommendations for promoting wellness based on an individual's genetic profile. With respect to diet, this field is known as nutritional genetics, or nutrigenetics. Some small studies (Arkadianos et al. 2007) have suggested that people provided with genotypic information lose weight more quickly than controls, although it has not been established that the specific genetic information, as opposed to the generic motivation that comes with participation in a study, is responsible. Nevertheless, over two dozen companies are already providing nutrigenetics services. One, LifeGenetics, offers a €249 "DNA test Slim" that assigns a person to one of four nutritional types—Mediterranean, fitness, high-fat, or low-fat—and provides corresponding personalized diet and exercise advice. Whether or not these categories are accurate, this approach does recognize that people vary not only in their metabolic profiles, but also in their response to dietary intervention: vegetarian, Mediterranean, South Beach, or pescetarian, one size does not fit all.

It is also well known that people have very different body shapes that reflect the distribution of fat, muscle, and bone mass. The canonical hourglass figure for women and the triangular upper body of athletic men are in fact the minority body types: most people fit better into pear-shaped (more lower-body mass), banana (also called barrel or tube shaped, more unitary waist-to-hip ratio), or apple-shaped (larger waist than hips) types. A related categorization recognizes ectomorphs (tall, thin, and delicate with a fast metabolism), mesomorphs (athletic and muscular but prone to fat gain), and endomorphs (stocky, round, soft musculature with a slow metabolism; **Figure 12.8**). The distribution of body fat in subcutaneous (thighs, buttocks, skin folds) versus visceral (abdominal, packed between the organs) deposits is as important to health outcomes as total fat mass, and the genetic and environmental factors underlying it are an active area of research (Heid et al. 2010). Fat distribution also clearly differs between the sexes and among populations, and changes with age almost certainly influence the rate of decline of metabolism-related health.

Nutrigenetics also concerns the ability to metabolize and respond to nutrients and stimulants and to synthesize essential vitamins. For instance, people vary fivefold in the speed with which they metabolize caffeine: some can readily fall asleep half an hour after a cup of coffee, while others would lie awake for hours. Some of this difference is due to variation in the gene that encodes the primary enzyme that metabolizes caffeine, CYP1A2, as well as in the AHR xenobiotic receptor that activates expression of this enzyme (Sulem et al. 2011). Both loci are associated with a difference in coffee consumption of around 0.2 cups per day per allele, with implications for heart disease risk.

Epidemiological studies suggest that variation in one-carbon metabolism, centered around folate and vitamin B_{12} function, has an important role in metabolic syndrome and aging through its effect on purine synthesis and subsequently on many aspects of cellular health. GWAS of 45,000 Icelanders

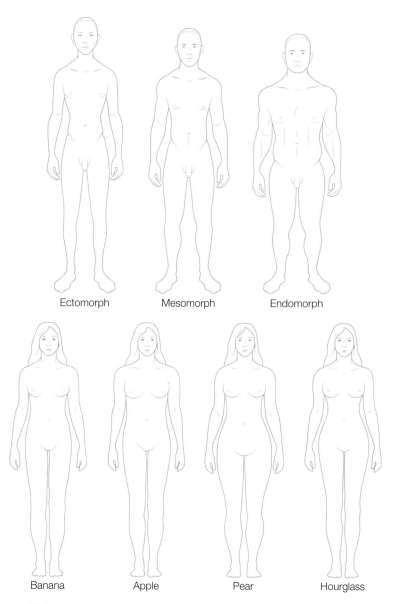

Figure 12.8 Body Types.   A classification that applies to both sexes is that of ecto-, meso-, and endomorphs. Women are often characterized as having banana-shaped (or tube), apple-shaped, pear-shaped, or hourglass body types.

and Danes (the former using imputation from whole-genome sequences of 1176 individuals) uncovered eleven loci that influence vitamin B_{12} levels in serum and two loci (*MTHFR* and *FOLR3*) that influence folate levels. These loci explain as much as 20% of the heritability of vitamin B_{12} levels, but just 6% of that of folate levels. Perhaps because the heritability of both traits is low (just 27% and 17%), these variants are not directly associated with cardiovascular disease, cancer, or Alzheimer's disease onset (Grarup et al. 2013). An alternative interpretation is that the genetics calls into question the causal connection between these metabolites and disease, a topic taken up in the next chapter in relation to HDL cholesterol and atherosclerosis. Genetic analysis not only illustrates the complexity of the mechanistic basis of the regulation of metabolic traits, but also has the potential to help distinguish correlation from causation in nutrition research.

Summary

1. Metabolic syndrome is a combination of central obesity and any two of the following: elevated triglycerides, reduced HDL cholesterol, high blood pressure, or elevated fasting blood glucose.

2. Primary dyslipidemias are a series of familial disorders of serum lipid metabolism involving abnormal levels of the circulating lipid vesicles VLDL, LDL, and HDL.

3. Lipoprotein vesicles vary in their proportions of proteins, triglycerides, phospholipids, and cholesterol, and the relative abundance of vesicle types is associated with risk of metabolic disease, including type 2 diabetes.

4. Familial dyslipidemias are generally due to major effect mutations of major effect with high penetrance.

5. GWAS meta-analysis has identified close to 100 loci that influence lipid traits. Many of these loci also harbor monogenic causes of dyslipidemia, and several have multiple causal variants.

6. The adipokine hormones, including leptin, adiponectin, and resistin, vary in expression and influence metabolic traits as well as obesity.

7. GWAS for type 2 diabetes (T2D) has uncovered 65 loci, but it is estimated that at least 500 contribute to disease risk.

8. There are three major documented sources of genetic variance for T2D risk: hypothalamic function that regulates food consumption and obesity, pancreatic β cell division and development, and development of peripheral insulin resistance.

9. Genes also regulate fasting levels of serum glucose and insulin as well as the ratios of these two factors, which give rise to measures of insulin resistance (HOMA-IR) and sensitivity (HOMA-B).

10. The genetic basis of diabetes risk varies among populations primarily due to differences in allele frequencies, and secondarily due to differences in allele affect sizes. Asians appear to have reduced genetic risk relative to Europeans and Africans. Several of the protective variants are derived in humans.

11. A low-diversity gut microbiome is associated with increased inflammation, obesity, and increased risk of metabolic disease, while

individuals with high microbial diversity may respond better to weight reduction programs.

12. Nutrigenetics is the study of whether and how genetic variation can be used to predict how individuals will respond to different diets.

References

Arkadianos, I., Valdes, A. M., Marinos, E., Florou, A., Gill, R. D., and Grimaldi, K. A. 2007. Improved weight management using genetic information to personalize a calorie controlled diet. *Nutrition J.* 6: 29.

Brown, M. S. and Goldstein, J. L. 1986. A receptor-mediated pathway for cholesterol homeostasis. *Science* 232: 34–47.

Chen, R., Corona, E., Sikora, M., Dudley, J. T., Morgan, A. A., et al. 2012. Type 2 diabetes risk alleles demonstrate extreme directional differentiation among Human populations, compared to other diseases. *PLoS Genet.* 8: e1002621.

Cotillard, A., Kennedy, S. P., Kong, L. C., Prifti, E., Pons, N., et al. 2013. Dietary intervention impact on gut microbial gene richness. *Nature* 500: 585–588.

Dastani, Z., Hivert M-F., Timpson, N., Perry, J. R. B., Yuan, X., et al. 2012. Novel loci for adiponectin levels and their influence on Type 2 Diabetes and metabolic traits: A multi-ethnic meta-analysis of 45,891 individuals. *PLoS Genet.* 8: e1002607.

Flannick, J., Thorleifsson, G., Beer, N. L., Jacobs, S. B. R., Grarup, N., et al. 2014. Loss-of-function mutations in SLC30A8 protect against type 2 diabetes. *Nat. Genet.* 46: 357–363.

Gibson, G. 2009. Decanalization and the origin of complex disease. *Nat. Rev. Genet.* 10: 134–140.

Grarup, N., Sulem, P., Sandholt, C. H., Thorleifsson, G., Ahluwalia, T. S., et al. 2013. Genetic architecture of vitamin B12 and folate levels uncovered applying deeply sequenced large datasets. *PLoS Genet.* 9: e1003530.

Haiman, C. A., Fesinmeyer, M. D., Spencer, K. L., Buzková, P., Voruganti, V. S., et al. 2012. Consistent directions of effect for established type 2 diabetes risk variants across populations: The Population Architecture using Genomics and Epidemiology (PAGE) Consortium. *Diabetes* 61: 1642–1647.

Heid, I. M., Jackson, A. U., Randall, J. C., Winkler, T. W., Qi, L., et al. 2010. Meta-analysis identifies 13 new loci associated with waist-hip ratio and reveals sexual dimorphism in the genetic basis of fat distribution. *Nat. Genet.* 42: 949–960.

Johansen, C. T. and Hegele, R. A. 2012. The complex genetic basis of plasma triglycerides. *Curr. Atheroscler. Rep.* 14: 227–234.

Kodama, K., Toda, K., Tojjar, D., Patel, C. J., Yamada, S., and Butte, A. J. 2013. Ethnic differences in the relationship between insulin sensitivity and insulin response. *Diabetes Care* 36: 1789–1796.

Le Chatelier, E., Nielsen, T., Qin, J., Prifti, E., Hildebrand, F., et al. 2013. Richness of human gut microbiome correlates with metabolic markers. *Nature* 500: 541–546.

Ley, R. E., Turnbaugh, P. J., Klein, S., and Gordon, J. I. 2006. Microbial ecology: Human gut microbes associated with obesity. *Nature* 444: 1022–1023.

Morris, A. P., Voight, B. F., Teslovich, T. M., Ferrreira, T., Segrè, A. V., et al. 2012. Large scale association analysis provides insight into the genetic architecture and pathophysiology of type 2 diabetes. *Nat. Genet.* 44: 981–990.

Musunuru, K., Strong, A., Frank-Kamenetsky, M., Lee, N. E., Ahfeldt, T., et al. 2010. From non-coding variant to phenotype via *SORT1* at the 1p13 cholesterol locus. *Nature* 466: 714–719.

Neel, J. V. 1962. Diabetes mellitus: A "thrifty" genotype rendered detrimental by "progress"? *Am. J. Hum. Genet.* 14: 353–362.

Pal, A. and McCarthy, M. I. 2013. The genetics of type 2 diabetes and its clinical relevance. *Clin. Genet.* 83: 297–306.

Patel, C. J., Bhattacharya, J., and Butte, A. J. 2010. An environment-wide association study (EWAS) on type 2 diabetes mellitus. *PLoS ONE* 5: e10746.

Patel, C. J., Chen, R., Kodama, K., Ioannidis, J. P., and Butte, A. J. 2013. Systematic identification of interaction effects between genome- and environment-wide associations in type 2 diabetes mellitus. *Hum. Genet.* 132: 495–508.

Scott, R. A., Lagou, V., Welch, R. P., Wheeler, E., Montasser, M. E., et al. 2012. Large-scale association analyses identify new loci influencing glycemic traits and provide insight into the underlying pathways. *Nat. Genet.* 44: 991–1005.

Soranzo, N., Sanna, S., Wheeler, E., Gieger, C., Radke, D., et al. 2010. Common variants at 10 genomic loci influence Hemoglobin A_{1C} levels via glycemic and non-glycemic pathways. *Diabetes* 59: 3229–3239.

Soutar, A. K. and Naoumova, R. P. 2007. Mechanisms of disease: genetic causes of familial hypercholesterolemia. *Nat. Clin. Pract. Cardiovasc. Med.* 4: 214–225.

Speakman, J. R. 2008. Thrifty genes for obesity, an attractive but flawed idea, and an alternative perspective: The "drifty gene" hypothesis. *Int. J. Obes. (Lond).* 32: 1611–1617.

Subramanian, S., Huq, S., Yatsunenko, T., Haque, R., Mahfuz, M., et al. 2014. Persistent gut microbiota immaturity in malnourished Bangladeshi children. *Nature* 510: 417-421.

Sulem, P., Gudbjartsson, D. F., Geller, F., Prokopenko, I., Feenstra, B., et al. 2011. Sequence variants at *CYP1A1-CYP1A2* and *AHR* associate with coffee consumption. *Hum. Mol. Genet.* 20: 2071–2077.

Teslovich, T., Musunuru, K., Smith, A. V., Edmondson, A. C., Stylianou, I. M., et al. 2010. Biological, clinical and population relevance of 95 loci for blood lipids. *Nature* 466: 707–713.

Turnbaugh, P. J., Ley, R. E., Mahowald, M. A., Magrini, V., Mardis, E. R., and Gordon, J. I. 2006. An obesity-associated gut microbiome with increased capacity for energy harvest. *Nature* 444: 1027–1031.

Utzschneider, K. M., Prigeon, R. L., Faulenbach, M. V., Tong, J., Carr, D. B., et al. 2009. Oral disposition index predicts the development of future diabetes above and beyond fasting and 2-h glucose levels. *Diabetes Care* 32: 335–341.

Voight, B. F., Scott, L. J., Steinthorsdottir, V., Morris, A. P., Dina, C., et al. 2010. Twelve type 2 diabetes susceptibility loci identified through large-scale association analysis. *Nat. Genet.* 42: 579–589.

Wang-Sattler, R., Yu, Z., Herder, C., Messias, A. C., Floegel, A., et al. 2012. Novel biomarkers for pre-diabetes identified by metabolomics. *Mol. Sys. Biol.* 8: 615.

Wild, C. 2005. Complementing the genome with an "exposome": The outstanding challenge of environmental exposure measurement in molecular epidemiology. *Cancer Epid. Biomarkers Prevent.* 14: 1847–1850.

13

Cardiovascular Disease

Cardiovascular disease (CVD) is the number one source of mortality worldwide. Globally, 23 million people now die from heart failure each year, the majority in low- and middle-income countries, where the incidence of CVD is rising just as it has begun to decline in high-income countries. Aside from genetics, the major risk factors for CVD are associated with contemporary lifestyles: smoking, lack of activity, and poor diet.

Technically, CVD includes disease of any organ affected by vascular function (kidneys, brain, peripheral veins), but this chapter will focus mainly on coronary disease. Heart failure takes a wide variety of forms, from sudden cardiac death to gradual decline. We will start with consideration of the genetic basis of hypertension and of the function of nonwhite blood cells (erythrocytes and platelets), then move to the three major sources of heart-related mortality: atherosclerosis (or coronary artery disease, CAD), arrhythmia, and cardiomyopathy (cardiac muscle failure; Kathiresan and Srivastava 2012). Congenital heart defects are a major source of infant morbidity, as nearly 1% of newborn babies are recommended for surgery to repair malformed vessels or chambers; the genetic bases of these defects are surveyed at the end of the chapter. Many other ailments that affect the heart, such as Marfan syndrome (primarily a disorder of the extracellular matrix protein fibrillin), also have genetic contributors, but are not considered here due to space limitations.

Stroke is not discussed in detail here because only a small number of monogenic causes are known, while the literature on polygenic factors has not yet produced consistent findings or novel associations that point conclusively to biology independent of coronary artery disease (Markus 2012). This situation may reflect the heterogeneity of the disease. A primary distinction is made between hemorrhagic and ischemic stroke: the former refers to localized bleeding in the brain, the latter to acute restriction of blood flow. Ischemic stroke, the much more common form, is further subdivided into large-artery stenosis, or narrowing of the arteries that supply the brain; cardioembolic stroke, in which ruptured plaque migrates to the brain where it causes embolism; and small-vessel disease, which leads to deep infarcts in the white and gray matter and is also responsible for vascular dementia. Undoubtedly, large GWAS meta-analyses will eventually illuminate the genetic contributions to each of these subtypes of stroke.

Hypertension

Hypertension (high blood pressure) affects as many as 1 billion people, many of whom are unaware that they have the condition, even though it more than doubles the risk of ultimate heart failure. Optimal blood pressure (BP) is below 120 millimeters of mercury (mm Hg) systolic and below 80 mm Hg diastolic. Stage I hypertension is in the range of 140–160 mm Hg systolic and 90–100 mm Hg diastolic, and stage II is yet more severe. People with diabetes or kidney disease generally require lower blood pressure. Hypertension is more prevalent in African Americans than in European Americans, and although age of onset is earlier in men, the condition has similar prevalence in both sexes, and women are slightly more likely to take medication to control it. WebMD lists over 200 prescription drugs used for control of high blood pressure and their side effects. These drugs include diuretics, beta blockers (of the adrenergic receptor that regulates blood flow), agonists of the renin-angiotensin pathway (ARBs), and antagonists of the angiotensin-converting enzyme (ACE inhibitors), among others.

The largest GWAS for hypertension netted just 28 loci, despite the inclusion of over 200,000 people and a relatively high heritability of approximately 30%. No variants were discovered in the initial WTCCC study of 2000 cases, and studies of Asian populations have not uncovered novel associations either. Thus, despite its strong genetic basis, there are no common variants of major effect affecting blood pressure (Agarwal et al. 2005). The discovered variants explain less than 1% of the phenotypic variance for BP. One-half of the variants are candidate eQTL, but fewer than ten are in the vicinity of genes that were considered candidate BP genes a priori. A genetic risk score including all 28 variants is significantly associated with both systolic (SBP) and diastolic (DBP) blood pressure in East and South Asians, but only modestly in individuals of African ancestry. These differences in the score's predictive ability may reflect variation in allele frequencies, variation in effect sizes, or linkage disequilibrium between tagging and causal polymorphisms.

The International Consortium for Blood Pressure Genome-Wide Association Studies (2011) also investigated whether the aggregate genetic risk score predicts heart disease. An increase of 1 standard deviation unit of the score was associated with a 21% increase in risk of hypertension, translating to a differential prevalence of 29% and 16% in the highest and lowest deciles, respectively, of an independent sample of 23,000 women. The corresponding difference between the top and bottom quintiles approximates the effect of a single antihypertensive agent. This same risk score also associates with left ventricular wall thickness, with occurrence of stroke, and very strongly with CAD (Lieb et al. 2013), but not with measures of kidney function or disease.

More research needs to be done to establish which specific genes are affected at each locus and to affirm the identity, and thence the mechanism of action, of the causal variant(s). However, three plausible modes of action are implicated by nine of the candidate genes. Four of the loci near *ADM*, *NPPB*, *NPR3*, and *GUCY1A3* are linked to natriuretic peptide function in homeostatic regulation of body water, sodium, and fat: *NPPB* encodes one of the atrial forms of the peptide, and adrenomedullin (encoded by *ADM*) also has natriuretic properties, while the other two genes regulate natriuretic peptide expression. Two of the genes, which encode the bicarbonate transporter SLC4A7 and the phospholipase component PLCE1, are expressed in the kidney and influence renal physiology and disease. Metal ion transport also seems to be affected by missense variants encoded by the transporter gene *SLC39A8* and by *HFE*, which is known to be a low-penetrance source of hereditary hemochromatosis, or iron overload. The only GWAS locus previously implicated in hypertension by genetic linkage analysis of rare familial cases is *CYP17A1*, which was first identified in Mendelian congenital adrenal hyperplasia and encodes an enzyme involved in mineralocorticoid and other steroid biogenesis.

Monogenic hypertension, which accounts for only 1% of cases, almost always involves increased sodium transport in the distal nephron of the kidney, which results in increased plasma volume and high blood pressure (Armani et al. 2011). For example, 11βHSD deficiency prevents metabolism of cortisol, which activates the mineralocorticoid receptor, and in turn the epithelial sodium channel (licorice also inhibits this enzyme and when consumed in excess upsets electrolyte balance). The most common form of congenital hypertension is Liddle's syndrome, or pseudohyperaldosteronism, which is due to a series of nonsense mutations that truncate the C-terminal portion of the renal tubular sodium channel subunits. These conditions also alter the renin-aldosterone balance, but more direct disruption of this pathway gives rise to another class of congenital renal abnormality, called glucocorticoid-remediable aldosteronism (GRA), that can be corrected by dexamethasone therapy. Unusual chimeric fusions of *CYPB11A* and *CYPB11B* result in overactive aldosterone synthase that has an autosomal dominant mode of inheritance.

Platelet and Erythrocyte Function

Platelets (also called thrombocytes) are anuclear disc-shaped cell fragments that play an essential role in blood clotting and release growth factors that promote vessel health and wound healing. They are derived from a progenitor cell type, known as the megakaryocyte, that is distinct from the lymphoid and myeloid white blood cell (leukocyte) lineages (**Figure 13.1**). Typically, they circulate at concentrations between 150,000 and 400,000 "cells" per milliliter of blood. Low platelet counts are associated with abnormal bleeding; high counts are associated with predisposition to thrombosis, or clotting, which can contribute to myocardial infarction (MI, or heart attack) or stroke. A large GWAS of almost 70,000 people uncovered 68 loci associated with

(A)

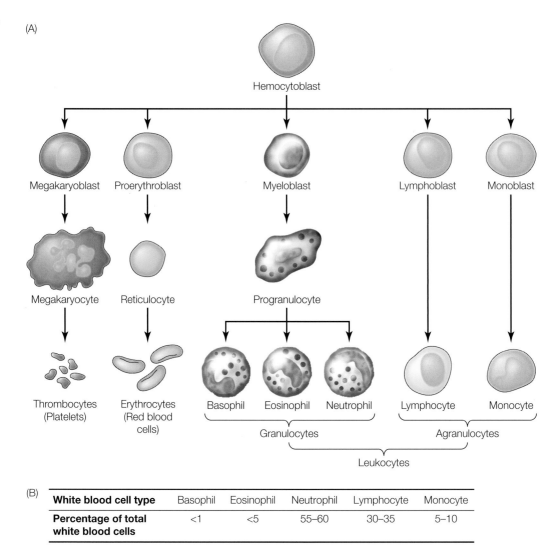

(B)

White blood cell type	Basophil	Eosinophil	Neutrophil	Lymphocyte	Monocyte
Percentage of total white blood cells	<1	<5	55–60	30–35	5–10

Figure 13.1 Blood Cell Development. Red blood cells (erythrocytes), platelets (thrombocytes), and white blood cells (leukocytes) derive from different progenitor cells. The two most common classes of white blood cells are the neutrophils and lymphocytes, representing the myeloid and lymphoid lineages, respectively. Neither platelets nor red blood cells contain nuclei, though they do carry a small representation of mRNAs held over from their progenitors.

either platelet counts or platelet volume, explaining 5% and 10% of these traits, respectively (Gieger et al. 2011). As usual, individual effect sizes are generally small, but there seems to be good replication across ethnicities. Several of the loci had been previously implicated in hemostasis (*ITGA2B*, *F2R*, *GP1BA*), megakaryopoiesis (*THPO*, *MEF2C*), and platelet life span (*BAK1*), but a particularly interesting aspect of this study was the assembly of an expanded interaction network of 633 genes connected to a core set of 34 of the GWAS genes (**Figure 13.2**). All of these genes are much more likely to be upregulated during in vitro megakaryopoiesis than erythropoiesis (red blood cell development). Furthermore, knockdown of 5 out of 6 of the GWAS genes, which were tested by morpholinos in zebrafish, resulted in loss of platelet development, and knockdown of more of the genes by RNAi even disrupted blood cell development in *Drosophila*. These results connect cellular differentiation to variation in a physiologically relevant trait. It is likely that other loci not in the core set have roles in platelet activation and function.

The international consortium behind these studies carried out a parallel GWAS meta-analysis of red blood cell (erythrocyte, RBC) traits, including hemoglobin concentration, RBC volume, and RBC counts, all of which are positively correlated (Van der Harst et al. 2012). Variation in these parameters is associated with measures of cardiovascular health, as well as with pregnancy and cognitive decline. Seventy-five loci correlate with these RBC traits, including variants in genes that had previously been

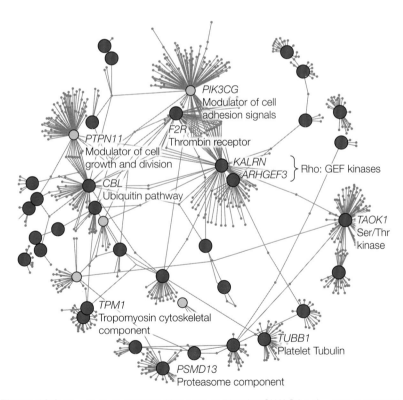

Figure 13.2 The Protein-Protein Network of Platelet GWAS Loci. The expanded version of the network shown here has 39 core genes identified by GWAS connecting with 747 other proteins via 1085 edges. It was assembled from two different databases of protein-protein interactions (Reactome and IntAct). The 34 genes represented by the large nodes (green) lie within 10 kb of the GWAS association signal. (After Gieger et al. 2011.)

linked to red blood cell disorders (*ANK1, SLC4A1, SPTA1*), red blood cell proliferation (*IKZF1, KIT, SH2B3, SH3GL1,* and *TAL1*), hemolytic anemia (*HK1*), and iron deficiency or overload (*TMPRSS6, HFE, TFR2*). Half of the candidate genes produce hematological phenotypes when knocked out in mice or flies, and they also have annotated functions in regulation of the cell cycle and cytokine activity. The utility of a genetic risk score based on these variants remains to be evaluated, but it certainly differentiates individuals at the ends of the distribution for a range of traits that are well known to correlate with health.

Ongoing genetic analyses are exploring other blood endophenotypes relevant to heart disease (Sabater-Lleal et al. 2013). Fibrinogen (also known as factor I) is the precursor of the fibrin protein matrix in blood clots, and 23 loci explain less than 5% of its variability. They are not, however, associated with stroke or venous thromboembolism. Circulating plasminogen activator protein 1, von Willebrand factor, and factor VIII are all genetically regulated, but GWAS to date have identified only a small number of loci that capture a minor fraction of the inferred SNP contributions to heritability of their serum levels and subsequent risk of thrombosis.

Arteriosclerosis

Arteriosclerosis, or hardening of the arteries, is now the major cause of heart disease, due to the combination of substantial genetic risk with very high levels of LDL cholesterol that is observed in contemporary humans. Arterial walls thicken with age, beginning in the third decade, as lipids begin to accumulate in the lumen as well as within cells. By the fourth decade, in some individuals, lipid deposits can begin to become fibrous as collagen and smooth muscle infiltrate along with macrophages, giving rise to atheroma, or plaque. Atherosclerosis refers to the subset of arteriosclerosis that is due to plaque buildup. Eventually, either condition will compromise heart function, but rupture of plaque gives rise to clogged arteries, especially after platelets initiate a coagulation cascade at the lesion. This condition can become a myocardial infarction within 5 minutes. Detached plaque material may also directly occlude smaller arteries downstream in the circulatory system, causing thromboembolism. If the major arteries also become clogged by accumulated plaque, stenosis (narrowing of the vessel) occurs. Coronary artery disease is sometimes defined as more than 50% stenosis of at least one artery, and it is the major risk factor for myocardial infarction and sudden cardiac death. Ischemia refers to the consequences of long-term loss of blood flow either to the extremities, or in this case, to the heart itself. Cardiac ischemia is often asymptomatic, and it may be the source of angina, but it is the number one cause of hospitalization and death in the West.

In 2011, the CARDIoGRAM (Coronary Artery Disease Genome-Wide Replication and Meta-analysis) Consortium carried out a meta-analysis of 14 studies involving 22,000 cases of people with coronary artery disease and 65,000 controls, which led to the discovery of 23 CAD-associated loci explaining 10% of the genetic component of risk (Schunkert et al. 2011). No more than

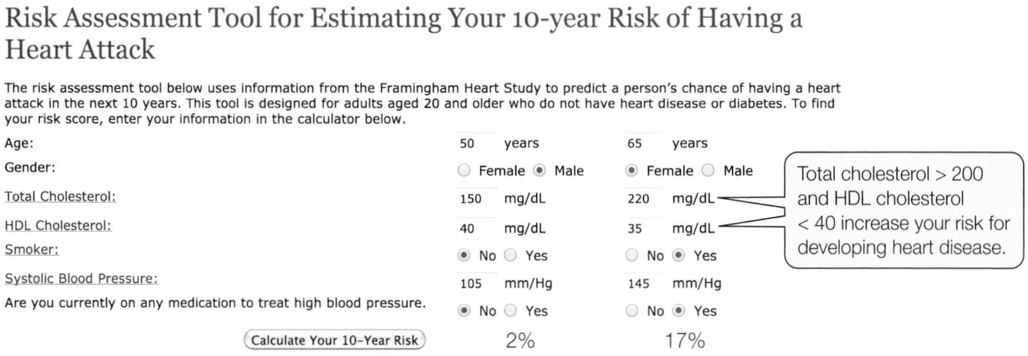

Risk Assessment Tool for Estimating Your 10-year Risk of Having a Heart Attack

The risk assessment tool below uses information from the Framingham Heart Study to predict a person's chance of having a heart attack in the next 10 years. This tool is designed for adults aged 20 and older who do not have heart disease or diabetes. To find your risk score, enter your information in the calculator below.

Age:	50 years	65 years
Gender:	○ Female ● Male	● Female ○ Male
Total Cholesterol:	150 mg/dL	220 mg/dL
HDL Cholesterol:	40 mg/dL	35 mg/dL
Smoker:	● No ○ Yes	○ No ● Yes
Systolic Blood Pressure:	105 mm/Hg	145 mm/Hg
Are you currently on any medication to treat high blood pressure.	● No ○ Yes	○ No ● Yes
(Calculate Your 10–Year Risk)	2%	17%

Total cholesterol > 200 and HDL cholesterol < 40 increase your risk for developing heart disease.

Figure 13.3 **The NHLBI's Cardiovascular Risk Calculator.**   Based on the Framingham Heart Study, this site takes a person's age, gender, total and HDL cholesterol levels, smoking status, and systolic blood pressure into account in assessing the 10-year risk of having a heart attack. Individuals taking medication for high blood pressure are also at elevated risk. (From cvdrisk.nhlbi.nih.gov)

6 of these genes clearly act in the same physiological pathway as the classical risk factors—high cholesterol, smoking, and high blood pressure—that have been identified by epidemiologists. Consequently, there is considerable new biology waiting to be discovered. Furthermore, the genetic risk score derived from the 23 SNPs is not strongly correlated with the Framingham cardiovascular risk score derived from the classical measures (**Figure 13.3**). Interestingly, a half-dozen of the variants have pleiotropic associations with the autoimmune diseases type 1 diabetes and celiac disease, with smoking and lung adenocarcinoma, and with cranial aneurysm. Risk allele frequencies in Europeans range across the full spectrum, from 0.13 to 0.91, with the largest effect increasing the odds of CAD by just 17% per allele. Effect sizes tend to be larger in early-onset cases, but whereas an early age of onset is more common in men (but with more severe disease in women), no gender-biased genetic effects have been documented.

A second large-scale GWAS conducted by the CHARGE (Cohorts for Heart and Aging Research in Genomic Epidemiology) Consortium (Bis et al. 2011) focused specifically on atherosclerosis, which they defined by non-invasive ultrasound imaging of the thickness of the carotid intima—namely, the innermost layers of the arteries that supply the head and neck with oxygenated blood—and by the existence of plaque in those arteries. They found just five loci, three of which were also discovered by CARDIoGRAM. One of those loci, *APOC1*, encodes one of the apolipoproteins that regulate lipid metabolism, and along with the nearly significant *LDLR* locus, it establishes the expected link between CAD and hyperlipidemia. *PIK3CG* was one of the loci associated with platelet volume, and along with the endothelin receptor encoded by *EDNRA*, it establishes links to plaque establishment and hypertension. The two other loci, *PINX1* and a gene desert at 8q24, have unknown biological connections to arteriosclerosis (**Figure 13.4**).

(A)

Stages

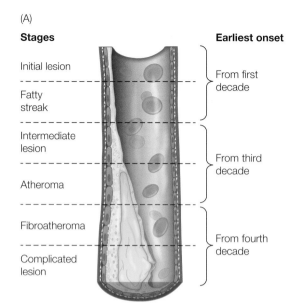

Initial lesion

Fatty streak

Intermediate lesion

Atheroma

Fibroatheroma

Complicated lesion

Earliest onset

From first decade

From third decade

From fourth decade

(B)

Loci associated with atherosclerosis	
Genes	**Functions**
PINX1 8q24	Unknown
APOC1 LDLR	Lipid metabolism
PIK3CG EDNRA	Platelet volume, plaque establishment

Figure 13.4 Atherosclerosis. This schematic of an artery illustrates the phases of progression of thickening of the arteries and narrowing of the arterial lumen (stenosis). Accumulation of lipids typically accelerates in a person's late 30s. Some of the genes that link lipid metabolism and thrombosis to coronary artery disease are indicated.

A third large GWAS by the National Heart, Lung, and Blood Institute (NHLBI) CARe (Candidate Gene Association Resource) Project, also published in 2011 (Lettre et al. 2011), focused explicitly on CAD-related phenotypes in almost 8100 African Americans. Evidence for replication was found for 9 loci associated with HDL-C and 6 loci associated with LDL-C, but not for loci associated with hypertension or smoking. In several of these cases, the authors discussed how the reduced linkage disequilibrium in Africans can actually help to refine the association signal, though they also pointed out that in some cases, different variants may be responsible, and that differences in allele frequency alter the power of studies across populations. Since the vast majority of genetic variants affecting CAD remain undetected in any population, effect sizes are small, and the failure to replicate many loci is not informative about possible differences in the architecture of atherosclerosis risk among races.

Myocardial Infarction

The ability to predict the likelihood of sudden death due to a heart attack in the CAD population would represent an important advance in cardiology. Currently, the risk factors for atherosclerosis are well known, but those that predispose people with atherosclerosis to progression to MI have not been defined well enough to support any type of personalized intervention.

Most cases of MI occur in adults over the age of 65, but early-onset MI actually has higher heritability. Already by 2009, an international consortium had identified 8 loci that contribute common SNPs to risk of MI in men younger than 50 and women younger than 60 (MIGC 2009). Three of the loci, *SORT1*, *LDLR*, and *PCSK9*, are associated with LDL-C, and thus presumably reflect the role of plaque accumulation in MI, but the remaining five loci are in the vicinity of genes encoding proteins with diverse molecular functions, such as phosphatases (*PHACTR1*), solute transporters (*SLC5A3*), and even a mitochondrial ribosomal protein (*MRPS6*), whose roles in MI, if any, remain to be clarified. The consortium also evaluated the potential effect of 554 common copy number polymorphisms and 8065 rare copy number variants, finding neither any association of MI with the common copy number variants (CNVs) nor an excess burden of rare CNVs in almost 3000 cases.

The most replicated allele is in a gene-poor region at chromosome 9p21.3, which is homozygous in one-fourth of all Europeans, on whom it confers twice the risk of MI relative to non-carriers. Deep sequencing of this region has identified more than 100 SNPs, spread over 100 kb, that are in high enough linkage disequilibrium for them all to be candidate contributors to the risk haplotype (**Figure 13.5**). By contrast, the immediately adjacent region of 9p21.3 that is associated with type 2 diabetes is less then 11 kb in length. Deep chromatin profiling for histone marks, DNase hypersensitive sites, and chromatin immunoprecipitation with key transcription factors revealed 33 candidate enhancers, making this one of the regions of the genome most enriched for regulatory sequences. Two of the MI risk polymorphisms disrupt a binding site in vascular endothelium for the interferon-inducible transcription factor STAT-1, and they have been shown to regulate the expression of the most proximal genes, specifically inducing transcription of the *ANRIL* antisense RNA (*CDKN2BAS*), which then represses abundance of the adjacent *CDKN2B* transcript (Shea et al. 2011). However, Harismendy et al. (2011) went on to show by chromatin conformation capture that the so-called *ECAD9* enhancer region surrounding these candidate SNPs also makes contact with at least four other promoters in a 1 Mb region, so the 9p21.3 locus could have multiple modes by which it influences risk of MI.

Related research is also exploring arterial features that may relate to plaque formation and rupture. These features include artery calcification and ventricular fibrillation following MI. Recurrent MI is also an interesting phenotype that may identify the highest-risk patients. Independent of GWAS, physiologists have identified several other pathways or mechanisms that likely influence arterial health, including inflammatory cytokines, which destabilize plaque; nitric oxide synthesis, which relaxes smooth muscle and increases blood flow; shear stress within the arterial wall, which alters gene expression, thereby contributing to likelihood of rupture; and the possible contribution of cardiac progenitor cells to repair of arterial walls. Kim et al. (2014) identified a gene expression profile in peripheral blood that was predictive of impending cardiovascular death

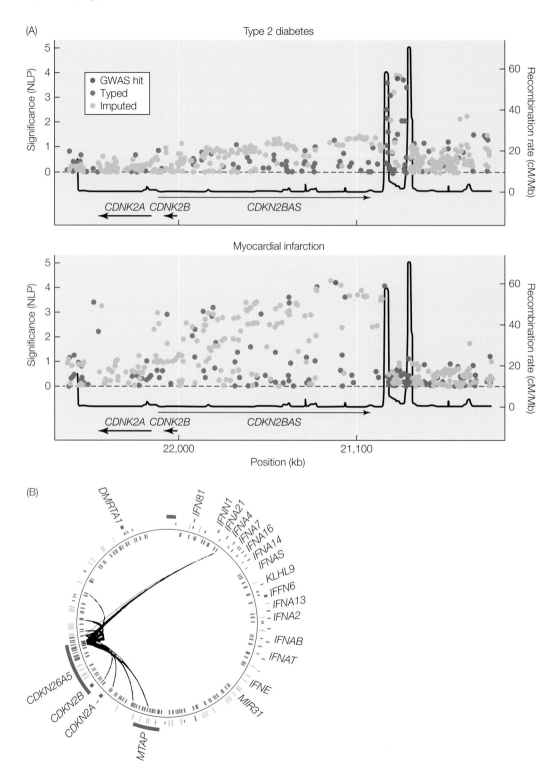

◀ **Figure 13.5 Fine Mapping of the 9p21.3 Locus.** (A) Regional Manhattan plots of the 100 kb gene-poor region that is rich in enhancers and is associated with both type 2 diabetes (T2D) and myocardial infarction (MI), showing how two adjacent intervals are responsible for the risk of the two diseases. Note that due to high LD, the MI association extends across hundreds of possible SNPs. (B) Long-range chromatin interactions in vascular endothelial HUVAC cells extend over 1.5 Mb in the vicinity of the association peak, which lies within the antisense RNA at the *CDKN2B* locus, but cross-link to promoters of four known genes (*CDKN2B*, *CDKN2A*, *MTAP*, and *IFNA21*) and several other sites (red tick marks, inner circle), any of which could explain the association with MI. On the outside of the circle, blue ticks are T2D associated, red ticks are coronary artery disease associated, and yellow ticks are other enhancers. (A, after Shea et al. 2011; B, after Harismendy et al. 2011.)

within 5 years for almost 10% of a small cohort of around 300 CAD patients, but the underlying mechanism is as yet unclear, and independent replication is needed.

Mendelian Randomization

It has recently been recognized that genetics provides an opportunity to evaluate the causal relationships between physiological risk factors and disease phenotypes. Observational epidemiology typically establishes only that a factor, such as an increased level of homocysteine in serum, is correlated with an outcome, such as elevated incidence of coronary disease. Biochemical arguments might be used to bolster the correlation, in this case, the knowledge that dietary folate facilitates conversion of homocysteine to methionine by virtue of a reaction involving the MTHFR enzyme. However, it is always possible that the correlated factors are both downstream consequences of the same underlying agent without there being a causal relationship. That is, if something causes both A and B, then A and B may be correlated, but this does not demonstrate that A causes B. If, though, there is a way to perturb A independently of other sources of correlation, then it is possible to demonstrate a likely causal relationship. For example, an amino acid polymorphism in the MTHFR protein is associated with reduced enzyme activity, mimicking low dietary folate intake (Jacques et al. 1996). Individuals carrying this polymorphism show the same degree of elevated coronary disease risk on average, relative to that of homozygotes for the wild-type allele, as individuals who naturally have the equivalent difference in homocysteine levels within either class of genotype alone. That is, the effect of homocysteine on heart disease is fully explained by the genotype. This would not be the case in the absence of a causal relationship because the genotype would be random with respect to normal dietary or other influences on folate levels.

This mode of analysis is known as **Mendelian randomization** (Davey Smith and Ebrahim 2003). It is formally an example of the statistical method of instrumental variable analysis, in which the genotype is the instrument. **Figure 13.6** illustrates the principle in relation to a case in which it falsifies the previously inferred causal relationship of HDL to protection against heart disease.

(A)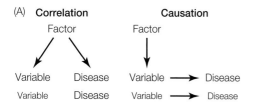

(B)

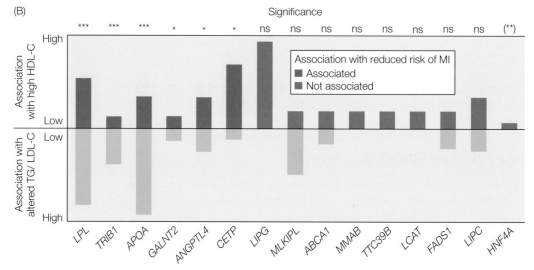

Genes with variants associated with increased HDL-C

Figure 13.6 Mendelian Randomization. (A) Whenever a variable is associated with a disease, it can either be because both the variable and the disease are correlated with another factor (left), or because the variable causes the disease (right). Perturbation of the variable (e.g., a genetic polymorphism) changes the disease prevalence in the expected direction if the relationship is causal, but has no effect if it is simple correlation. (B) Of 15 SNPs that are associated with elevated HDL cholesterol, only 6 are also associated with risk of myocardial infarction (MI) in the expected direction (*HNF4A* is in the opposite direction; ***, $p < 10^{-5}$; *, $0.01 < p < 0.05$; ns, nonsignificant). All of those 6 SNPs are also associated with reduced triglycerides (or LDL cholesterol for *CETP*). *MLXIPL* is also associated with triglycerides (TG) and LDL, but in the opposite direction. These data are consistent with a causal effect of the variants on TG/LDL and subsequently risk of MI (blue), but simple correlation explains the association of HDL with MI in the general population (red). (B, data from Voight et al. 2012.)

HDL cholesterol is so widely regarded as a protective factor against heart disease that it is often called "good cholesterol." Surprisingly, though, an extensive Mendelian randomization study published in 2012 has questioned the causal relationship, at least with respect to myocardial infarction (Voight et al. 2012). First, as a positive control, the authors showed in a meta-analysis of 21,000 cases and 95,000 controls that 9 out of 10 SNPs that were associated with elevated LDL cholesterol were also associated with increased risk of MI, at least at nominal levels (the tenth SNP had the smallest effect and was likely a false negative). In parallel, only 6 of 15 SNPs associated with HDL also decreased MI risk, while one actually increased the risk significantly. It turns out that all of those 6 SNPs are also associated with either triglycerides (TG) or LDL, which could explain the effect. By contrast, the amino acid substitution Asn396Ser in the endothelial lipase LIPG had the strongest association with HDL, and the genotype was not correlated with any other potential mediators of CAD (LDL, triglycerides, BMI, blood pressure), yet it showed absolutely no association with risk of MI. Furthermore, a polygenic risk score derived from 14 similar SNPs associated only with HDL was not associated with risk of MI, whereas a corresponding LDL polygenic risk score was very strongly so. Formally, the Mendelian randomization analysis contrasts the magnitude of the genotypic effect with the magnitude of the continuous variable effect. Whereas a 1 standard deviation rise in HDL correlates with an odds ratio of 0.62 for MI from epidemiology, the odds ratio for a 1 standard deviation increase in HDL due to the genotypes was not significantly different from 1. The clinically important inference to be drawn is that it is unlikely that interventions designed specifically to increase HDL will lower MI risk. Indeed, this conclusion is consistent with observations of the effect of hormone replacement therapy, which raises HDL but does not reduce MI incidence.

For Mendelian randomization to be informative, a number of conditions must be met. Most importantly, the analysis must be sufficiently well powered to detect associations between the genotype and the disease, which does not necessarily follow even where associations between the genotype, the intermediate phenotype, and the disease have been established. Second, the genotype cannot be associated with any other potential intermediate phenotypes that could cause the association—though these phenotypes may be unknown and lead to false positive conclusions, as pointed out by Clarke et al. (2012) in a meta-analysis of the MTHFR example. Third, the genotype should have a plausible connection to the intermediate phenotype, preferably lying in the gene that directly causes it. Similar analyses have also been used to question whether C-reactive protein (CRP), a known marker of inflammation, is causal in promoting coronary disease. Plenge et al. (2013) have further called for expansion of the approach as a general mode of validating therapeutic targets as an adjunct to randomized clinical trials.

Arrhythmia

The second major source of adult heart disease is cardiac arrhythmia: an aberrant heartbeat (which may be too fast, too slow, or more unstable than usual). Research in the 1990s and 2000s established that multiple arrhythmia syndromes trace to abnormal function of cardiac ion channels, which is very often due to mutations of major effect that most often have dominant, but sometimes recessive, modes of inheritance (Priori et al. 1999; Wilde and Bezzina 2005). Such syndromes are thus also called cardiac "channelopathies." Channelopathies affecting noncardiac tissue also give rise to other classes of disease, including epilepsy, myotonia, and pain insensitivity.

Cardiac arrhythmias are diagnosed by abnormal electrocardiograms (ECGs), though it is also recognized that the resting heart in some patients will appear normal and that many instances of sudden death may be due to a stress-induced abnormality that was never diagnosed. A human heartbeat consists of five waves, P through T. Abnormality typically occurs in the length of the QT interval, which may be long or short, reflecting characteristic effects on action potentials depending on which channel is affected, whether the current is inward or outward flowing, and which layer of the ventricular cardiac wall is affected.

The most common form of arrhythmia is long-QT syndrome, which affects approximately 1 in 7000 people. The dominant form is also known as Romano-Ward syndrome, whereas a recessive form that is often accompanied by deafness is known as Jervell and Lange-Nielsen syndrome (JLNS). Thirteen causal loci have now been recognized from family-based linkage mapping. These loci provide an alternative classification scheme for long-QT arrhythmias, since the pharmacology and symptomatology of these arrhythmias differs according to which channel is disrupted (**Table 13.1**). Six of the loci affect potassium channels, two affect sodium

TABLE 13.1 Classification scheme for long-QT syndrome subtypes based on 13 causal loci

Locus	Gene	Function	Property
LQT1	KCNQ1/KvLQT1	Potassium channel	Both Romano-Ward syndrome and Jervell and Lange-Nielsen syndrome; >30% of cases
LQT2	hERG/KCNH2	Potassium channel	Prone to drug-induced long-QT syndrome; >25%
LQT3	SCN5A	Sodium channel	Also causes Brugada syndrome
LQT4	Ankyrin	Muscle protein	
LQT5	KCNE1/MinK	Potassium channel with LQT1 proteins	
LQT6	KCNE2/MiRP1	Potassium channel with LQT2 proteins	
LQT7	KCNJ2	Potassium channel	Andersen-Tarwil syndrome
LQT8	CACNA1c	Calcium channel	Timothy's syndrome
LQT9	Caveolin 3	Muscle protein	Similar to LQT3, affects sodium current
LQT10	SCN4B	Sodium channel	
LQT11	AKAP9	Muscle protein	
LQT12	SNTA1	Muscle protein	
LQT13	KCNJ5/GIRK4	Potassium channel	

channels, and one a calcium channel, while four less commonly mutated loci affect other aspects of cardiac muscle. *LQT1* and *LQT2* account for more than half of all known cases of long-QT arrhythmias.

Each of the more common LQT genes is known to harbor a variety of mutations that have variable penetrance and expressivity, reflecting both intrinsic differences in the severity of the effect of the amino acid substitution and the modifier effects of other genes (Splawski et al. 2000). In addition, the environment can affect a patient's disease: it is thought that LQT2 and LQT3 proteins are prone to adverse interaction with beta or sodium blocker drugs. The two major potassium channels, KvLQT1 and hERG, act in a dominant negative fashion because the mutations are loss-of-function variants but reduce total cellular protein activity by more than half by disrupting the complexing of the heterotetrameric protein with the normal protein encoded by the other allele. The SCN5A mutations in *LQT3*, by contrast, are dominant gain-of-function variants that prolong the sodium current. People with *LQT1* mutations are more likely to have exercise-induced cardiac arrest, whereas in those with *LQT3* mutations, episodes typically occur during sleep or rest. This heterogeneity of phenotypes is something of an impediment to clinical intervention based solely on knowledge of the genotype, but once a mutation is detected, all family members can be screened for it. Screening offers the potential benefits of diagnosis of undetected cases as well as consideration of treatment options.

Short-QT syndrome is very rare, but interestingly, seems to be caused by gain-of-function mutations in either the LQT1 or LQT2 potassium channels. It is a more severe disease, with a high rate of sudden death. Similarly, loss-of-function mutations of LQT3 give rise to the abnormality of the ST interval that defines Brugada syndrome, a disorder that is more common in Southeast Asia, where it contributes to sudden unexplained death syndrome. Other disorders of cardiac rhythmicity that have less clearly defined genetics include idiopathic sick sinus syndrome (SSS), catecholaminergic polymorphic ventricular tachycardia, and familial atrial fibrillation.

Genome-wide association studies have also been pursued to investigate the contribution of common variants to nonfamilial risk of atrial fibrillation and sudden cardiac death. Analysis of resting heart rate, which is a predictor of heart disease at the extremes of the distribution, in 180,000 Europeans, identified 21 loci spread throughout the genome (den Hoed et al. 2013). Individuals in the upper and lower deciles of the polygenic risk score derived from these loci differ on average by more than 4 heartbeats per minute. Three of the heart rate–increasing alleles in *SLC35F1*, *LINC00477*, and *NKX2-5* are also associated with increased risk of atrial fibrillation, while two in *GJA1* and *HCN4* are protective against it. Collectively, the 21 loci also mildly predict SSS and the need for a pacemaker, but they are not associated with blood pressure, CAD, or MI. The causal mechanisms of most of the variants have yet to be defined, and in most cases the affected gene can only be speculated upon, but there is strong pathway enrichment for roles in muscle contraction and cardiomyopathy as well as in heart development. Mutations of *NKX2-5* also lead to atrial septal defects; interestingly, this gene is the human homolog of the critical *Drosophila* heart-determining gene *tinman*. In fact, knockdown of 20 of 31 candidate genes in the vicinity of the GWAS signals in either *Drosophila* or zebrafish resulted in specific modulation of heart rate. Given the strong conservation of both the development and physiology of the heart across animal species, it can be expected that more of the hundreds of genes identified in flies and fish may contribute to susceptibility to abnormal cardiac function (Bodmer and Venkatesh 1998).

Cardiomyopathy

Cardiomyopathy is progressive deterioration of the heart muscle, ultimately leading to heart failure. It can have extrinsic causes, such as ischemia due to prolonged coronary disease, or it may be due to intrinsic factors that induce primary pathology of the myocardium itself, whether environmental agents, such as viral infection or alcohol toxicity, or genetic factors. Idiopathic cardiomyopathy has no known cause, but an increasing proportion of the disease can be traced to mutations of large effect affecting the sarcomeres, cytoskeleton, or desmosomes of heart fibers (**Figure 13.7**).

Hypertrophic cardiomyopathy (HCM), in which the heart muscle thickens, is a leading cause of sudden death in young adults, notably elite athletes, in whom the disease is often asymptomatic. Thirteen genes encoding

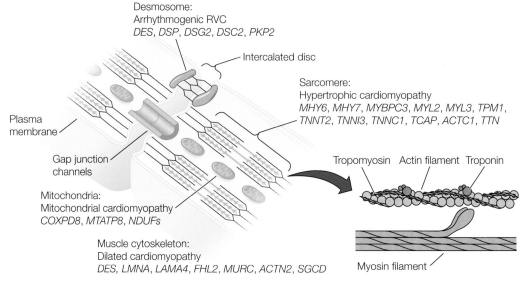

Desmosome:
Arrhythmogenic RVC
DES, DSP, DSG2, DSC2, PKP2

Intercalated disc

Sarcomere:
Hypertrophic cardiomyopathy
MHY6, MHY7, MYBPC3, MYL2, MYL3, TPM1, TNNT2, TNNI3, TNNC1, TCAP, ACTC1, TTN

Plasma membrane

Gap junction channels

Tropomyosin   Actin filament   Troponin

Mitochondria:
Mitochondrial cardiomyopathy
COXPD8, MTATP8, NDUFs

Muscle cytoskeleton:
Dilated cardiomyopathy
DES, LMNA, LAMA4, FHL2, MURC, ACTN2, SGCD

Myosin filament

Figure 13.7 Cardiomyopathies. Genes that are involved in cardiomyopathy fall into four main groups: those that affect the function of the sarcomeres (myosins, actins, and troponins), the muscle cytoskeleton, components of desmosomes, and mitochondria. RVC, right ventricular cardiomyopathy.

sarcomeric proteins are implicated in over half the cases (Cahill et al. 2013). Sarcomeres are the basic building units of muscle fibers, in which the actin and myosin filaments slide by one another during contraction and relaxation. The most common mutations have a dominant mode of familial inheritance and affect either the β myosin chain, encoded by *MHY7* or *MHY6*, or the cardiac myosin binding protein C, encoded by *MYBPC3*. Three other mutations affect other myosin chains (*MYL2*, *MYL3*, and *TPM1*), four affect troponins (*TNNT2*, *TNNI3*, *TNNC1*, and *TCAP*), and two affect cardiac actin (*ACTC1*) and titin (*TTN*, muscle's giant spring), while *CSRP3* may have a regulatory rather than a structural role. Hypertrophy of the left ventricle leads to obstructed blood flow, angina, and shortness of breath (dsypnea), eventually causing ischemia and possibly arrhythmia. The first-line treatment is beta blockers that slow down the heart rate, but heart surgery or transplantation is required in severe cases.

In dilated cardiomyopathy (DCM), the heart enlarges and weakens, losing its capacity to pump blood as systolic blood pressure drops. One-fourth of cases of DCM are also attributable to familial dominant inheritance that traces to genes that encode muscle proteins, in this case affecting connective scaffold proteins, such as desmin (*DES*) and laminin (*LMNA* and *LAMA4*), as well as *FHL2*, *MURC*, *ACTN2* and *SGCD* (Parvari and Levitas 2012). Aberrant calcium and iron regulation have also been implicated, and OMIM has more than 40 entries associated primarily with the term "dilated cardiomyopathy." It is possible that many idiopathic cases are due to unrecognized mutations in these or other components of the muscle cytoskeleton, many of which may arise de novo in affected individuals, given the large mutational target size.

Arrhythmogenic right ventricular cardiomyopathy (ARVC) is due to defective function of the desmosomes, which are the adhesive junctions that link muscle cells. Incidence is only around 1 in 10,000, once again often tracing to familial inheritance of dominant mutations in up to a dozen genes. The affected genes include *DES*, desmoplakin (*DSP*), desmoglein (*DSG2*), desmocollin (*DSC2*), and plakophilin (*PKP2*), and some regulatory genes are also implicated.

Mitochondrial myopathy is also a major source of disease, particularly for children born with some severe heart conditions that are usually fatal early in life. The heart (along with the brain) is particularly sensitive to reduced energy production by respiration in the mitochondria; thus, mutations affecting mitochondrial electron transport are well-known sources of cardiomyopathy. Examples include mutations in *COXPD8*, *MTATP8*, and at least 15 nuclear mitochondrial complex I deficiencies affecting the family of NADH-ubiquinone oxidoreductases encoded by the *NDUF* genes.

Congenital Heart Defects

Congenital heart defects (CHDs) are the most common survivable birth defects. With improved imaging, the documented incidence has risen to 0.8% of live births, up from 0.5% a few decades ago, when only the most severe defects were detected. A wide range of defects are included, ranging from general heterotaxy (abnormal positioning or left-right symmetry, including situs invertus), to atrioventricular canal defects (AVCs, also known as a hole in the heart, since the walls between the chambers are incomplete) and atrial septal defects (ASDs, in which the left and right atria are not properly separated), to hypoplastic left heart syndrome (HLHS, in which the left ventricle is underdeveloped), various types of valvular defects, and conditions involving multiple abnormalities, such as tetralogy of Fallot (**Figure 13.8A**). Many of these defects can be corrected by surgery soon after birth or, in rare cases, in utero.

CHDs are characterized by both marked genetic heterogeneity and highly variable expressivity (Yuan et al. 2012). In other words, the mapping of genotype onto phenotype is not simple, even though it is clear that many instances are largely attributable to a single mutation. Several

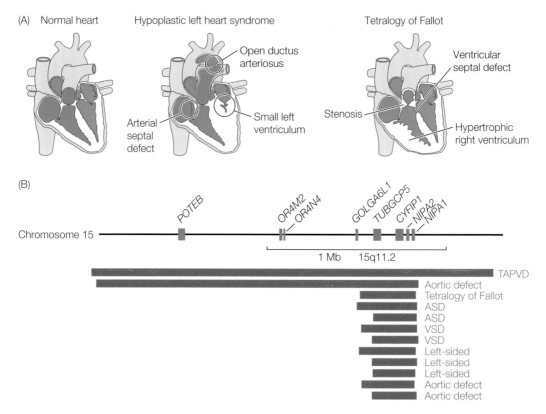

Figure 13.8 Congenital Heart Defects.   (A) Congenital heart defects include a number of phenotypes, such as atrial and ventricular septal defects (ASD and VSD, respectively), tetralogy of Fallot, and other complex heart syndromes, which are often restricted to one side of the heart. (B) These defects are often due to de novo mutations or copy number variations, such as deletions covering the indicated intervals in the 1 Mb 15q11.2 region. (See Soemedi et al. 2012 for details.)

dozen genes have been identified, and whole-genome profiling suggests that there are likely to be hundreds, if not thousands, of contributing genes. To date, three major types of gene function have been implicated. The first type encompasses genes involved in specification of cardiac tissue and developmental patterning, which includes the homeobox transcription factor *NKX5.2*, a half-dozen T-box transcription factors (*TBX1*, *TBX2*, etc.), and components of several of the major growth factor signaling pathways (Wnt, Hedgehog, TGF-β, Notch). The functions of these genes are being well defined by model organism studies in *Drosophila*, zebrafish, chick, and mouse. The second type encompasses genes involved in ciliopathies, defects in the biogenesis of the cilia—small hairlike projections on cells—which are now recognized to disrupt the establishment of left-right asymmetry as the heart chambers first form. Examples include *DNAH5* and *DNAH11*, which encode the dynein heavy chains, and genes involved in the syndromic NPHP ciliopathies. Third, genes involved in chromatin remodeling, particularly through histone H3K4 methylation, also cause CHD. Mutations in *MLL2*, a histone methyltransferase gene, give rise to Kabuki syndrome, and mutations in *CHD7*, a chromodomain-DNA helicase gene, give rise to another craniofacial/cardiac syndrome known as CHARGE.

The nonspecific nature of chromatin remodeling adds to the perception that heart development is highly sensitive to changes in the copy numbers of many genes. This idea first gained traction with the observation that half of all individuals with trisomy 21 (Down syndrome) have heart defects, and it was reinforced by the discovery of multiple common CNVs associated with certain CHDs. The best known of these is DiGeorge, or velo-cardio-facial, syndrome, which is due to a deletion of one copy of the 22q11.2 interval. This condition is modeled in mice by hemizygosity for *TBX1*. Interestingly, the phenotypic heterogeneity among littermates is quite large, indicating that other factors influence the pathology. That pathology also has a large sporadic component, which may be attributed to the ability of developmental processes to correct defects naturally. Two groups of researchers have followed up by assessing the burden of rare, including de novo, CNVs in genetically undiagnosed cases of CHD, and have found a significant—almost twofold—excess relative to unaffected controls (Fakhro et al. 2011; Soemedi et al. 2012). Almost 8% of cases have a rare deletion that affects a gene, compared with just over 4% of controls (**Figure 13.8B**), whereas duplications or deletions that do not cover a gene are not so enriched. Rare CNVs may explain as many as 5% of congenital heart defects and an even larger proportion of syndromic cases.

As with several other types of birth defects, whole-exome sequencing of trios of the affected infant and both biological parents is now being used in the attempt to track down causal mutations. The first large-scale study (Zaidi et al. 2013) compared 362 individuals with nonfamilial CHDs with 264 controls, finding an average of 0.87 de novo coding mutations per individual regardless of disease status. However, when the researchers focused on a set of 4169 genes that are in the upper quartile for expression in the developing heart, they found a twofold enrichment of de novo mutations in individuals with heart defects. Furthermore, missense mutations in these genes at conserved positions showed even higher, 7.5-fold, enrichment, strongly suggesting that new mutations in genes involved in heart development are likely to be major contributors to individual cases of heart defects. These results alone do not establish the cause of individual cases, since 10% of controls, compared with 22% of cases, have mutations in the heart-expressed genes. In each case, more research is required to establish whether a mutation is causal. It is noteworthy that in several instances of new mutations in a gene known to be involved in heart defects, the observed abnormalities were quite different from those typically observed in mutations of the same gene. Furthermore, 90% of cases remain unexplained, presumably resulting from mutations that are inherited, are polygenic and/or influenced by the environment, or affect gene expression rather than protein sequence. Nevertheless, clinical sequencing of newborn infants, or even of fetal DNA, is now feasible and is set to transform the genetic diagnosis of tens of thousands of cases of congenital heart defects each year.

Summary

1. There are essentially four major types of cardiovascular disease: coronary artery disease, arrhythmia, cardiac muscle failure, and pediatric developmental heart defects.

2. Hypertension, or high blood pressure, affects a substantial fraction of the world's population and is a major risk factor for coronary disease.

3. Only 28 genes associated with blood pressure have been identified from studies of over 200,000 people. An aggregate risk score from these variants distinguishes the top and bottom quintiles by approximately the same amount as typical antihypertensive drugs.

4. Platelet count and volume, which are affected by hundreds of genes, may influence risk of heart attack through the effect of platelets on blood clotting and vascular repair.

5. Oxygen-carrying capacity is tightly correlated with red blood cell count and volume, which are influenced by 75 loci found by GWAS, many of which have been confirmed in animal models.

6. Arteriosclerosis is thickening of the arteries largely due to lipid deposition associated with metabolic disease; it progresses to atherosclerosis when plaques become established on artery walls.

7. GWAS for coronary artery disease (CAD) have uncovered 23 loci, while an additional few loci were found by association with a more specific phenotype, carotid intima thickness.

8. Mendelian randomization is a method for evaluating the causality of an intermediate phenotype by conditioning the effect on genotypes that jointly influence that phenotype and the disease. It casts doubt on the role of HDL cholesterol as a protective factor against myocardial infarction.

9. Cardiac arrhythmia is usually due to channelopathies; specifically, mutations affecting cardiac potassium, sodium, or calcium channels. The most prevalent type of arrhythmia is long-QT syndrome.

10. Cardiomyopathy has multiple causes, but familial cases generally trace to dominant mutations affecting proteins of the sarcomeres, muscle cytoskeleton, or desmosomes.

11. Several hundred genes can mutate to yield congenital heart defects that used to be lethal for newborn children, but can now often be surgically corrected.

12. Gene dosage appears to be tightly controlled during normal heart development as rare and common large deletions as well as de novo mutations are observed in up to 20% of cases

References

Agarwal, A., Williams, G. H., and Fisher, N. D. 2005. Genetics of human hypertension. *Trends Endocrinol. Metab.* 16: 127–133.

Armani, C., Botto, N., and Andreassi, M. G. 2011. Susceptibility genes in hypertension. *Curr. Pharm. Des.* 17: 2973–2986.

Bis, J. C., Kavousi, M., Franceschini, N., Isaacs, A., Abecasis, G. R., et al. 2011. Meta-analysis of genome-wide association studies from the CHARGE consortium identifies common variants associated with carotid intima media thickness and plaque *Nat. Genet.* 43: 940–947

Bodmer, R. and Venkatesh, T. V. 1998. Heart development in *Drosophila* and Vertebrates: Conservation of molecular mechanisms. *Dev. Genet.* 22: 181–188.

Cahill, T. J., Ashrafian, H., and Watkins H. 2013. Genetic cardiomyopathies causing heart failure. *Circ. Res.* 113: 660–675.

Clarke, R., Bennett, D. A., Parish, S., Verhoef, P., Dötsch-Klerk, M., et al. 2012. Homocysteine and coronary heart disease: Meta-analysis of MTHFR case-control studies, avoiding publication bias. *PLoS Med.* 9: 1001177.

Davey Smith, G. and Ebrahim, S. 2003. "Mendelian randomization": Can genetic epidemiology contribute to understanding environmental determinants of disease? *Int. J. Epidemiol.* 32: 1–22.

den Hoed, M., Eijgelsheim, M., Esko, T., Brundel, B. J., Peal, D. S., et al. 2013. Identification of heart rate-associated loci and their effects on cardiac conduction and rhythm disorders. *Nat. Genet.* 45: 621–631.

Fakhro, K. A., Choi, M., Ware, S. M., Belmont, J. W., Towbin, J. A., et al. 2011. Rare copy number variations in congenital heart disease patients identify unique genes in left-right patterning. *Proc. Natl. Acad. Sci. U.S.A.* 108: 2915–2920.

Gieger, C., Radhakrishnan, A., Cvejic, A., Tang, W., Porcu, E., et al. 2011. New gene functions in megakaryopoiesis and platelet formation. *Nature* 480: 201–208.

Harismendy, O., Notani, D., Song, X., Rahim, N. G., Tanasa, G., et al. 2011. 9p21 DNA variants associated with coronary artery disease impair IFNγ signaling response. *Nature* 470: 264–268.

International Consortium for Blood Pressure Genome-Wide Association Studies. 2011. Genetic variants in novel pathways influence blood pressure and cardiovascular disease risk. *Nature* 478: 103–109.

Jacques, P. F., Bostom, A. G., Williams, R. R., Ellison, R. C., Eckfeldt, J. H., et al. 1996. Relation between folate status, a common mutation in methylenetetrahydrofolate reductase, and plasma homocysteine concentrations. *Circulation* 93: 7–9.

Kathiresan, S. and Srivastava, D. 2012. Genetics of human cardiovascular disease. *Cell* 148: 1242–1247.

Kim, J., Ghasemzadeh, N., Eapen, D. J., Chung, N. C., Storey, J. D., et al. 2014. Gene expression profiles associated with acute myocardial infarction and risk of cardiovascular death. *Genome Med.* 6: 40.

Lettre, G., Palmer, C. D., Young, T., Ejebe, K. G., Allayee, H., et al. 2011. Genome-wide association study of coronary heart disease and its risk factors in 8,090 African Americans: The NHLBI CARe Project. *PLoS Genet.* 7: e1001300.

Lieb, W., Jansen, H., Loley, C., Pencina, M. J., Nelson, C. P., et al. 2013. Genetic predisposition to higher blood pressure increases coronary artery disease risk. *Hypertension* 61: 995–1001.

Markus, H. S. 2012. Stroke genetics. *Hum. Mol. Genet.* 20: R123–R131.

Myocardial Infarction Genetics Consortium (MIGC). 2009. Genome-wide association of early-onset myocardial infarction with single nucleotide polymorphisms and copy number variants. *Nat. Genet.* 41: 334–341.

Parvari, R. and Levitas, A. 2012. The mutations associated with dilated cardiomyopathy. *Biochem. Res. Intl.* 2012: 639250.

Plenge, R. M., Scolnick, E. M., and Altshuler, D. 2013. Validating therapeutic targets through human genetics. *Nat. Rev. Drug Discov.* 12: 581–594.

Priori, S. G., Barhanin, J., Hauer, R. N. W., Haverkamp, W., Jongsma, H. J., et al. 1999. Genetic and molecular basis of cardiac arrhythmias: Impact on clinical management Parts I and II. *Circulation* 99: 518–528.

Sabater-Lleal, M., Huang, J., Chasman, D., Naitza, S., Dehghan, A., et al. 2013. Multiethnic meta-analysis of genome-wide association studies in >100 000 subjects identifies 23 fibrinogen-associated loci but no strong evidence of a causal association between circulating fibrinogen and cardiovascular disease. *Circulation* 128: 1310–1324.

Schunkert, H., König, I. R., Kathiresan, S., Reilly, M. P., Assimes, T. L., et al. 2011. Large-scale association analysis identifies 13 new susceptibility loci for coronary artery disease. *Nat. Genet.* 43: 333–338.

Shea, J., Agarwala, V., Philippakis, A. A., Maguire, J., Banks, E., et al. 2011. Comparing strategies to fine-map the association of common SNPs at chromosome 9p21 with type 2 diabetes and myocardial infarction. *Nat. Genet.* 43: 801–805.

Soemedi, R., Wilson, I. J., Bentham, J., Darlay, R., Töpf, A., et al. 2012. Contribution of global rare copy-number variants to the risk of sporadic congenital heart disease. *Am. J. Hum. Genet.* 91: 489–501.

Splawski, I., Shen, J., Timothy, K. W., Lehmann, M. H., Priori, S., et al. 2000. Spectrum of mutations in Long-QT Syndrome genes *KVLQT1*, *HERG*, *SCN5A*, *KCNE1*, and *KCNE2*. *Circulation* 102: 1178–1185.

Van der Harst, P., Zhang, W., Leach, I. M., Rendon, A., Verweij, M., et al. 2012. Seventy-five genetic loci influencing the human red blood cell. *Nature* 492: 369–375.

Voight, B. F., Peloso, G. M., Orho-Melander, M., Frikke-Schmidt, R., Barbalic, M., et al. 2012. Plasma HDL cholesterol and risk of myocardial infarction: A Mendelian randomisation study. *Lancet* 380: 572–580.

Wilde, A. A. M. and Bezzina, C. R. 2005. Genetics of cardiac arrhythmias. *Heart* 91: 1352–1358.

Yuan, S., Zaidi, S., and Brueckner M. 2012. Congenital heart disease: Emerging themes linking genetics and development. *Curr. Op. Genet. Dev.* 23: 352–359.

Zaidi, S., Choi, M., Wakimoto, H., Ma, L., Jiang, J., et al. 2013. De novo mutations in histone-modifying genes in congenital heart disease. *Nature* 448: 220–223.

14

Cancer Genetics

C ancer genetics is at the leading edge of the adoption of genomic methods for personalized medical care. Whole-genome sequencing and gene expression profiling are becoming routine components of the identification of pharmacological targets and have revolutionized our understanding of cancer progression. Surgery and radiation therapy remain the primary treatments, but genomic profiling can provide insight into survival probabilities and assist in designing further treatment options. Rather than treating cancers solely according to the cell type of origin, oncologists are now directing chemotherapy against specific mutations that indicate the likelihood that the tumor will respond to a particular drug, even one that may have initially been approved for another cancer type. This chapter, however, will only touch on cancer therapy; instead, it aims to provide an overview of the concepts and methods used in genomic analyses across all cancer types while also providing some insights that are specific to each of seven tissues: breast, prostate, lung, colon, blood, skin, and brain. There is simply insufficient space to consider the thousands of other types of carcinomas, many of which have unique genomic attributes that are being intensely studied. Cancer incidence and mortality rates in the United States are listed in **Table 14.1**.

Unlike the other diseases considered in this book, cancer is not a heritable disease, in the sense that each case has its own spectrum of mutations that arise in the DNA of somatic cells. These mutations are not transmitted from parents or to children. However, cancer predisposition is certainly heritable and has a clear basis in the distributions of rare and common variants. These variants include inherited mutations in cancer-driving genes themselves (that is, in genes that promote cancer when mutated) and in modifier genes that affect the likelihood that cancer will proliferate. It is highly probable that genetic variation also affects the ability of a tumor to evade the immune system or to metastasize (migrate to another location in the body).

All cancers probably start with a single somatic mutation. Some may go through a long period of relatively benign slow and undetected growth before growing into a visible tumor or beginning to spread. Metastasis may

TABLE 14.1   Cancer incidence and deaths in the United States

Cancer	New Cases	Deaths	Incidence[a]	Death rate[a]	Lifetime risk
Breast (women only)	230,000	40,000	124	23	13%
Prostate (men only)	240,000	30,000	152	23	15%
Lung	230,000	160,000	61	50	7%
Colon	143,000	50,000	45	16	5%
Leukemia	48,500	24,000	13	7	1%
Melanoma	77,000	9,500	21	3	2%
Brain	23,000	14,000	7	4	<1%
Pancreatic	45,000	38,000	12	11	2%
Liver	30,000	22,000	8	6	1%
Ovarian	22,000	14,000	12	8	1%
Stomach	21,500	11,000	8	4	1%

Source: seer.cancer.gov/statfacts; data represents estimates for 2013.
Note: Incidence and death rates are age-adjusted, and lifetime risk is based mostly on 2008–2010 data.
[a]Number per 100,000 individuals.

require the accumulation of secondary mutations. By the time an advanced tumor is detected, the DNA may include hundreds or thousands of mutations and chromosomal rearrangements. We start with a quick introduction to some of the key concepts of cancer genetics.

Conceptual Foundation

All cancer-promoting genes have normal roles in development and physiology, but have the potential to mutate to a form that is pathogenic. The regulation of cell division is so complex—involving well over 10% of the genes in the genome—that there is plenty of scope for mutations to lead to the uncontrolled cell division and growth that defines cancer (Balmain et al. 2003). There are three major classes of cancer-promoting genes: oncogenes, tumor suppressors, and DNA repair genes (**Figure 14.1**). These genes can be thought of as the accelerators, brakes, and mechanics of cell division, respectively.

Proto-oncogenes are normal genes that encode proteins that sense the context of each cell and signal whether or not it is appropriate for it to enter a cycle of division. Some authors prefer to restrict the term "proto-oncogene" to cellular homologs of transforming genes that were historically identified in viruses, but I use it here to distinguish the normal copy from mutated versions that become active oncogenes. They may have roles as secreted growth factors, signal-transducing kinases, or transcription factors; examples

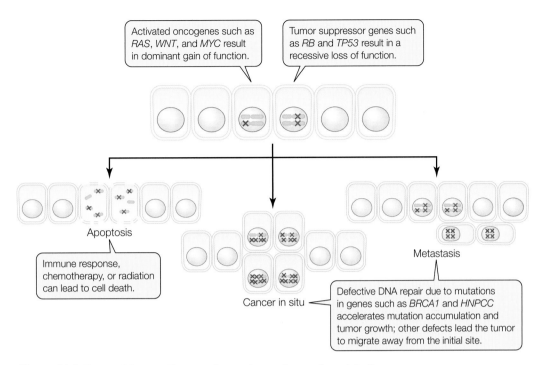

Figure 14.1 Types of Cancer Genes. Genes that are frequently mutated in cancer generally fall into three broad categories as described in the text. These are (1) oncogenes such as *RAS*, which typically have a dominant gain of function mode of action that causes cells to progress through cell division abnormally; (2) tumor suppressors such as *RB*, which typically have a recessive loss of function mode of action since the proteins normally inhibit cell division; and (3) DNA repair enzymes or regulators such as *BRCA1*, loss of which causes mutation accumulation. Other genes are involved in the progression to metastasis later in cancer progression. The figure shows three possible outcomes once a cell becomes cancerous: apoptosis or chemo/radiation induced cell death; tumor growth in situ associated with mutation accumulation; and metastasis, or spread of the tumor away from the initial clone.

are *WNT*, *RAS*, and *MYC*, respectively. A single activating mutation is sufficient to turn a proto-oncogene into a cancer-causing **oncogene**. Activating mutations may affect the protein sequence, activate overexpression, or in some cases create a new fusion protein (such as BCR-ABL, generated from a chromosomal translocation that produces the so-called Philadelphia chromosome). Since just a single activating mutation is necessary, activated oncogenes act in a dominant manner.

Tumor suppressor genes, by contrast, usually act in a recessive manner: one mutation can be tolerated because the normal copy on the other chromosome provides the needed protein activity. The function of these genes is to prevent cells from moving through the cell cycle, either because it is not warranted by the developmental program or because DNA damage has been detected and it is "unsafe to proceed" (in which case apoptosis is usually initiated). Typical examples include two genes that encode nuclear proteins, *TP53*, which is mutated in more than 50% of advanced breast, ovarian, colon, lung, and blood cancers, and *RB*, which is most often associated with retinoblastoma when one mutation is inherited, but can cause other tumor types as well. Many of the genes that facilitate metastasis can also be regarded as tumor suppressors because they regulate cell adhesion or contact inhibition, and because a single copy of the gene is sufficient for normal function.

DNA repair is an ongoing process in every cell of every individual every day, correcting point mutations and DNA breaks that are induced by radiation or toxic chemicals or just happen by chance. More than 30 **DNA repair genes** have been associated with cancer, including those with roles in the enzymology of mismatch repair (e.g., those involved in hereditary nonpolyposis colorectal cancer, HNPCC genes) and those that sense DNA damage and regulate DNA repair pathways and homologous recombination (e.g., *BRCA1* in breast cancer). Some of these genes are specific for particular cancer types, and others are generically found to be mutated across types. Loss of DNA repair capacity leads to the accelerated accumulation of mutations that occurs in cancer cells, allowing new mutations that promote faster cell division or migration to be selected as cancer progresses. Catastrophic loss of repair capacity results in chromothrypsis, in which thousands of rearrangements and deletions build up in the genome of an advanced cancer cell.

MicroRNAs (miRNAs) are increasingly implicated in cancer as well. Aberrant miRNA expression may drive some cancers, and normal miRNA activity can modify the effects of oncogenes by regulating expression of accessory proteins or by modulating therapeutic responses. Profiling of miRNA also provides a novel means for classifying the origin of the cancer (Lu et al. 2005). Similarly, long noncoding RNAs (lncRNAs) with diverse cellular functions are sometimes disrupted in cancer (Cheetham et al. 2013).

One of the important concepts in oncology is the **two-hit hypothesis**, formulated by Alfred Knudson in 1971. He noted that the age of onset of inherited forms of retinoblastoma is typically younger than in sporadic cases, and that affected children often develop independent tumors in both eyes. This observation implied that an inherited mutation alone is insufficient to promote the cancer, but that it may predispose the individual to cancer if a second mutation occurs. In sporadic cases, two separate mutations are required. While the hypothesis was initially proposed to explain the specific case of retinoblastoma, it is now widely accepted that most cancers require at least two, if not multiple, mutational "hits." These mutations do not have to affect the same gene, although the predisposition to certain familial cancers due to inheritance of a mutation of one copy usually involves a second mutation in the same tumor suppressor gene. **Box 14.1** discusses some of the concepts that are specific to **familial cancers**: those cancers that have such a strong genetic component that they can be said to run in families.

Loss of heterozygosity (**LOH**) refers to the situation in which a gene that is heterozygous in the germ line is represented by only one of the two alleles in cancer cells because those cells have become hemizygous—that

BOX 14.1
Familial Cancer Syndromes

Whereas 90% of cancer is sporadic, 10% of cases have a strong hereditary component attributable to a single Mendelian gene (Garber and Offit 2005). As a result, half a million Americans and Europeans each year are diagnosed with cancers that are also observed in family members. These types of cancer, known as **familial cancer susceptibility syndromes**, originate in most of the tissue types discussed in the text, including breast, colorectal, skin, blood, and neuroendocrine tissues. The National Cancer Institute provides information about many of these conditions at www.cancer.gov/cancertopics/pdq/genetics/overview/health-professional/page3, and a Dutch website with comprehensive information about hundreds of familial syndromes for genetic counselors and clinicians can be referenced at www.familialcancer-database.nl. Genetic testing is recommended where a strong family history of the cancer exists, an unambiguous genetic diagnosis is possible, and the results may guide effective medical care.

Since one-half of all men and one-third of all women will contract cancer at some point in their lives, recognition of familial cases is not straight-forward (Riegert-Johnson et al. 2009). The most important tool in identifying patients who have a familial cancer syndrome is a thorough family history in which incidence of the cancer is placed on the family pedigree, from which dominant or recessive transmission can be inferred. Familial clustering can also appear to occur if family members share an environmental exposure (such as smoking or proximity to a radiation source) or by chance in large families. The cancer is more likely to have a relatively simple genetic etiology if the primary tumors are the same type of malignancy, occur relatively early in life, or are comorbid with other congenital abnormalities. Some cancers are observed in conjunction with multiple benign tumors, such as the thousands of polyps in FAP, and only progress due to secondary mutations appearing somatically in individuals who inherit an oncogenic mutation in the *APC* gene.

It is important to note that inheritance of a predisposing mutation does not guarantee that someone will eventually get the cancer. As in other diseases attributed to rare alleles of major effect, penetrance varies quite widely and may be modified both by the genetic background and by lifestyle. Nevertheless, genetic tests are available for approximately 50 familial cancer syndromes, and genome sequencing will increase both the number of genes that can be screened for and the number of cancer susceptibility variants found segregating in different families.

As discussed in the text, familial breast cancer is usually associated with mutations in the *BRCA1* or *BRCA2* genes, and Lynch syndrome (hereditary nonpolyposis colorectal cancer) is associated with a set of mismatch repair genes. Four other well-known familial cancers are Cowden, Li-Fraumeni, and von Hippel-Lindau syndromes and retinoblastoma. In each of these cases, heritable susceptibility is due to transmission from a carrier parent with a mutation in a tumor suppressor gene; namely, *PTEN*, *TP53*, *VHL*, or *RB*, respectively. Cowden syndrome is characterized by tumorlike growths, called hamartomas, of the skin, thyroid, and intestine. These growths arise from a primary defect in dephosphorylation of cell cycle signaling proteins and typically give rise to cancers of the breast, thyroid, and endometrium. Li-Fraumeni syndrome is a similarly rare autosomal dominant condition resulting from defective DNA repair and apoptosis when the p53 protein is inactive, which gives rise to a 25-fold elevated risk of malignancies relative to the general population, particularly of the breast, brain, bone, blood, and adrenal gland. The causal gene in von Hippel-Lindau syndrome encodes an E3 ubiquitin ligase, a protein that regulates degradation of specific target proteins; inheritance predisposes individuals to central nervous system and pancreatic tumors, among others, with 90% penetrance by the age of 70, but a suite of other symptoms, including problems with balance, dizziness, headaches, and vision, is observed. Retinoblastoma is a malignant tumor of the eye with early onset, usually before the age of 3, that leads to deteriorating vision, one-third

(Continued)

BOX 14.1 (continued)

of the time manifesting in both eyes, but one-half of the cases are not obviously inherited. Tumor growth results from defective recruitment of E2F family transcription factors by the RB protein during the cell cycle. Hereditary leukemias are also well known, but do not fall into a single syndrome, although a handful of transcription factors involved in hematopoiesis have been identified from mutations in the *RUNX1*, *CEBPA*, *GATA2*, and *PAX5* genes (Benson and Horwitz 2006).

---

is to say, the copy of the gene on one of the two diploid chromosomes has been lost. Loss of heterozygosity is one type of **somatic copy number alteration (SCNA)**; the other is the gain of a third copy of the gene. If loss and gain occur in the same cell, the LOH is said to be copy-neutral. Somatic copy number alteration is similar to the copy number variation (CNV) observed at hundreds of locations throughout every individual's genome, except that it arises by de novo mutation only in the tumor cells.

Another important concept in contemporary cancer genetics is that of "driver" and "passenger" mutations. When cancer genomes are sequenced and compared with the genome of the individual's normal healthy cells, dozens to hundreds of novel variants may be seen. By comparing the prevalence of mutations in normal cells and tumor cells, then adjusting for attributes such as gene length and tolerance of the sequence to disruption, it is possible to identify genes that are "significantly disrupted (mutated) in cancer," which are inferred to be **driver mutations (Figure 14.2)**. Only a handful of the mutations in a tumor are actually likely to be causing the loss of control of cell division or the other attributes of cancer. Many are **passenger mutations**, simply going along for the ride, having arisen in cells that have gained a selective advantage due to a coincident driver mutation. Some authors also now recognize the notion of "backseat driver" mutations, which while not sufficient to promote cancer, nevertheless modify cell behavior. These distinctions are easier to define in theory than in practice, as it is difficult to ascertain what the effect of any given mutation may be, but recurrent observation of mutations in the same gene in many different cases of cancer is a good sign that a gene is capable of acting as a driver when mutated.

Widespread use of cancer genome sequencing is clarifying the incidence and nature of driver mutations. Reports often refer to four tiers of mutation. Tier 1 involves amino acid replacement substitutions in annotated exons, disrupted consensus splice sites, or mutant miRNAs and lncRNAs. Most of these variants may also be captured by whole-exome sequencing. Tier 2 begins to prioritize potential regulatory variants by focusing on highly conserved regions and DNA elements annotated by ENCODE data as potentially regulatory. Tier 3 refers to alterations in other non-repetitive intergenic regions, or in regions less likely to be in regulatory portions of the genome, while tier 4 refers to changes in the remainder of the genome.

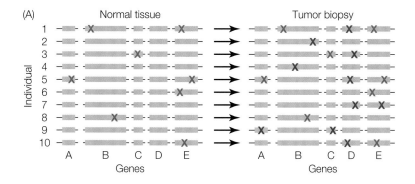

(B)

	Number of mutations in each gene				
Mutation type	**A**	**B**	**C**	**D**	**E**
Rare alleles in normal tissue	1	2	1	0	4
All rare alleles in tumor biopsy	2	4	2	5	5
Inferred de novo in tumor biopsy	1	2	1	5	1

The one de novo mutation in the tumor biopsy of this gene is likely a passenger mutation.

The higher mutation count in this gene is likely due to its greater length.

The excess of de novo mutations in the tumor biopsy means this gene is a candidate driver gene.

Figure 14.2 Detecting Genes Significantly Mutated in Cancer. (A) Suppose that five hypothetical genes, A–E, are sequenced in ten pairs of samples, each of which consists of normal healthy tissue and a tumor biopsy from the same individual. Rare mutations are indicated by red Xs if they appear in the normal tissue (as well as in the tumor) and green Xs if they appear only in the tumor. (B) Simply counting the total number of mutations observed in the tumors suggests that genes B, D, and E are enriched for rare mutations, but after accounting for mutations already present in the progenitor cells, it turns out that only gene D has an excess of de novo mutations. This gene is "significantly mutated in cancer" and is a candidate driver. Gene B has an extra tumor mutation relative to genes A, C, and E, but it is also more than twice as long. Gene E appears to be tolerant of rare mutations, so the new variant in individual 7 is likely to be a passenger riding along with the driver mutation in gene D.

There is a wide variety of software available for detecting and annotating cancer mutations (Wang et al. 2013a), including the popular VarScan and MuTect, as well as increasingly sophisticated tools that incorporate biological knowledge into the algorithm, such as MuSiC and DrGaP (Dees et al. 2012; Hua et al. 2013).

Cancer genetics research is big science, highly coordinated both in the United States, by the National Cancer Institute, and internationally. Three major initiatives have supported much of the research that the surveys in this chapter are based on. The Cancer Genome Anatomy Project (CGAP;

cgap.nci.nih.gov) profiled gene expression in cancer cells during the course of cancer progression. The Cancer Genetic Markers of Susceptibility (CGEMS) Project (dceg.cancer.gov) focuses on GWAS to detect polymorphic genetic risk factors. The Cancer Genome Atlas (TCGA; cancergenome.nih.gov) takes an integrative approach centered around whole-genome and transcriptome sequencing of the hundreds of different cancer types, building on the foundation laid by CGAP. All of these activities are coordinated at the NIH by the Office of Cancer Genomics (ocg.cancer.gov). Of course, the European Union also has major investments in cancer genetics and genomics, as do the health research agencies of most developed countries. An important international initiative is the International Cancer Genetics Consortium (ICGC; icgc.org; **Figure 14.3**), which, as of October 2014, had committed to 71 projects. A few examples will provide a sense of the diversity of cancer types these projects address and their international scope: ovarian cancer in Australia, pancreatic cancer in Canada, blood cancer in South Korea, gastric cancer in China, liver cancer in Japan, oral cancer in India, head and neck cancer in Mexico, renal cancer in France, chronic lymphocytic leukemia in Spain, rare tumors in Italy, and thyroid cancer in Saudi Arabia. Note that these foci do not imply that the cancers are specific to those countries, they are just being studied intensively in those locations.

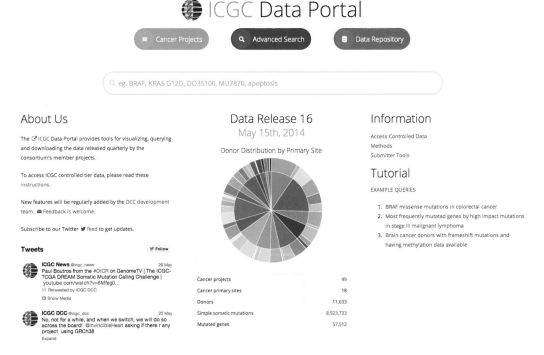

Figure 14.3 International Cancer Project Resources.   The International Cancer Genome Consortium is only one organization that provides critical resources for researchers. (From dcc.icgc.org)

Breast Cancer

Breast cancer is the most common invasive cancer in women, affecting 1 in 8 women over their lifetimes. In women it accounts for 16% of all cancers and a slightly lower percentage of cancer deaths. It also affects men, but with 100 times lower incidence. The environmental risk factors for breast cancer are well defined: age of menarche and menopause (hence lifetime estrogen exposure), alcohol intake, and being overweight. It is less prevalent in African American than in European American women, but has an earlier age of diagnosis, and tends to be more aggressive, in African American women.

Breast cancer is a highly heterogeneous condition because cancers can arise from different breast tissues and respond to different growth factors. Most breast cancers derive from the epithelial lining of either the milk ducts or the lobules that produce milk, where they remain in situ (stage 0), invade the surrounding tissues (stages 1–3), or metastasize through the lymphatic system (stage 4). **Stage** refers to the location and size of a tumor, whereas **grade** is a description of a tumor's histology that is used to predict its likely rates of growth and spread. Low-grade tumors have almost normal cellular structure, but progress to high-grade tumors as the nuclei lose their shape and the cells become disorganized. A receptor classification system for breast cancer has also been developed, in which HER2/Neu (epidermal growth factor)–positive and estrogen receptor– or progesterone receptor–negative status, for example, are associated with a poor prognosis (Slamon et al. 1987). Two of the major objectives of genomic profiling of tumors are to provide better predictions of likely survival time in the absence of treatment and of likely responses to chemotherapy, which ideally would be personalized for each patient on the basis of her or his molecular profile.

The future of unbiased genomic profiling in breast cancer research was revealed in five papers published in the June 21, 2012, edition of *Nature*. Four of these papers (Ellis et al. 2012; Stephens et al. 2012; Shah et al. 2012; Banerji et al. 2012) examined the somatic mutational spectrum in approximately 100 individuals with breast cancer by comparing the whole-exome sequences of tumors with those of the individual's normal genome. The average tumor exhibited over 1700 point mutations, at least 50 of which were tier 1, as well as 17 SCNA. A core set of nearly 50 breast cancer driver genes was identified. Most tumors possessed a mutation in at least one of these driver genes, while some tumors had as many as six such mutations. The seven most prevalent breast cancer drivers were each mutated in at least 10% of tumors, and three (*MAP3K1*, *PIK3CA*, and *TP53*) were even more prevalent, as has been confirmed by targeted sequencing of more samples. The most affected pathway seems to be ERK/JNK/JUN kinase signaling: Ellis et al. (2012) specifically noted an excess of mutations in this cluster in those tumors that had become resistant to treatment with aromatase inhibitors, which are used to block estrogen production. There is some indication that different mutation types are also diagnostic of tumor subtypes identified by histology or transcription profiling, notably A and B luminal types, which have

Figure 14.4 Breast Cancer Susceptibility Loci. (A) The cumulative lifetime risk of breast cancer increases from about 10% at the age of 70 for most women to 50% for women who are carriers of a *BRCA2* mutation. This risk is further increased in the 1% of women who are homozygous for all four risk alleles at the *FGFR2* and *TOX3* loci (high risk group), but decreases in women without any such alleles (no risk group). (B) This forest plot shows the estimated effect sizes (odds ratios) for 13 breast cancer susceptibility genes identified by GWAS. The three genes highlighted in the blue area seem to make different contributions to the risk of estrogen receptor-positive and estrogen receptor-negative risk tumors, whereas odds ratios for women who carry *BRCA1* or *BRCA2* mutations are for the most part constant. (After Mavaddat et al. 2010.)

different prognoses. Shah et al. (2012) focused on triple negative tumors (lacking receptors for estrogen, progesterone, and epidermal growth factor) and found wide variation in the overall mutation frequency at diagnosis. They also provided evidence that the polyclonal nature of advanced tumors is reflected in the proportion of reads that have each mutation as well as in their co-occurrence. They then inferred that mutations affecting cell motility and shape are more likely to appear late in tumor progression.

The other mode of genomic profiling for tumor classification that has received a lot of attention is gene expression profiling. The fifth of the June 2012 *Nature* studies (Curtis et al. 2012) jointly considered transcript abundance and copy number aberration in almost 2000 individuals with diverse types of breast cancer. They argued that in both the discovery and validation subsets, ten subcategories of tumors could be identified. These subcategories showed marked differences in 15-year survival probability, which reflected the nature of the drivers that were either overexpressed or deleted by somatic mutations. These results extended a series of smaller early studies (Perou et al. 2000; van't Veer et al. 2002) that did not produce sufficiently reproducible results to support clinical use, perhaps reflecting technological and analytic differences as well as the heterogeneity of the cancers. Nevertheless, Fan et al. (2013) showed that a 70-gene poor-outcome profile and/or a classification based on two transcriptional profiles of luminal tumors, as well as expression differences associated with estrogen and HER2 receptor status, have prognostic potential beyond classical histological grades. Individual genes alone are not predictive, but whole-transcriptome profiling shows how multiple genes tend to be co-expressed. Collectively co-expressed genes provide strong prognostic indicators, much as polygenic genotype risk scores distinguish different levels of disease susceptibility.

Although as many as 13% of all women will experience breast cancer in their lifetimes, approximately 10% of the disease is believed to be clustered in families. One-fifth of the risk in this group of women is explained by rare variants in two genes, *BRCA1* and *BRCA2*, which were first cloned using fine-structure linkage mapping (Friedman et al. 1994; King et al. 2003). Both of these genes are involved in DNA repair and chromosome stability. The lesions tend to lead to truncated proteins. Collectively, they represent less than 1% of *BRCA* alleles, so they make a very small contribution to overall risk in the population. However, their effect sizes are so large that carriers

(A)

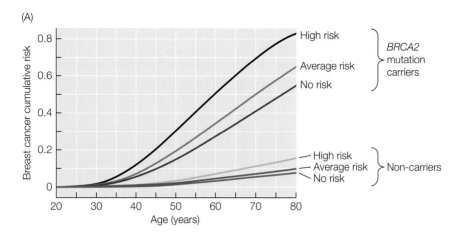

(B)

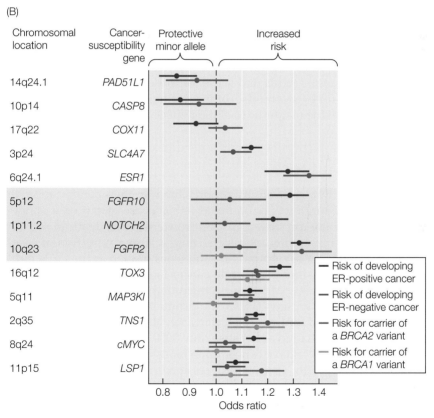

have a fivefold to tenfold increased lifetime risk, generally over 50% but up to 80% in some families. This risk leads many women to opt for mastectomy (breast removal) and sometimes oophorectomy (ovary removal), since ovarian cancer risk is also strongly increased by these variants. **Figure 14.4A** shows how risk increases with age as well as with genetic background at the *FGFR2* and *TOX3* loci. Thus, while *BRCA1* and *BRCA2* are popularly regarded as "the breast cancer genes," they are each responsible for at most 1% of all cases of the disease, although carriers are very clearly at elevated risk that justifies close monitoring. Risk is higher in certain populations, such as Ashkenazi Jews, since up to 2% of individuals carry mutations in these genes.

Several medium-sized GWAS have sought common variants that modify breast cancer risk, finding 13 loci that each confer modest relative risks of less than 1.3 (Easton et al. 2007; Mavaddat et al. 2010). Another 6 loci have larger effects, but are too rare to account for an appreciable amount of the total population risk. These loci appear to have diverse molecular functions; the most clearly defined is *FGFR2*, which encodes a fibroblast growth factor receptor. A variant in a gene desert at 8q24 is also associated with several other cancer types and probably affects expression of the *MYC* oncogene, while a common variant in *MAP3K1* links normal variation in signal transduction to cancer risk. An interesting variant, rs1045485, leads to the D302H substitution on CASP8, a key protein in the apoptosis pathway, but is common only in Europeans, for whom the derived minor allele offers significant protection. **Figure 14.4B** shows the estimated odds ratios (odds of disease for those with the genotype relative to the odds for everyone else) of the documented genetic risk factors and suggests that the magnitudes of their effects may differ among breast cancer subtypes.

Prostate Cancer

Prostate cancer in men is slightly more common than breast cancer in women and, although it is not quite as lethal as breast cancer, it is nevertheless one of the top three causes of cancer-related death in men. It is difficult to estimate lifetime risk because a large proportion of cases are benign and even asymptomatic; autopsies of men in their 70s indicate that 80% have signs of prostate cancer in situ, compared with just 30% of men in their 50s. Most cases are accompanied by elevated levels of prostate-specific antigen (PSA), the product of the *KLK3* kalleikrein gene that is required for liquefying ejaculate. However, PSA is no longer recommended for diagnosis because only 30% of individuals with high PSA are determined to have cancer upon biopsy, so the cost-benefit ratio is considered too high. Disease progression is often slow, but eventual mortality is high. Prevalence is slightly lower in Hispanic than in non-Hispanic Caucasian Americans, but more than 50% higher in African Americans, with an even higher mortality rate. It is not known to what extent this difference is due to genetic or environmental factors, neither of which is particularly well defined.

If there is a *BRCA* equivalent for prostate cancer, it is the *HOXB13* gene. When Ewing et al. (2012) sequenced 200 genes across the 17q21–22 interval that linkage studies had implicated in familial prostate cancer, they found a glycine to glutamate replacement at amino acid 84 of the encoded protein

(G84E) in 4 of 94 families. The mutation was further observed in all 14 individuals with prostate cancer in these families, and it was subsequently found at a frequency of 1.4% in just over 5000 unrelated patients, a 20-fold excess compared with the general population. This mutation had an incidence as high as 3.1% in familial cases and was also associated with early onset. Members of the HOX family of proteins are regulatory transcription factors that are critical throughout development for patterning of the body, and *HOXB13* is particularly important for prostate development, so its role in cancer promotion makes biological sense.

Genome-wide association studies have identified 14 loci, 2 of which have multiple independent variants (Eeles et al. 2008): the 8q24 locus, which is in a gene desert in the vicinity of *MYC* among other candidate genes, and *HNF1B*, which encodes a homeodomain transcription factor. The International Consortium for Prostate Cancer Genetics (ICPCG) confirmed that 8 of these 14 loci were replicated in family-based association testing (Jin et al. 2012). This experimental design involves simply asking whether heterozygous parents transmit one or the other allele to a single affected child more often than the 50% of the time expected by chance. For example, if 60% of probands receive the A from A/G parents at a particular site, then, with a sufficiently large sample, this observation would imply that the A allele confers risk of cancer. This design avoids any potential influence of population structure that may produce false positives and thus confirms the role of the locus. It also establishes that the same variants that contribute to sporadic cases in the general population also contribute to cases in which prostate cancer is enriched in the almost 2000 families studied.

One of the primary treatments for prostate cancer is androgen deprivation therapy, which starves the prostate-derived cancer cells of the hormone they need to grow. Although over 70% of patients respond with partial remission, most experience recurrence after several months or years. Already in 1995, a likely molecular mechanism of recurrence had been revealed by comparative genome hybridization (CGH), which showed specific amplification of the Xq11–13 region that includes the androgen receptor (*AR*) gene (Scher and Sawyers 2005). Quantitative fluorescent in situ hybridization to biopsy sections confirmed high-level amplification of the androgen receptor in 7 out of 23 recurrent tumors, but this phenomenon was not observed in the primary tumors. Other studies have suggested that point mutations at this locus that allow the cells to continue to grow in the presence of low levels of androgen may be selected for. Furthermore, there appear to be multiple modes by which some tumors escape androgen withdrawal or inhibition (also known as chemical castration for example, with Depo-Provera); the term "castration-resistant prostate cancer" is used to describe these advanced tumors (Berger et al. 2011). Whole-exome sequencing has identified multiple mutations in prostate cancer linking androgen signaling to chromatin modeling through histone methylation and through the ETS and FOXA1 transcription complexes (Grasso et al. 2012).

Lung Cancer

Worldwide, lung cancer is the number one cause of cancer deaths, and in the United States alone it takes the lives of more than twice as many men and women as breast and prostate cancer combined. Its major cause is tobacco smoking, but incidence remains high despite the recent decline in cigarette use. Other environmental risk factors include exposure to asbestos, radon gas, air pollution, and secondhand smoke, all of which increase the burden of mutation that airway cells are exposed to. The heritability of lung cancer, as estimated from a large Swedish cancer database, is quite low, less than 10%, but that of nicotine dependence is very high, and so predisposition to heavy smoking contributes much of the genetic risk for lung cancer (Czene et al. 2002).

The majority of lung cancers are of the non-small cell variety (NSCLC), primarily adenocarcinomas but also including squamous cell and large cell carcinomas. Lung cancer was one of the first cancer types for which high-throughput sequencing was used to characterize the mutational landscape. Ding et al. (2008) screened 623 candidate genes in 188 tumor biopsies and identified an average of five somatic mutations per sample. Seventeen of those genes were enriched for multiple mutations, including suspected tumor suppressors (*TP53, CDKN2A, STK11, NF1, ATM, RB1, APC*) and oncogenes (*KRAS, NRAS, EGFR, ERBB2, ERBB4, EPHA3, NTRK, AKT1*). The tumor suppressors tended to be disrupted by deletions and/or reduced gene expression, the oncogenes by duplications and/or increased gene expression—a finding that highlights the role of SCNA alongside point mutations in tumor promotion. Mutation frequency increased with tumor grade, and across all samples, several signaling pathways seemed more likely to be targets than expected by chance, including Wnt, Ras (targeted in almost three-fourths of the samples), and the insulin/mTOR growth pathway.

Depending on exposure rates, which vary among different countries, between 10% and 40% of cases are observed in never-smokers. Govindan et al. (2012) compared the mutational landscapes of adenocarcinomas from 6 nonsmokers and 11 smokers (**Figure 14.5**), finding a more than tenfold elevation of the somatic mutation rate in the smokers as well as a bias toward transitions due to the mechanism of action of carcinogens in cigarette smoke. Deep targeted sequencing of 24 candidate driver genes that are significantly mutated in lung cancer allowed estimation of variant allele frequencies within tumors, supporting the inference that more than half of

Figure 14.5 Mutational Landscapes of Lung Cancers of Smokers and ▶
Never-Smokers.   (A) Tumors from 17 individuals with lung cancer, 6 of whom were never-smokers, all carried a mutation in at least one of 24 genes that are significantly mutated in lung cancer. (B) The frequency of mutations in these genes and in general, however, differed between the smokers and the never-smokers. (C) The prevalence of transitions, from C to A in particular, also differed between the smokers and the never-smokers. (After Govindan et al. 2012.)

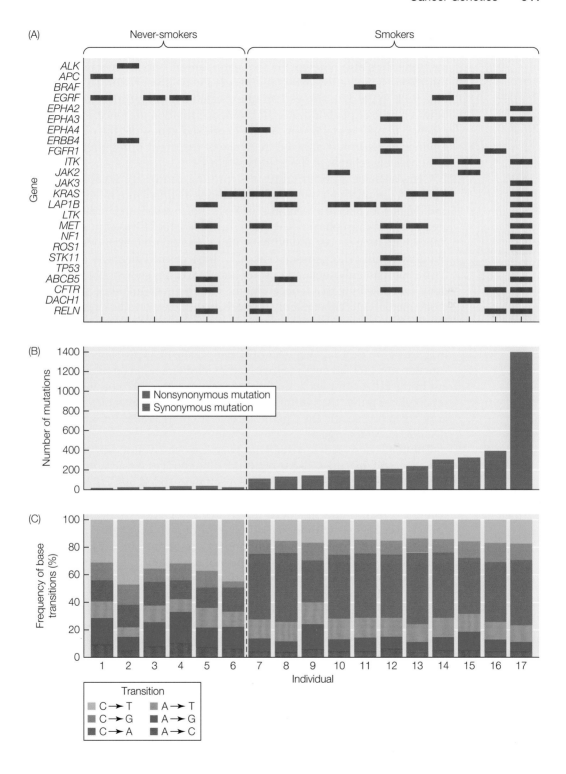

the tumors were polyclonal. The therapeutic implication is that multiple drugs may be required to target different mutations in the various cell subtypes. It is also noteworthy that each sample had an average of three mutations in so-called druggable target genes. Integration of RNA-Seq data with WGS data showed an unexpected overall negative correlation between transcript abundance and somatic mutation presence, suggesting a deficit of new mutations in highly expressed genes. On the other hand, a minority of drivers showed evidence of recurrent upregulation of the mutant allele relative to the wild-type copy.

Genome-wide association results for lung cancer are notable for two findings: the relative paucity of associations despite large meta-analyses, and the major effect of a locus at 15q24 that is also associated with aspects of cigarette-smoking behavior. In fact, in April 2008, 15q24 was simultaneously linked with lung cancer and with nicotine dependence in separate analyses of the two traits (Thorgeirsson et al. 2010; Tobacco and Genetics Consortium 2010). Subsequently, meta-analysis of lung cancer has shown that the region does not contribute to adenocarcinoma risk in never-smokers, suggesting that it mediates its effect mainly through an influence on smoking behavior, though it may also have a direct effect on lung function in smokers (Timofeeva et al. 2012). Two additional regions have been robustly associated with lung cancer in general. A cluster of variants at 5p15.33 probably affects the function of TERT, an enzyme involved in telomere replication, and a variant in the HLA complex is thought to influence the *BAT3* or *MSH5* DNA repair genes rather than immunity. An independent association at 5p15.33 appears to be specific for squamous cell carcinoma, as is variation at the *RAD52* DNA repair gene.

The 15q24 association, which has been intensively studied, is probably due primarily to a missense substitution that reduces the activity of the α5 subunit of the nicotinic acetylcholine receptor encoded by *CHRNA5* (Ware et al. 2012). Knockout of this gene in mice leads to inability to self-regulate nicotine consumption, and accordingly, the association of the minor, lung cancer–risk allele is with increased nicotine dependence rather than with age of onset of smoking or likelihood of cessation. The polymorphism explains only 0.5% of the variance in smoking quantity, but this variance corresponds to one extra cigarette a day, if not more: self-reporting of all addictive behaviors is notoriously inaccurate, and *CHRNA5* SNPs explain a lot more of the variation for nicotine metabolite levels in smokers. The lead SNP, rs16969968, is also associated with opioid and alcohol dependence, as well as with chronic obstructive pulmonary disease, coronary artery disease, and schizophrenia. All of these effects are attributed to activation of the reward pathway in a specific nucleus of the brain. Two other loci are also associated with nicotine dependence, one a possible lncRNA at 10q25, the other near the gene encoding the enzyme that metabolizes nicotine to cotinine, *CYP2A6*. Smoking initiation and cessation are associated independently with two loci that are plausibly related to dopaminergic neuron function—*BDNF* and *DBH*, respectively—but do not make a measurable contribution to lung cancer susceptibility (TGC 2010).

A unique feature of cancer genomes is that they often accumulate gene fusions that create novel oncogenes, and this is particularly true of lung cancer. Software such as BreakFusion, BreakDancer, and ChimeraScan (reviewed in Wang et al. 2013b) has found recurrent fusions of the *ROS*, *ALK1*, and *RET* oncogenes. *ALK* fusions, along with *EGFR* activating mutations, are common targets of lung cancer chemotherapy, but resistance commonly emerges within a couple of years. As with resistance to androgen inhibition in prostate cancer, various molecular mechanisms are implicated, from mutation of receptors and downstream signaling proteins to emergence of novel signaling pathways that allow the cells to bypass the inhibition of the growth factor they require (Niederst and Engelman 2013).

Colorectal Cancer

Colorectal cancer is another prevalent cancer (with a lifetime risk of approximately 5%) that has modest heritability but high mortality. Screening programs are encouraged because survival rates are high if the disease is caught early, and progression through the early stages of hyperplasia to benign adenoma, then carcinoma in situ, and eventually metastatic disease takes several years. As in breast cancer, the contribution of rare, highly penetrant alleles is substantial, but because they are seen in only a small number of families, their overall contribution to risk is small. To date, GWAS have identified only a dozen common variants that contribute to sporadic cases, and even these have rather small effects that have not been replicated robustly in different studies.

The causal gene is known for a handful of familial colorectal syndromes. Lynch syndrome, also called hereditary nonpolyposis colorectal cancer (HNPCC), and familial adenomatous polyposis (FAP), which jointly explain 5% of familial clustering, are due to mutations in the MMR mismatch repair pathway and *APC* gene, respectively. The *APC* gene, which encodes a tumor suppressor, provides a classic illustration of the two-hit hypothesis. Children born with one defective copy are almost sure to develop FAP cancer at some point in their lives due to independent secondary mutations in the hundreds or thousands of polyps they have. There is a strong genotype-to-phenotype relationship, as mutations affecting different parts of the APC protein phosphatase have variable expressivity, including one variant that introduces a premature stop codon that is partially rescued by translation from an internal ribosome initiation site. Four mismatch repair genes—*MLH1*, *MSH2*, *MSH6*, and *PMS2*—contribute to Lynch syndrome with high penetrance, but have lower penetrance for induction of other classes of cancer, notably endometrial and ovarian cancer. The second hit in Lynch syndrome sometimes occurs as a result of aberrant methylation of the promoter region of *MLH1*, providing an instance of epigenetic onset of cancer. One type of *MSH2* mutation is unusual in that it has a high germ line recurrence rate, since the A→T transition, which affects a splice donor site, occurs in a highly mutable string of 26 A nucleotides in a row. Age of onset of Lynch syndrome is generally postreproductive, so selection against

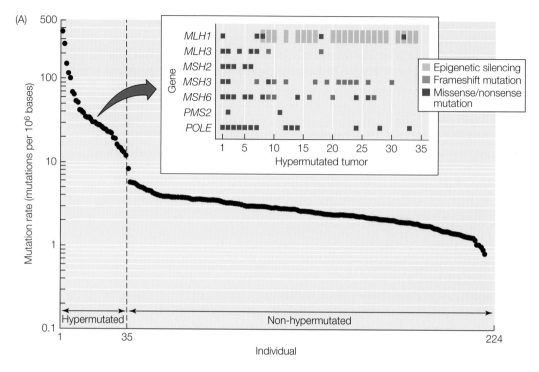

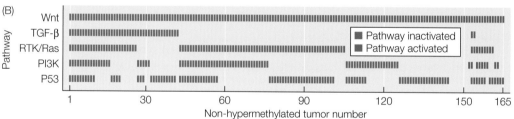

Figure 14.6 The Mutational Landscape of Colorectal Cancer (CRC). (A) Rank-ordered plot of the mutation rate per million bases on the log scale for 224 colorectal adenocarcinomas, showing that 16% of the tumors were hypermutated. The inset shows which of these tumors carried missense (green) or frameshift (purple) mutations or epigenetic silencing (yellow) of seven mismatch repair genes. (B) Each of the five signaling pathways is commonly mutated in CRC. Red or blue bars indicate which pathway is affected in each of 165 tumors. (After CGAN 2012.)

the mutations is expected to be weak; for this reason, some of the variants are thought to have arisen hundreds to thousands of years ago and affect unique isolated populations.

Comprehensive exome sequencing and transcriptome, methylation, and DNA copy number profiling of 224 tumor–normal tissue pairs has established the existence of two broad classes of colorectal tumors (**Figure 14.6A**; CGAN 2012). The minority, around 16%, are hypermutated, with a lesion every 10 kb throughout the genome in some cases, and tend to show both extensive *microsatellite instability* (MSI) and a peculiar *CpG island methylator phenotype* (CIMP). The remainder generally have high chromosomal instability, resulting in SCNA as well as up to ten interchromosomal translocations per tumor, which sometimes generate novel fusion proteins. At least 15 genes are recurrently mutated in the hypermutation type, as are a largely distinct set of 17 genes in the majority type, though *APC* is somatically altered in one-half to three-fourths of tumors of both types. **Figure 14.6B** shows how five major signaling pathways are disrupted: the Wnt pathway in 92% of tumors, the PI3K-Ras pathway in 55% of tumors with apparent mutually exclusive patterns of mutation, and the TGF-β pathway in one-third of tumors. Efforts are under way to integrate all these data types to identify the molecular basis of invasive and drug-resistant tumors, as well as to find a core set of potential drug targets that may correct aberrant activity in each of these signaling pathways.

Leukemia

Cancers of the blood, bone marrow, or lymph nodes are collectively known as hematological malignancies. The most common is lymphoma, a solid tumor of white blood cells often found in the lymph nodes, which is subcategorized as Hodgkin's or non-Hodgkin's lymphoma on the basis of visible features of the tumor. Leukemias originate in the bone marrow but typically present in the blood after patients complain of bruising, frequent infections, or anemia, all of which result from disrupted hematopoiesis and hence deficiencies of platelets, leukocytes, and erythrocytes. Four major subtypes of leukemia are distinguished by whether the lymphoid or myeloid lineages of blood cells (see Figure 13.1) are affected and by whether the disease is acute or chronic. The vast majority of leukemias have an adult onset, but childhood leukemia tends to be acute, requiring aggressive therapies that, encouragingly, result in survival rates of over 75%.

Acute myeloid leukemia (AML) presents much simpler mutation profiles than other cancer types, though all of the typical features are observed. A survey of 200 tumors (Cancer Genome Atlas Research Network 2013) revealed an average of just 13 somatic mutations per case, only 5 of which affected one of 23 recurrently mutated driver genes. Each tumor, on average, had only a single SCNA event, while gene fusions were even rarer, but where present, chromosomal aberrations typically led to a poor prognosis. This relative simplicity of the genomic DNA profiles has led to a nine-pathway classification that includes almost all AML tumors. The following fractions of tumors carry a mutation in each of the nine classes, illustrating the diversity of cancer mechanisms in a single cell type: activated signaling genes (59%), DNA methylation–related genes (44%), chromatin-modifying genes (30%), the nucleophosmin gene *NPM1* (27%), transcription factor genes that regulate myeloid blood cell differentiation (22%), transcription factor fusions

(18% of cases), tumor suppressor genes (16%), spliceosome complex genes (14%), and cohesin complex genes (13%).

Many studies have explored the efficacy of subtype classification using genetic profiling, starting with chromosomal aberrations observed with classical karyotyping. Leukemia was one of the first cancer types for which gene expression profiling was used to discriminate tumor subtypes with different survival probabilities. Golub et al. (1999), who used an early Affymetrix array to profile just 38 childhood leukemia samples, showed how a variety of classification approaches extract transcripts whose abundances discriminate among AML and acute lymphoblastic (ALL) subtypes principally, and between B and T lineage ALL types secondarily. Similarly, Alizadeh et al. (2000) identified two novel subtypes of B cell lymphomas that have significantly different survival rates (**Figure 14.7**). More recently, methylation and miRNA profiles have been employed for the same purpose. It is clear that some translocation or fusion events are associated with combined profiles across these omic domains (notably the Philadelphia chromosome, which fuses *BCR* and *ABL*, and another relatively common *PMR-RARA* fusion involving the retinoic acid receptor). Other profiles correlate in some cases with specific tumor morphologies known as the FAB (French, American, British) types, but more generally, the relationship between transcription, methylation, and miRNA abundance is complex.

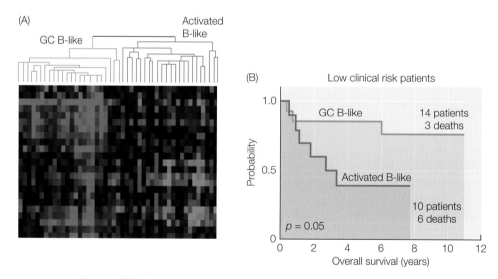

Figure 14.7 Molecular Classification of Cancer Subtypes. (A) Heat map of the expression of 22 genes that differentiate two subtypes of B cell lymphomas, named GC B-like and activated B-like. Red shading indicates high expression and green shading, low expression. (B) Kaplan-Meier survival curve showing the cumulative probability of survival for 12 years after diagnosis for 14 individuals with GC B-like and 10 with activated B-like tumors. After 8 years, only 3 of the individuals with GC B-like tumors had died, whereas the prognosis was much worse for individuals with the activated subtype, of whom 6 had died. (After Alizadeh et al. 2000.)

Consequently, the clinical potential of combined profiles has yet to be realized, at least to the extent that specific mutations are just beginning to be used to guide therapeutic options.

One of the most pressing questions in current leukemia research is whether the mutational spectra of the different classes of acute and chronic leukemia are different. A relatively small study of 88 individuals with chronic lymphocytic leukemia (CLL) found enrichment for mutations affecting DNA damage repair, cell cycle control, Wnt signaling, and Notch signaling, the last possibly defining a lymphoid-specific pathway (Wang et al. 2013c). A related question is whether the mutation profile for pediatric acute leukemia is distinct from that for adult-onset leukemia, or whether age of onset is more influenced by genetic background.

Genome-wide association studies have identified 23 loci that explain 17% of the familial risk of chronic lymphocytic leukemia, confirming that there is a strong common variant component to leukemia risk (Berndt et al. 2013). Six of these loci are located near genes annotated to functions in apoptosis. They include an intronic variant in the *FAS* gene, which encodes one of the tumor necrosis factor receptors that initiates the cell death cascade, variants in the *CASP8* caspase locus, and multiple members of the *BCL2* (B cell CLL/lymphoma) pathway that regulates apoptosis. Leukemia thus provides a new perspective by adding aberrant regulation of cell death to control of cell division as a key component of cancer. Childhood ALL has also yielded some hits to GWAS, although published studies are not yet sufficiently large to facilitate comparative genomics of leukemia risk.

Melanoma

Skin cancer can derive from different layers of the skin. The most common type is basal cell carcinoma, which originates in the epidermis, whereas squamous cell carcinoma and melanoma are less common, but more malignant. Melanoma, cancer of the pigmented melanocytes, is typically diagnosed due to asymmetrical growth of a freckle or nevus, including the appearance of rough edges or discoloration (**Figure 14.8A,B**). If caught early, it can be successfully dealt with by surgical excision; however, malignant melanoma is among the most aggressive cancers and has a survival time measured in months. The most important risk factor is exposure to ultraviolet radiation, either from bright sunlight or tanning beds. As a consequence, melanoma has an unusual mutation profile characterized by a very high basal rate of C→T transitions at dipyrimidines that result from failure to repair UV-induced damage. Individuals with fair skin and freckles are particularly susceptible, and the combination of British ancestry with burning sun and an outdoor lifestyle leads to a greatly elevated rate of melanoma in Australia.

Three noteworthy features of melanoma have emerged from genetic analysis so far (Chin et al. 2006). The first is the unusual nature of the gene that is responsible for a familial form of the disease, *CDKN2A*. At this locus,

(A)

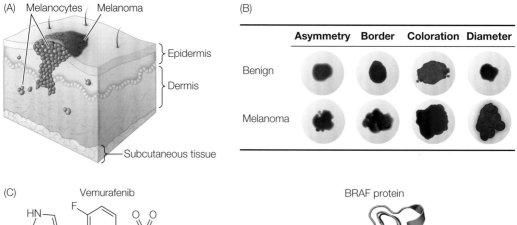

(B)

	Asymmetry	Border	Coloration	Diameter
Benign				
Melanoma				

(C)

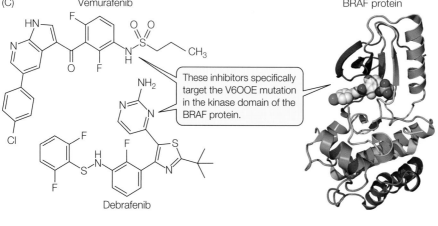

These inhibitors specifically target the V600E mutation in the kinase domain of the BRAF protein.

Vemurafenib

Debrafenib

BRAF protein

(D)

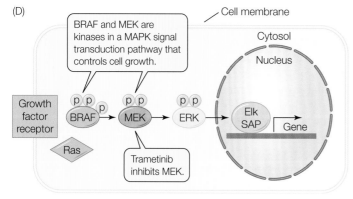

BRAF and MEK are kinases in a MAPK signal transduction pathway that controls cell growth.

Trametinib inhibits MEK.

two different promoters and first exons are spliced to the same second exon, but in a different reading frame, resulting in the production of two different proteins, INK4A and ARF, that have distinct roles in cell cycle regulation through the RB and P53 tumor suppressors. The second is the frequent amplification of a novel oncogene, *MITF*, which turns out to encode a key transcription factor involved in melanocyte stem cell development.

◀ **Figure 14.8 Melanoma and BRAF.**   (A) Melanoma is cancer of the melanocytes. It is extremely invasive, and the prognosis is generally poor once metastasis has commenced. (B) There are five "ABCDE" signs of conversion of freckles or nevi into melanoma: *asymme*try of the pigmented region, irregular *b*orders, abnormal *c*oloration, including mixing of dark and light spots, enlarged *d*iameter greater than 5 mm, and *e*volution of the shape and morphology of the nevus over time (not shown). (C) The *BRAF* oncogene is commonly mutated in melanoma. The drug vemurafenib has been licensed for treatment, as it specifically binds the active site of the V600E mutation only in the activated kinase. (D) Combination therapy with trametinib and BRAF inhibitors disrupts consecutive steps in the signal transduction pathway and can be more effective than either drug alone.

Congenital mutation of this gene leads to type 2a Waardenburg syndrome, which includes pigmentation defects and a white forelock, and it is also associated with premature graying. Furthermore, it is the downstream target of phosphorylation by the MAPK kinase signaling pathway, which is mutated in the majority of high-grade melanomas.

The third feature is mutation of *BRAF*, one of the members of the MAPK pathway, whose discovery has led to the paradigmatic example of the development of personalized cancer therapeutics (Davies et al. 2002). Around 50% of melanomas (and a smaller percentage of numerous other cancer types) have mutations in the kinase domain of BRAF, the most common being V600E, which constitutively activates the oncogene. Depletion of the *BRAF* message by small interfering RNA knockdown not only inhibits activation of the ERK phosphorylation cascade, but can trigger apoptosis in the presence of the mutation. This observation has led to the development of multiple small-molecule BRAF inhibitors, initially sorafenib, which was subsequently approved for use with other cancer types carrying mutations in the kinase, but more recently inhibitors such as vemurafenib and debrafenib, which are selective for the mutated active site (Chapman et al. 2011; **Figure 14.8C**). These drugs are not effective against all BRAF-positive tumors, but can buy time. Combination therapies for targeting additional mutated loci or other genes in the pathway (e.g., the MEK inhibitor trametinib; **Figure 14.8D**) hold promise for overcoming resistance and increasing survival times and remission (Luke and Hodi 2013).

Glioblastoma

Brain tumors, perhaps surprisingly, are rarely neuronal in origin: the two most common forms are glioblastoma, derived from the glial support cells that hold the brain together, and meningioma, derived from the meninges that encase the brain. Lifetime risk is low, and incidence is generally sporadic. Epidemiologists have implicated risk factors associated with modern life, namely radiation from cell phone use and one of the chemicals in artificial sweeteners, but these need to be further validated. Men are at greater risk for glioblastoma than women for unknown reasons.

Glioblastoma was actually the first cancer type to be studied by the Cancer Genome Atlas Research Network (2008; see also Frattini et al. 2013).

Analysis of 91 tumor–normal tissue pairs for SCNA and nucleotide mutations in 600 candidate genes found a very consistent pattern whereby over 80% of tumors show inactivation of the tumor suppressor p53 and RB pathways as well as activation of receptor tyrosine kinase growth factor (EGF, ERBB2) signaling. One-fourth of the tumors were also affected by inactivation of *NF1*, which is also responsible for noncancerous neurofibroma tumors. A hypermutation phenotype was documented in a half dozen cases in which a mismatch repair gene was also mutated, and methylation of the promoter of the *MGMT* DNA repair gene (Hegi et al. 2005), which is required for removal of alky adducts from G residues, was associated with a divergent pattern of nucleotide substitutions genome-wide. Since one of the major drugs used to treat glioblastoma, temozolomide, is an alkylating agent, this finding also has implications for personalized cancer treatment. Indeed, patients with the epigenetic modification of the *MGMT* promoter survive, on average, for 6 months after drug treatment, whereas those without it do not see a significant benefit.

Pan-Cancer Analysis

With a catalog of genomic profiles for over 5000 tumors, researchers have begun to perform comparative analyses across the panoply of cancer types. Two papers published in September 2013 (Zack et al. 2013; Ciriello et al. 2013) provide some first insights. One considers in detail the pan-cancer distribution of SCNA, and the other divides tumors into subclasses based on joint analysis of copy number and mutational variance as well as methylation and gene expression. Both papers conclude that all cancer types are highly heterogeneous at the molecular level, reflecting the independent mutational history of each tumor, but also note the extensive sharing not only of classes of lesions, but of specific mutational events. This research is leading to the notion that from a therapeutic perspective, cancers may be better classified according to genomic features than by traditional tissue lineage or histological grades. Subsequently, Lawrence et al. (2014) catalogued the Cancer5000 set of 254 genes that are regarded as highly significant drivers, 21 of which are found in at least three cancer types. The numbers of these genes found in each of 19 tumor types are listed in **Table 14.2**, and the online representation of the mutation spectrum for one gene (*BRAF*) from TumorPortal is shown in **Figure 14.9**.

Zack et al.'s (2013) consideration of 4934 tumors of 11 types found over 200,000 SCNA, a median of 39 per tumor, including focal deletions and duplications smaller than the length of a chromosome arm, longer events at least as long as a chromosome arm, and copy number–neutral losses of heterozygosity where the deleted region was replaced with a duplicate of the remaining chromosome. Just over a third of the tumors were found to involve whole-genome duplication events, and these tumors also exhibited an excess of focal SCNA. Recurrent gains and losses were observed at 70 regions each, accounting for one-fifth of all events, the former including 24 known oncogenes and the latter 12 known tumor

TABLE 14.2 Cancer driver genes

Cancer type	Tumor–normal tissue pairs studied	Number of highly significant drivers mutated	Mutated genes from list of 21 most commonly mutated cancer genes[a]
Acute myeloid leukemia	196	27	IDH1, NRAS, TP53, KRAS
Bladder cancer	99	34	TP53, RB1, PIK3CA, ARID1A, MLL2, ERBB3, FBXW7, HRAS
Breast cancer	892	37	PIK3CA, TP53, PTEN, PIK3R1, CTCF, ARID1A, RB1, KRAS
Chronic lymphocytic leukemia	159	15	TP53
Colorectal cancer	233	35	TP53, FBXW7, NRAS, BRAF, KRAS, PIK3CA, ARID1A, ERBB3, CASP8
Diffuse large B cell lymphoma	58	23	TP53, MLL2, HLA-A, BRAF
Endometrial cancer	248	73	PTEN, PIK3CA, TP53, KRAS, FBXW7, CTCF, ARID1A, ERBB3, NRAS
Esophageal adenocarcinoma	141	15	TP53, CDKN2A, PIK3CA
Glioblastoma multiforme	291	26	PTEN, TP53, PIK3R1, NF1, RB1, IDH1, BRAF
Head and neck carcinoma	384	34	TP53, CDKN2A, CASP8, MLL2, HRAS, PIK3CA, PTEN, HLA-A, CTCF
Kidney clear cell cancer	417	21	TP53, MLL2, HLA-A, BRAF
Lung adenocarcinoma	405	32	TP53, CDKN2A, KRAS, NF1, BRAF, PIK3CA, ARID1A, ATM, RB1
Lung squamous cell carcinoma	178	24	TP53, CDKN2A, MLL2, PIK3CA, NFE2L2, RB1, HRAS, HLA-A
Medulloblastoma	92	3	MLL2, TP53
Melanoma	118	28	BRAF, TP53, CDKN2A, PTEN
Multiple myeloma	207	14	KRAS, TP53, BRAF, IDH1
Neuroblastoma	81	1	
Ovarian cancer	316	10	TP53, RB1, NF1
Prostate cancer	138	6	TP53

Source: Lawrence et al. 2014.

[a]The 21 genes found to be significantly mutated in three or more cancer types were ARID1A, ATM, BRAF, CASP8, CDKN2A, CTCF, ERBB3, FBXW7, HLA-A, HRAS, IDH1, KRAS, MLL2, NF1, NFE2L2, NRAS, PIK3CA, PIK3R1, PTEN, RB1, and TP53.

suppressors. This observation implies that many more driver genes of both types remain to be defined, many of which are candidates due to their significantly elevated somatic point mutation rates. Analysis of the

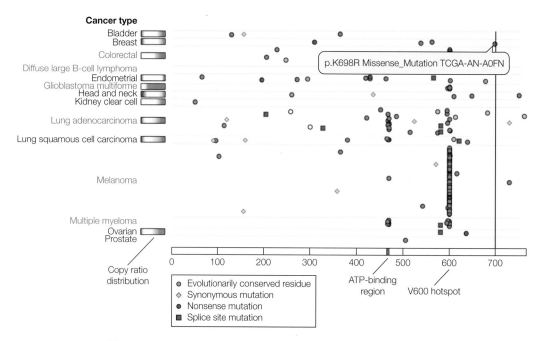

Cancer type

Bladder
Breast
Colorectal
Diffuse large B-cell lymphoma
Endometrial
Glioblastoma multiforme
Head and neck
Kidney clear cell

Lung adenocarcinoma

Lung squamous cell carcinoma

Melanoma

Multiple myeloma
Ovarian
Prostate

Copy ratio
distribution

p.K698R Missense_Mutation TCGA-AN-A0FN

0   100   200   300   400   500   600   700

ATP-binding
region   V600 hotspot

○ Evolutionarily conserved residue
◇ Synonymous mutation
● Nonsense mutation
■ Splice site mutation

Figure 14.9 TumorPortal Representation of Cancer Mutations.   This screenshot shows the locations for all mutations within the coding regions of the *BRAF* gene detected in the Cancer5000 project's exome sequencing of normal tissue–tumor pairs, ordered by tumor type. Evolutionarily conserved residues are shown in green, synonymous mutations in gray, and splice site and nonsense mutations in red. The location of mutational hotspot at valine 600 (V600; which mutates to glutamate or lysine) and the predicted ATP-binding region are shown. Mousing over each mutation gives its details, as illustrated here for one breast cancer mutation. The blue/red shaded boxes to the left reflect the incidence of somatic copy number alterations. Data for other genes and tumors can also be accessed from www.tumorportal.org. (From cancergenome.broadinstitute.org/index.php?gene=BRAF)

point mutation data by Lawrence et al. (2014) confirmed this inference that the cancer gene list is far from complete. This scale of analysis is also sufficient to identify significant anticorrelations: instances in which one SCNA is observed less commonly than expected when another is present. Most of these anticorrelations involve proteins that form interaction networks, suggesting that once a pathway is severely disrupted, it is less likely that another event affecting the same pathway will be selected during tumor progression.

Ciriello et al. (2013) overlaid point mutations on SCNA in 3299 tumor data sets including 12 cancer types and found an unexpected inverse relationship between these two types of mutations, leading to the notion of the cancer genome hyperbola. M-class (mutation) tumors, which are dominated by point mutations, were most prevalent in cancers of the kidney, colon, and uterus and in glioblastomas and AML. C-class (copy-number) tumors, by contrast, which are enriched for SCNA, were most prevalent in breast, ovarian, lung, head and neck, and endometrial cancers. Both of these classes were further divided into more than two dozen subtypes, which for the most part included samples from multiple tissue lineages, but captured the effects of recurrent patterns of 199 significantly mutated genes, 267 copy number aberrations, and 13 epigenetically silenced genes. Whether and how these categories will emerge as defining features for therapeutic intervention remains to be seen. All indications are that combinations of lesions will suggest combinatorial drug treatment, all personalized to the specifics of the tumor (Watson et al. 2013).

Summary

1. Cancer is both a disease of genetic predisposition due to rare and common variants and of somatic mutations that ultimately lead to a loss of cell cycle control and gain of invasiveness.

2. Three major classes of genes are recognized in cancer genetics: onco-genes, tumor suppressor genes, and DNA repair genes.

3. Advanced tumors accumulate a very large number of mutations, including point mutations and somatic copy number alterations (SCNA). Among these mutations are drivers, which are essential for malignancy as well as passengers, which are simply present in the growing tumor cells.

4. Genome-wide association studies have identified dozens of risk fac-tors for many of the major cancer types. Highly penetrant rare alleles are also known to promote familial forms of cancer.

5. Knudson's two-hit hypothesis, derived from statistical analysis of age of onset of cancer, is that all cancers require at least two muta-tional events. This is clearly seen in the case of the tumor suppressors that are responsible for cancer syndromes such as retinoblastoma.

6. The two well-known breast cancer genes, *BRCA1* and *BRCA2*, each actually account for no more than 1% of cases, but women with muta-tions of these genes are at very high lifetime risk of breast and/or ovar-ian cancer.

7. Prostate cancer is just as prevalent as breast cancer and has a higher mortality rate, in part due to the evolution of resistance to hormone therapy as a result of amplification of the androgen receptor.

8. Heritable lung cancer risk is largely attributable to genetic suscepti-bility to nicotine addiction in smokers with a common variant in the nicotinic acetylcholine receptor.

9. Leukemia is a highly heterogeneous cancer of the blood with acute and chronic classes affecting both the myeloid and lymphocytic lineages.

Gene expression and methylation profiling have been employed to categorize subtypes, predict survival probabilities, and suggest therapies.

10. Skin cancer, including melanoma, is attributable to a unique pattern of somatic mutations induced by UV radiation. The *BRAF* oncogene is mutated in over half of melanomas and is a major target of chemotherapy for melanoma and other cancer types.

11. Glioblastoma is the most common form of brain cancer, and while it shows some unique features, it is also remarkably similar to other cancers with regard to its spectrum of mutational mechanisms.

12. Some cancers, notably colorectal cancer, are associated with aberrant methylation of the promoters of DNA repair enzymes and other cancer drivers, implicating epigenetics in cancer progression.

References

Alizadeh, A. A., Eisen, M. B., Davis, R. E., Ma, C., Lossos, I. S., et al. 2000. Distinct types of diffuse large B-cell lymphoma identified by gene expression profiling. *Nature* 403: 503–511.

Balmain, A., Gray, J., and Ponder, B. 2003. The genetics and genomics of cancer. *Nat. Genet.* 33: 238–244.

Banerji, S., Cibulskis, K., Rangel-Escareno, C., Brown, K. K., Carter, S. L., et al. 2012. Sequence analysis of mutations and translocations across breast cancer subtypes. *Nature* 486: 405–409.

Benson, K. F. and Horwitz, M. 2006. Familial leukemia. *Best Pract. Res. Clin. Haematol.* 19: 269–279.

Berger, M. F., Lawrence, M. S., Demichelis, F., Drier, Y., Cibulskis, K., et al. 2011. The genomic complexity of primary human prostate cancer. *Nature* 470:214–220.

Berndt, S. I., Skibola, C. F., Joseph, V., Camp, N. J., Nieters, A., et al. 2013. Genome-wide association study identifies multiple risk loci for chronic lymphocytic leukemia. *Nat. Genet.* 45: 868–876.

Cancer Genome Atlas Network (CGAN). 2012. Comprehensive molecular characterization of human colon and rectal cancer. *Nature* 487: 330–337

Cancer Genome Atlas Research Network. 2008. Comprehensive genomic characterization defines human glioblastoma genes and core pathways. *Nature* 487: 330–337.

Cancer Genome Atlas Research Network. 2013. Genomic and epigenomic landscapes of adult de novo acute myeloid leukemia. *N. Engl. J. Med.* 368: 2059–2074.

Chapman, P. B., Hauschild, A., Robert, C., Haanen, J. B., et al., BRIM-3 Study Group. 2011. Improved survival with vemurafenib in melanoma with *BRAF* V600E mutation. *N. Engl. J. Med.* 364: 2507–2516.

Cheetham, S. W., Gruhl, F., Mattick, J. S., and Dinger, M. E. 2013. Long noncoding RNAs and the genetics of cancer. *Brit. J. Cancer* 108: 2419–2425.

Chin, L., Garraway, L. A., and Fisher, D. E. 2006. Malignant melanoma: Genetics and therapeutics in the genomic era. *Genes Dev.* 20: 2149–2182.

Ciriello, G., Miller, M. L., Aksoy, B. A., Senbabaoglu, Y., Schultz, N., and Sander, C. 2013. Emerging landscape of oncogenic signatures across human cancers. *Nat. Genet.* 45: 1127–1133.

Curtis, C., Shah, S. P., Chin, S-F., Turashvili, G., Rueda, O. M., et al. 2012. The genomic and transcriptomic architecture of 2,000 breast tumours reveals novel subgroups. *Nature* 486: 346–352.

Czene, K., Lichtenstein, P., and Hemminki, K. 2002. Environmental and heritable causes of cancer among 9.6 million individuals in the Swedish Family-Cancer Database. *Int. J. Cancer* 99: 260–266.

Davies, H., Bignell, G. R., Cox, C., Stephens, P., Edkins, S., et al. 2002. Mutations of the *BRAF* gene in human cancer. *Nature* 417: 949–954.

Dees, N. D., Zhang, Q., Kandoth, C., Wendl, M. C., Schierding, W., et al. 2012. MuSiC: Identifying mutational significance in cancer genomes. *Genome Res.* 22: 1589–1598.

Ding, L., Getz, G., Wheeler, D. A., Mardis, E. R., McLellan, M. D., et al. 2008. Somatic mutations affect key pathways in lung adenocarcinoma. *Nature* 455: 1069–1075.

Easton, D. F., Pooley, K. A., Dunning, A. M., Pharoah, P. D. P., Thompson, D., et al. 2007. Genome-wide association study identifies novel breast cancer susceptibility loci. *Nature* 447: 1087–1093.

Eeles, R. A., Kote-Jarai, Z., Giles, G. G., Olama, A. A., Guy, M., et al. 2008. Multiple newly identified loci associated with prostate cancer susceptibility. *Nat. Genet.* 40: 316–321.

Ellis, M. J., Ding, L., Shen, D., Luo, J., Suman, V., et al. 2012. Whole-genome analysis informs breast cancer response to aromatase inhibition. *Nature* 486: 353–360.

Ewing, C. M., Ray, A. M., Lange, E. M., Zuhlke, K. A., Robbins, C. M., et al. 2012. Germline mutations in *HOXB13* and prostate-cancer risk. *N. Engl. J. Med.* 366: 141–149.

Fan, C., Oh, D. S., Wessels, L., Weigelt, B., Nuyten, D. S. A., et al. 2013. Concordance among gene-expression based predictors for breast cancer. *N. Engl. J. Med.* 355: 560–569.

Frattini, V., Trifonov, V., Chan, J. M., Castano, A., Lia, M., et al. 2013. The integrated landscape of driver genomic alterations in glioblastoma. *Nat. Genet.* 45: 1141–1149.

Friedman, L. S., Ostermeyer, E. A., Szabo, C. I., Dowd, P., Lynch, E. D., et al. 1994. Confirmation of *BRCA1* by analysis of germline mutations linked to breast and ovarian cancer in ten families. *Nat. Genet.* 8: 399–404.

Garber, J. and Offit, K. 2005. Hereditary cancer predisposition syndromes. *J. Clin. Oncol.* 23: 276–292.

Golub, T. R., Slonim, D. K., Tamayo, P., Huard, C., Gaasenbeek, M., et al. 1999. Molecular classification of cancer: Class discovery and class prediction by gene expression monitoring. *Science* 286: 531–537.

Govindan, R., Ding, L., Griffith, M., Subramanian, J., Dees, N. D., et al. 2012. Genomic landscape of non-small lung cancer in smokers and never-smokers. *Cell* 150: 1121–1134.

Grasso, C. S., Wu Y-M, Robinson, D. R., Cao, X., Dhanasekaran, S. M., et al. 2012. The mutational landscape of lethal castration-resistant prostate cancer. *Nature* 487: 239–243.

Hegi, M. E., Diserens, A. C., Gorlia, T., Hamou, M. F., de Tribolet, N., et al. 2005. *MGMT* gene silencing and benefit from temozolomide in glioblastoma. *N. Engl. J. Med.* 352: 997–1003.

Hua, X., Xu, H., Yang, Y., Zhu, J., Liu, P., and Lu, Y. 2013. DrGaP: A powerful tool for identifying driver genes and pathways in cancer sequencing studies. *Am. J. Hum. Genet.* 93: 439–451.

Jin, G., Lu, L., Cooney, K. A., Ray, A. M., Zuhlke, K. A., et al. 2012. Validation of prostate cancer risk-related loci identified from genome-wide association studies using family-based association analysis: Evidence from the International Consortium for Prostate Cancer Genetics (ICPCG). *Hum. Genet.* 131: 1095–1103.

King, M.-C., Marks, J. H., Mandell, J. B., and The New York Breast Cancer Study Group. 2003. Breast and ovarian cancer risks due to inherited mutations in *BRCA1* and *BRCA2*. *Science* 302: 643–646.

Knudson, A. 1971. Mutation and cancer: Statistical study of retinoblastoma. *Proc. Natl. Acad. Sci. U.S.A.* 68: 820–823.

Lawrence, M. S., Stojanov, P., Mermel, C. H., Robinson, J. T., Garraway, L. A., et al. 2014. Discovery and saturation analysis of cancer genes across 21 tumour types. *Nature* 505: 495–501.

Lu, J., Getz, G., Miska, E. A., Alvarez-Saavdera, E., Lamb, J., et al. 2005. MicroRNA expression profiles classify human cancers. *Nature* 435: 834–838.

Luke, J. J. and Hodi, F. S. 2013. Ipilimumab, vemurafenib, dabrafenib, and trametinib: Synergistic competitors in the clinical management of BRAF mutant malignant melanoma. *Oncologist* 18: 717–725.

Mavaddat, N., Antoniou, A. C., Easton, D. F., and Garcia-Closas, M. 2010. Genetic susceptibility to breast cancer. *Mol. Oncol.* 4: 174–191.

Niederst, M. J. and Engelman, J. A. 2013. Bypass mechanisms of resistance to receptor tyrosine kinase inhibition in lung cancer. *Sci. Signal.* 6: re6.

Perou, C. M., Sorlie, T., Eisen, M. B., van de Reijn, M., Jeffrey, S. S., et al. 2000. Molecular portraits of human breast tumours. *Nature* 406: 747–752

Riegert-Johnson, D. L., Boardman, L. A., Hefferon, T., and Roberts, M. (eds.). 2009. Cancer Syndromes. National Center for Biotechnology Information (US) Bookshelf: www.ncbi.nlm.nih.gov/books/NBK1825

Scher, H. I. and Sawyers, C. L. 2005. Biology of progressive, castration-resistant prostate cancer: Directed therapies targeting the androgen-receptor signaling axis. *J. Clin. Oncol.* 23: 8253–8261.

Shah, S. P., Roth, A., Goya, R., Oloumi, A., Ha, G., et al. 2012. The clonal and mutational evolution spectrum of primary triple-negative breast cancers. *Nature* 486: 395–399.

Slamon, D. J., Clark, G. M., Wong, S. G., Levin, W. J., Ullrich, A., and McGuire, W. L. 1987. Human breast cancer: Correlation of relapse and survival with amplification of the HER-2/neu oncogene. *Science* 235: 177–182.

Stephens, P. J., Tarpey, P. S., Davies, H., Van Loo, P., Greenman, C., et al. 2012. The landscape of cancer genes and mutational processes in breast cancer. *Nature* 486: 400–404.

Thorgeirsson, T. E., Gudbjartsson, D. F., Surakka, I., Vink, J. M., Amin, N., et al. 2010. Sequence variants at CHRNB3-CHRNA6 and CYP2A6 affect smoking behavior. *Nat. Genet.* 42: 448–453.

Timofeeva, M. N., Hung, R. J., Rafnar, T., Christiani, D. C., Field, J. K., et al. 2012. Influence of common genetic variation on lung cancer risk: Meta-analysis of 14 900 cases and 29 485 controls. *Hum. Mol. Genet.* 21: 4980–4995.

Tobacco and Genetics Consortium. 2010. Genome-wide meta-analyses identify multiple loci associated with smoking behavior. *Nat. Genet.* 42: 441–447.

van't Veer, L. J., Dai, H., van de Vijver, M. J., He, Y. D., Hart, A. A., et al. 2002. Gene expression profiling predicts clinical outcome of breast cancer. *Nature* 415: 530–536.

Wang, L., Lawrence, M. S., Wan, Y., Stojanov, P., Sougnez, C., et al. 2013c. SF3B1 and other novel cancer genes in chronic lymphocytic leukemia. *N. Engl. J. Med.* 365: 2497–2506.

Wang, Q., Jia, P., Li, F., Chen, H., Ji, H., et al. 2013a. Detecting somatic point mutations in cancer genome sequencing data: A comparison of mutation callers. *Genome Med.* 5: 91.

Wang, Q., Xia, J., Jia, P., Pao, W., and Zhao, Z. 2013b. Application of next generation sequencing to human gene fusion detection: Computational tools, features and perspectives. *Brief Bioinform.* 14: 506–519.

Ware, J. J., van den Bree, M., and Munafò, M. R. 2012. From men to mice: *CHRNA5/CHRNA3*, smoking behavior and disease. *Nicotine Tob. Res.* 14: 1291–1299.

Watson, I. R., Takahashi, K., Futreal, P. A., and Chin, L. 2013. Emerging patterns of somatic mutations in cancer. *Nat. Rev. Genet.* 14: 703–718.

Zack, T. I., Schumacher, S. E., Carter, S. L., Cherniack, A. D., Saksena, G., et al. 2013. Pan-cancer patterns of somatic copy number alteration. *Nat. Genet.* 45: 1134–1140.

Neurological and Psychiatric Disorders

This chapter surveys what has been learned since 2010 about the genetic contributions to five domains of neurological and psychiatric disease: autism spectrum disorder (ASD), schizophrenia (SCZ), intellectual disability (ID), major and bipolar depression (MDD and BPD), and epilepsy. Each of these conditions afflicts on the order of 1% of the population at any given time, but some of them have higher lifetime prevalence. With the exception of major depression, all have very high heritability, as estimated from twin and sibling recurrence rates, but there is considerable debate over the exact magnitude of the genetic contributions and the nature of the environmental influences. Perhaps because of this debate, as well as disappointing returns from early GWAS, most attention has focused on identification of rare alleles with major effects, and the RAME model has dominated thinking in psychiatric genetics (Sullivan et al. 2012; McCarroll and Hyman 2013). It is ironic, then, that the most spectacular advances have been made in relation to two sources of variation that do not cleanly explain the within-family heritability: de novo mutations and rare copy number variants (CNVs) that may be transmitted from unaffected parents and have variable penetrance and expressivity. As with immune-related disease, it has become unambiguous that the domains of brain disease share a large component of genetic risk involving both common and rare variants. This is particularly true of autism, BPD, and SCZ, but ID and epilepsy are also converging on some common genes.

Lamentably, little is known about how genes and environment interact in neuropsychiatric disorders, despite hundreds of papers on the topic. Individual genetic effects are either small if common or very rare if large, and the environmental triggers are likely to include specific stresses experienced by individuals. Autism prevalence seems to be increasing dramatically, perhaps reflecting new diagnostic practices as well as the tendency of men to father children at a later age, but some argue for a cultural contribution as well. Epidemiologists are well aware that schizophrenia is not evenly distributed between developing and developed countries, native-born and migrant populations, or southern and northern latitudes

(A)

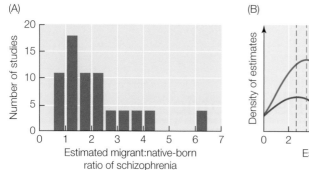

(B)

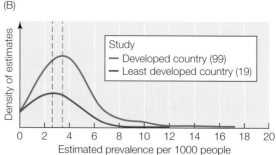

Figure 15.1 Epidemiology of Schizophrenia. (A) The histogram shows the ratio of schizophrenia incidence between migrants and native-born residents in 20 different locations globally, indicating a significant tendency for increased prevalence in migrants. (B) There also appears to be an increased burden of disease contrasting prevalence estimates in 99 studies in developed countries, and 19 studies from least developed countries ($p < 0.05$). Dashed lines indicate median estimate of prevalence. (Data from Saha et al. 2005.)

(**Figure 15.1**). Intellectual disability on the other hand is more common in developing countries, most likely because of malnutrition and infection rather than the cultural bias of IQ tests. Obviously, depression can be brought on by the stresses of life and changes in circumstance and season. All these observations highlight the fact that genetic explanations of neuropsychiatric disease that fail to account for lifestyle and both the biotic and abiotic environment are incomplete. Interesting gene expression profiling studies have established that these environmental factors, including socioeconomic factors, have a long-term effect on the immune system, but it remains to be seen whether and how they affect the brain (Cole 2013).

The five domains considered in this chapter by no means encompass all the neurological and psychiatric conditions of interest to geneticists. Attention deficit hyperactivity disorder shares genetic risk factors with them, but is not discussed here because relatively few robust associations have been reported. Multiple triplet repeat expansion diseases, typified by Huntington's chorea and cerebellar ataxias, have fascinating genetic features. Restless leg syndrome was recognized as a disease only in the past decade, but contributing genes have already been identified, and drugs are being developed around them. Resilience to post-traumatic stress has a genetic basis, and it is likely that generalized risk for acquiring severe phobias (of which there are hundreds) does as well. The line between normality and disease is very blurry for many conditions—paranoia, social function, memory performance—and much interesting genetic analysis focuses on "endophenotypes," which are variable biochemical or cellular attributes that likely contribute mechanistically to disease (Cannon and Keller 2006). Brain imaging is too expensive and time-consuming to perform on the scale of genome-wide association studies, but it is being incorporated into follow-up research. Finally, the pharmacogenetics of neuropsychiatric disease is a compelling topic of great translational importance, but one that is just developing.

Autism

Autism is, many believe, a largely genetic disorder, with typical onset in the first or second year of life followed by gradual progression. Several diagnostic symptoms of autism are also found to some extent in the general population, so it is difficult to be certain how much of its increasing prevalence can be attributed to changes in diagnostic practice. Estimates in the 1980s were that 0.1% of the U.S. population had autism, but the most recent estimates place the prevalence of the broader category of ASD at 1 in 68 children. Boys are four times more likely than girls to have autism. The major impairments involve social interaction, starting with unwillingness to make eye contact, impaired verbal and nonverbal communication, repetitive behaviors, and restriction of interests. A minority, no greater than 10%, of affected individuals have extraordinary abilities such as prolific memory (savantism), while a majority have eating and gastrointestinal problems. Asperger syndrome, one of the major ASDs, is distinguished by the relative absence of language and cognitive delays; PDD-NOS (pervasive developmental disorder not otherwise specified) is also known as atypical autism because its symptoms are more variable. Until quite recently, a strong social stigma was associated with having a child with autism as this disorder was attributed to cold or distant maternal parenting ("refrigerator mothers"), so elucidation of the genetic basis of disease has important implications for the entire family.

Published GWAS to date have been too small to provide much information about the contribution of common variants to autism risk (Wang et al. 2009; Weiss et al. 2009). Just a handful of candidate genes have been implicated, none of which contribute anything approaching 1% of the risk. Nevertheless, several important genes have been highlighted that give a first hint of which biochemical pathways might be perturbed. These genes include the neuronal cell adhesion neurexins *NRXN1* and *CTNTNAP2* as well as cadherin *CDH10*; they also include an organizer of neurotransmitter receptors, *SHANK3*, and the associated neuroligin *NLGN3* (both of which implicate synaptic regulation).

Numerous studies have established a role for copy number variation in autism, as well as in most other neuropsychiatric disorders (Cook and Scherer 2008). Array-based comparative genome hybridization (aCGH, described in Chapter 5), has shown at least six rare CNVs to be enriched in autism cases, though it should be emphasized that the penetrance is far from complete, and in 40% of identified cases the CNV was inherited from a parent who is asymptomatic. One of the CNV duplications, 15q11–13, has a strong maternal effect: inheritance from the mother is associated with ASD in up to 85% of cases, whereas transmission from the father only sometimes leads to mild cognitive impairment. Three of the CNV deletions cover the genes *SHANK3*, *NLGN4*, and *NRXN1*, and more recent whole-genome genotyping studies identifying rare CNVs have expanded the list to include dozens of candidate neurodevelopmental and neuronal plasticity genes. Pinto et al. (2010) did not observe an elevated frequency of rare CNVs larger than 30 kb (most of us have around 2.5 of them), or a difference in their size spectrum, in individuals with ASD, but rather found that the number of genes included in their CNVs was around 20% higher than in controls. This finding corresponds to a modest twofold enrichment in candidate ASD genes after various bioinformatic filters are applied. Furthermore, they argued that in 3.3% of 876 children with ASD, new CNVs are likely to be the primary cause of disease. Although the evidence pertaining to any individual variant is inconclusive, as a whole there is broad agreement that individuals with ASD in general carry a burden of copy number–altered genes that contribute substantively to the etiology of 5%–10% of cases.

Similarly, de novo mutations are very likely to contribute approximately 5% of the risk. Three studies published back-to-back in *Nature* in May 2012 argued for an excess of severe nonsynonymous mutations in probands. Each study considered approximately 200 cases, but they used different methodologies and arrived at subtly different conclusions. O'Roak et al. (2012), for example, identified a protein-protein interaction network among their set of 113 mutated genes that was much more interconnected than expected by chance and suggested a key role for β-catenin and Wnt signaling in neuronal development leading to autism. Sanders et al. (2012) did not find such a network, but they did suggest that over 40% of the de novo single-nucleotide variants (SNVs) they found in brain-expressed genes are likely to be pathogenic. Neale et al. (2012) generated an expanded network in conjunction with intellectual deficiency genes and prior candidates without remarking on functional enrichment. All three groups noted an increase in the mutation rate with parental age, most likely driven by the father's age, since there seems to be a male bias in the generation of mutations. Eighteen genes were found mutated in at least two cases of autism, and although this number is not greater than expected by chance, further sequencing of several thousand cases identified multiple hits in a few of the genes that were followed up on, notably *GRIN2B*, *SNC2A*, and *CHD8*. The first two studies estimated that the mutational target for ASD

Figure 15.2 Gene Expression Profiling of Autism.   (A) Voineagu et al. (2011) used ▶ Illumina microarrays to contrast gene expression in the frontal and temporal cortex between 19 individuals with autism and 17 controls. 444 transcripts showed differential expression, 79 of which were replicated in a second sample. The heat map shows that after two-way hierarchical clustering of the top 200 genes, most of the samples from affected individuals form a separate cluster from the controls, and they seem to form two subclasses. Green represents low expression, red high, and the bars below the heat map show the age, sex, and brain region of each sample. Few significant differences were seen in the cerebellum. (B) Some of these genes form a co-expression module, shown here, in which edges linking transcripts imply very high correlation, and eight hub genes are identified. (C) An expanded network of 150 genes is enriched for gene ontology terms related to synaptic and axonal function as well as alternative splicing.

may consist of between 400 and 1000 genes, but Neale and colleagues emphasized that it would be wrong to think of autism as a collection of very rare single-gene disorders; their modeling suggests that it is much more likely that there are a large number of neuronal development genes that can yield mutations that elevate risk as much as 10- to 20-fold and are thus important contributors in one-fifth of all cases of autism. The source of the reported very high heritability remains to be ascertained, but seems likely to involve complex interactions among multiple genes segregating in each family.

Gene expression profiling in relation to neuropsychiatric disorders is much more difficult than for immune and metabolic disease because brain tissue can only be obtained postmortem. Nevertheless, Voineagu et al. (2011) reported encouraging results from a comparison of just 19 cases and 17 age-matched controls. Microarray analysis of cortex (but not cerebellum) samples found 444 genes to be differentially expressed (**Figure 15.2**). Two modules of genes

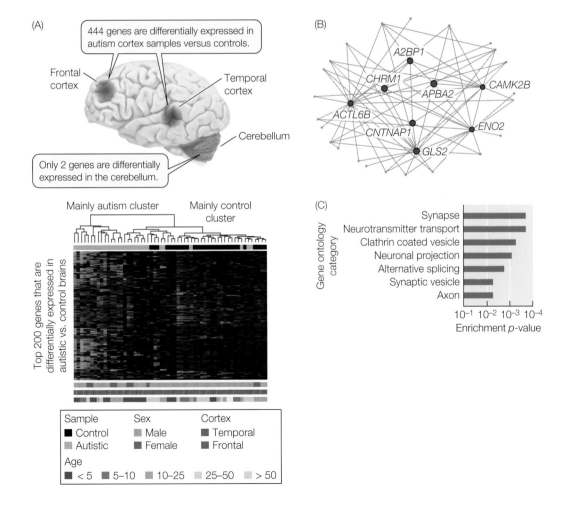

are particularly noteworthy: an upregulated module annotated to astrocyte and microglial function, which may indicate an inflammatory response to disease, and a downregulated module enriched for genes involved in synaptic and cytoskeletal functions. One of the hub genes of the latter module, *A2BP1*, encodes a splicing regulator, and follow-up RNA-Seq comparison of three cases, each with high or low expression of this gene, found differential splicing of 212 genes, many in the same module. Furthermore, the downregulated genes are enriched in the set of genes that approach genome-wide significance in GWAS, suggesting that synaptogenesis genes may be a major component of autism risk, due to both local regulatory variants and global factors that cause decreased expression or alter splicing.

More recently, two groups took a systems biology approach aiming to integrate gene co-expression information across a developmental time series as well as in different regions of the brain of neurologically healthy individuals (Parikshak et al. 2013; Willsey et al. 2013). Using different analytic strategies, both groups found several networks of transcripts that had different temporal and spatial trajectories and were significantly enriched for genes implicated in autism, but not in intellectual disability. An early- to midfetal network seemed to connect genes involved in neuronal migration and axon outgrowth that were disrupted by the de novo mutations identified by whole-exome sequencing, whereas networks expressed late in the fetal brain and soon after birth were more likely to be associated with common variants identified by GWAS and to have roles in synaptogenesis and neuronal signal transduction in particular layers of the cortex. Further analyses have demonstrated enrichment for transcription factor binding sites and chromatin features, providing hints as to how a common pathology can arise from disruption of different members of a core set of genes and how disease penetrance might be modified by gene activity in the networks.

Schizophrenia

Although schizophrenia is popularly characterized as "split personality," it is more commonly a disorder involving multiple psychiatric abnormalities. The symptoms include delusions, paranoia, hallucinations, and disorganized thinking on the one hand and loss of spontaneity, capacity to experience pleasure, or ability to take initiative on the other, all of which often lead people with schizophrenia to have problems maintaining employment and relationships and ultimately decrease their life expectancy. The age of onset is typically young adulthood. Prevalence is around 0.5%, but is modified by environmental factors, increasing in children born after parents move to cities, for example (McGrath et al. 2008; see Figure 15.1). The fact that antipsychotic medicines target dopamine and serotonin signaling naturally led to hundreds of studies investigating variation in the genes regulating the activity of these neurotransmitters, but association with the dopamine receptor gene *DRD2* only emerged after extremely large meta-analyes. Similarly, candidate gene-by-environment studies initially showed great hope for demonstrating how genes and behavioral or cultural variables may interact, but single-gene interactions have not proven to be robust (Duncan and Keller 2012).

The genetics of schizophrenia shares many features with that of autism. Contributions have been established for rare and de novo CNVs and for de novo point mutations (Xu et al. 2012; Fromer et al. 2014), but the penetrance of these variants is unknown, and they are unlikely to contribute to more than one-fourth of cases. Larger GWAS, including over 34,000 cases, have now been reported (Ripke et al. 2014), and although it seems that the effect sizes of the top hits are in general smaller than for other complex diseases, 128 independent polymorphisms with genome-wide significance have been discovered in 108 loci. The credible SNPs are enriched in enhancers that are active in numerous neuronal cell types, and individuals in the top decile for a composite risk score are approximately tenfold more likely to have schizophrenia than those in the bottom decile of the score. A method called approximate Bayesian analysis actually estimates that in excess of 8000 common variants will be found to explain half of the variation in susceptibility, and hence the majority of the genetic variance (see Figure 6.9A). Several interesting biological insights have emerged from these studies. One is an unexpected role for calcium signaling, with two channels and several associated genes at or near GWAS significance levels. Another is that the HLA complex, which is traditionally linked to immune function, is a major contributor, with variants spread across the entire HLA region. Third, the microRNA gene *MIR137* is predicted to target 15 of the 22 established loci and may also be a susceptibility locus itself. Similarly, long noncoding RNAs (lncRNAs) are present in the vicinity of more than half of those 22 loci.

Given the limited role for de novo variants, which in any case cannot explain the high heritability of schizophrenia because they are not usually shared by relatives (though they may elevate the recurrence rate for identical twins), Need et al. (2012) evaluated the potential contribution of Goldilocks alleles: low-frequency variants of modest effect that would not be tagged on genotyping arrays. They sequenced the exomes of 166 cases of treatment-refractory schizophrenia and found 5155 rare variants, none of which had effect sizes conferring a relative risk of greater than 5 that would be required to attain Bonferroni-adjusted significance with a minor allele frequency less than 0.05 (**Figure 15.3A**). They followed up 4028 of these variants by targeted genotyping in an additional 2600 cases and found a half-dozen individuals with multiple variants, but pointed out the low statistical power of even this scale of experiment. For example, 5 affected individuals had a rare coding mutation in the *KLOTHO* gene, compared with none of over 7400 controls, a seemingly convincing enrichment, but it would actually take almost 30 instances of the mutation, typed in something like 13,000 cases, to establish genome-wide significance. An interesting feature of this particular variant is that it affects vitamin D metabolism, which has previously been implicated in schizophrenia risk, which is greater at high latitudes and for children born in winter months (both of which relate to sunlight exposure and hence to vitamin D production).

(A)

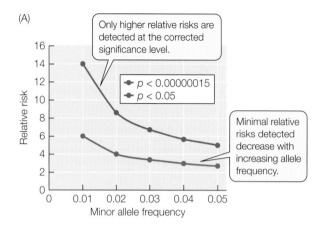

Only higher relative risks are detected at the corrected significance level.

● p < 0.00000015
● p < 0.05

Minimal relative risks detected decrease with increasing allele frequency.

(B)

Gene set	Number of genes	p-value
Disruptive		
All	2546	0.0001
Known de novo	87	0.0007
In de novo CNV	234	0.0039
Calcium channel	26	0.0214
ARC network	28	0.0014
NMDAR network	61	0.0251
PSD-95 genes	65	0.0009
Nonsynonymous		
All	2546	0.0015
Known de novo	611	0.0011

Figure 15.3 Power of Rare Variant Discovery.   (A) In a case-control whole-exome sequencing study of 166 people with schizophrenia and 307 controls, Need et al. (2012) computed the 99% power to detect variants with the indicated relative risks and minor allele frequencies. The two curves show power for nominal significance ($p < 0.05$) and after adjustment for the total number of rare variants discovered ($p < 1.5 \times 10^{-7}$). For example, an allele with a minor allele frequency of 0.01 will be confidently detected only if it increases the risk 6-fold at the nominal level and 14-fold at the adjusted level. (B) Statistical burden tests evaluate the significance of enrichment of all identified rare variants in subsets of genes. In the Purcell et al. (2014) study of 2536 affected individuals and 2543 controls, enrichment was observed for disruptive (that is, nonsense, frameshift, or splice site) mutations present in 10 or fewer individuals in general, as well as in those genes also disrupted by de novo mutations in affected individuals, or annotated to the activity-regulated cytoskeleton (ARC) or postsynaptic density (PSD), and less significantly to voltage-gated calcium signaling, or the N-methyl-D-aspartate receptor complex (NMDAR). Conserved nonsynonymous mutations also show some enrichment.

More recently, Purcell et al. (2014) sequenced the exomes of 2536 affected individuals (cases) and confirmed the expected enrichment for genes involved in synapse development (**Figure 15.3B**). They pointed out, though, that 46% of the cases, compared with 41% of controls, had mutations in a large set of candidate genes that encompasses one-tenth of the genome. The conclusion is that rare variants do contribute to schizophrenia risk, but most are very rare, and it will take extremely large samples to confirm individual genes. We are left with the impression that schizophrenia fits the infinitesimal model of genetic risk (ISC 2009), only with smaller common variant effect sizes, and possibly greater heterogeneity, than immune-related diseases.

A novel functional genomic approach was taken by Brennand et al. (2011). They caused human induced pluripotent stem cells (hiPSC) to form neurons by treating undifferentiated cells with neuronal growth factors. Cultures of cells derived from four controls and from four individuals with familial schizophrenia showed normal neuronal differentiation, but reduced synaptic connectivity and neurite outgrowth was observed in the cells derived from affected individuals. Microarray analysis then documented almost 650 differentially expressed genes that overlapped significantly with differences observed in the postmortem brain samples. Affected pathways included Wnt signaling, glutamate receptor activity, and cAMP signaling, the latter two of which are required for long-term potentiation and memory establishment. This observation is consistent with the deficit in short-term memory access that is commonly seen in people with schizophrenia. Furthermore, one of five antipsychotic drugs that was tested on the cell cultures, Clozapine, rescued the defect in connectivity and restored some of the gene expression, highlighting the potential utility of this ex vivo model of disease pathology. Genome engineering techniques are now being used to evaluate the function of rare variants in schizophrenia as well as other neuropsychiatric conditions.

Intellectual Disability

Intellectual disability is usually defined by an IQ of less than 70, and its prevalence in Western countries is around 0.5%. Lifetime costs to the healthcare system for each individual with ID are estimated to be in excess of $1 million, which, in addition to the emotional and financial burden placed on caregivers and families, makes ID one of the most significant public health concerns in developed countries. Some cases of ID are attributable to environmental factors such as fetal alcohol exposure, malnutrition, and infection, but it is thought that very rare mutations of large effect account for many, if not the majority, of cases. Incidence is approximately 30% higher in boys than in girls, in part due to the hemizygosity of the X chromosome, which harbors over 80 genes that are disrupted in X-linked intellectual disability (XLID), which accounts for 10% of cases (Tarpey et al. 2009). Another 15% of cases are associated with visible chromosome abnormalities, either aneuploidy or inversions, most famously trisomy 21 (Down syndrome). The advent of aCGH has led to the detection of smaller CNVs in up to 15% more cases. This leaves around 50% of cases of ID yet to be explained, and it is thought that many will be found to be due to autosomal dominant (ADID) and recessive (ARID) causes (Ropers 2010).

The most common single-gene cause of XLID is a massive expansion of the CGG trinucleotide repeat upstream of the fragile X syndrome gene, *FMR1*, that leads to transcriptional silencing and hence loss of function of a key protein component of synaptic reinforcement. Systematic X-exome sequencing of most of the exons on the X chromosomes of hundreds of affected boys has since identified candidate genes with a wide range of functions. Tarpey et al. (2009) identified probable pathogenic variants in 25% of 208 cases, with particularly high-confidence attribution to 6 genes (*AP1S2*, *CUL4B*, *BRWD3*, *UPF3B*, *ZDHHC9*, and *SLC9A6*) that exhibited 22 mutations leading to premature stop codons, accounting for 10% of XLID. In all, 30 genes had presumably deleterious truncation mutations,

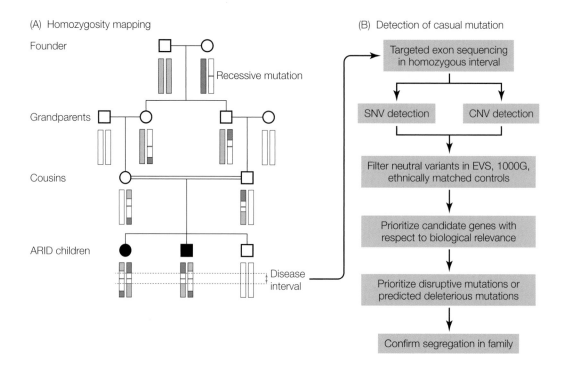

(A) Homozygosity mapping

Founder

Recessive mutation

Grandparents

Cousins

ARID children

Disease interval

(B) Detection of casual mutation

Targeted exon sequencing in homozygous interval

SNV detection

CNV detection

Filter neutral variants in EVS, 1000G, ethnically matched controls

Prioritize candidate genes with respect to biological relevance

Prioritize disruptive mutations or predicted deleterious mutations

Confirm segregation in family

but 19 of these mutations were observed in a single case, and 9 of those were also observed in unaffected family members. In follow-up analyses accounting for the presence of a large number of missense variants in the Exome Variant Server (EVS) database of 10,500 X chromosomes, Piton et al. (2013) argued that only 79 of 104 presumed causal XLID genes really are pathogenic. For example, assuming complete penetrance, and given that less than 1% of the 2500 EVS males are likely to have ID, any variant found twice would account for 8% of the expected 25 cases, much more than *FMR1*, which is highly unlikely. Nevertheless, it is clear that de novo X mutations and cases of transmission from normal-IQ mothers contribute substantially to the burden of disease.

The identification of autosomal genes involved in ID is particularly difficult in nonsyndromic cases, which account for around half the cases. Syndromic forms are easier to attribute because associated craniofacial or other abnormalities co-occur with the intellectual disability in family members. Thus, Kabuki syndrome and Schinzel- Giedion syndrome have been attributed to recurrent mutations in the *MLL2* and *SETBP1* genes, respectively (though the former is genetically heterogeneous). De Ligt et al. (2012) were among the first to carry out large-scale whole-exome sequencing in family trios, finding likely causal mutations in 16 of 100 children with ID. Ten of those children had de novo ADID mutations, one was a compound heterozygote with one inherited and one de novo variant, three had inherited XLID, and two had de novo XLID. Notably, no fully inherited

◀ **Figure 15.4** **Strategy for Mapping Rare Variants in Consanguineous Families.**
(A) Homozygosity mapping narrows the search space for rare variants to those regions
of the genome that are homozygous in the genome of a proband. Most such regions will
be identical by descent, for example, because two cousins who married both carried a
mutation that one of their parents inherited from a common progenitor. (B) The strategy for detecting likely causal mutations in such intervals is to identify all novel single
nucleotide variants (SNVs) or copy number variants (CNVs) in the exons covered by the
interval, filter out variants found in control individuals, ethnically matched where possible,
and then prioritize according to the known biological function of the gene and the likely
pathogenicity of the mutation. Subsequently, segregation of the mutation in affected and
unaffected family members is tracked, and if possible, a functional assay is developed.
Najmabadi et al. (2011) used this approach to screen 136 Iranian children with autosomal
recessive intellectual disability (ARID) and detected candidate mutations in 23 known and
50 novel genes.

ARID mutations were detected, which implied that in randomly mating
communities, recessive single-gene causes of ID are rare. By contrast, an
Iranian study of consanguineous families (40% of extended Iranian families
involve marriages between cousins, which are common in North Africa and
the Near East) did detect a high incidence of ARID genes. Najmabadi et al.
(2011) first identified stretches of homozygosity shared by family members
with ID, then sequenced the exons in just these regions (**Figure 15.4**). They
identified a single causal gene in 78 of 136 families. These causal genes
included 28 with truncation (nonsense) mutations, and each of 50 novel
genes was observed to be disrupted in a single family. The functions of the
affected proteins ranged from general housekeeping roles in chromatin and
metabolism to specific synaptic and neuronal differentiation, and specific
interventions were suggested in a few cases.

The picture that emerges is of a highly heterogeneous condition
attributable largely to very rare coding variants, most of which probably remain in the gene pool for only one or two generations. Conservative estimates imply that between 800 and 1000 genes may mutate to
cause ID, which means that average allele frequencies must be less than
0.001% (Topper et al. 2011). Yet these analyses still leave at least one-third
of the genetic contributions to ID unaccounted for by exonic variants,
suggesting either that rare regulatory variants that turn off gene expression may be involved or that interactions between two or more genes
are critical in some cases. Future studies will also address the role of
the genetic background in modifying penetrance and expressivity. Since
clinical whole-exome resequencing for discovery of likely causal variants
is now a reality, new resources that are helpful to parents and physicians
as they sift through the data are being developed. Particular note should
be taken of the Seattle-based Mendelian.org website and the UK-based
Decipher Developmental Disorders (DDD) website and database (www.
ddduk.org). The latter is an excellent example of a site that also provides
resources for family members of children with psychological and neurological disorders (**Figure 15.5**).

Figure 15.5 Websites for Associations Related to Psychological and Neurological Disorders.   Community-oriented websites that explain the genetics of schizophrenia (welivewithsz.com), autism (www.autismspeaks.org), intellectual disability (www.thearc.org), and depression (www.dailystrength.org or www.adaa.org for the American Depression and Anxiety Association) are available to individuals and families living with these disorders. The Epilepsy Foundation also has a similar site (www.epilepsyfoundation.org; not shown). DDD (www.ddduk.org) is a UK-based site that compiles information across a broad range of pediatric developmental disorders.

Major and Bipolar Depression

Depression encompasses two types of mood disorders, major depressive disorder and bipolar disorder. The former is an experience of anxiety, sadness, and a sense of worthlessness that can last for weeks or months, while the latter involves mood swings between major depression and episodes of elevated, excited mood (or mania) that can last for a week. The prevalence of BPD is as high as 4%, and the condition has very high heritability, on a par with that observed for schizophrenia and autism. Major depressive disorder is much more common, but has a modest heritability of less than 40%. It also has various environmental triggers, including loneliness, seasonal affect in gloomy latitudes, and childbirth (postpartum depression), so the incidence varies, but its lifetime prevalence approaches 20%. However, "only" 5% of men and 10% of women have a high likelihood of suffering long-term depression at any time. The age of onset is typically young adulthood, with a second peak in middle age. According to the "kindling theory," the threshold for triggers of depression is lowered after each episode, so early diagnosis and prevention is a priority (Kendler et al. 2000). Depression is associated with suicide attempts and suicidal ideation, both of which also have a heritable basis.

For many years, genetic analyses focused on the serotonin axis, since the two major classes of antidepressants (selective serotonin reuptake inhibitors, or SSRIs, and tricyclic antidepressants) target the length of time serotonin remains in the synaptic cleft or the activity of the serotonin receptor. A long polymorphic repeat in the promoter of the serotonin transporter gene, known as the *5HTT-LPR*, has been the subject of several thousand behavioral studies (Munafò 2012) because the short form of the repeat leads to lower transcription, which in turn has been associated with elevated anxiety and risk of depression, higher activity in the amygdala on MRI scans in response to aversive stimuli, and altered pharmacological responses.

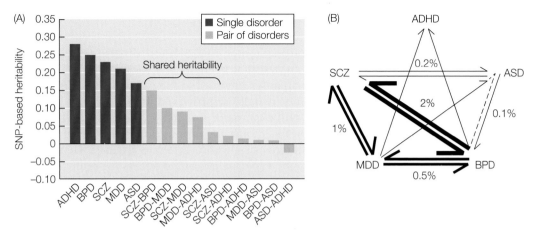

Figure 15.6 Cross-Disorder Shared Heritability.   (A) Estimates of the heritability explained by all common variants jointly for each of five neuropsychological disorders and of the shared heritability explained genome-wide for each pair of disorders. Common variants explain up to a quarter of the variance in disease liability, and a significant component of the risk for SCZ, BPD, and MDD is shared. ADHD is somewhat distinct. (B) An alternative demonstration of shared heritability was obtained by generating a genetic risk score from the sum of all SNPs at $p < 0.05$ for each of the five disorders and asking what proportion of the other four disorders is explained by the same score. As in (A), there is highly significant sharing of liability for MDD, BPD and SCZ. The width of each arrow is proportional to the percentage of variance explained by the genetic risk score. (A, after CDG-PGC 2013a; B, after CDG-PGC 2013b.)

However, meta-analyses are inconclusive, and it is now clear that *SLC6A4*, which encodes the serotonin transporter, is at best a minor genetic factor, one whose effect is less than that associated with most GWAS loci.

The largest meta-analysis of neuropsychiatric disease to date (as of June 2014) actually considered 33,000 cases encompassing all of the five major disorders—SCZ, ASD, MDD, BPD, and ADHD—simultaneously—that is, as if they were one disease (CDG-PGC 2013a, 2013b; Ripke et al. 2013). It confirmed that these five conditions have a shared genetic component (**Figure 15.6**). Genome-wide heritability estimated from all SNPs implies that the three adult-onset conditions, SCZ, BPD, and MDD, share the most variation, and in fact, the top four genes that exceeded the GWAS threshold of 5×10^{-8} all had effects in the same direction for each condition. Polygenic risk scores derived from all significant SNPs at even nominal levels predicted each disease. Intriguingly, MDD and BPD were both more closely correlated with SCZ than with each another. Among the top 14 loci, 3 may be specific for bipolar disorder and 4 for schizophrenia. A clear biological conclusion that has emerged from focused analysis of BPD is that calcium channels are contributors to this disorder, two of which (*CACNA1C* and *CACNAB2*) have genome-wide significance, while another five are implicated.

Epilepsy

Epilepsy is another not uncommon neurological disease for which the major focus of genetic analysis has been on rare variants and on familial cases in which linkage analyses have led to the causal genes. Epilepsy is characterized by seizures, usually unprovoked and recurring. While as many as 3% of people will experience seizures in their lifetimes, annual incidence is less than one in a thousand. Various different types of epilepsy are recognized, and different classification schemes are used. Generalized epilepsy, the most common type, is distributed across the brain. Absence epilepsy refers to disease that is accompanied by loss of consciousness for up to 20 seconds. Many other syndromes, defined by electroenceph-alogram, age of onset, and behavioral symptoms, are recognized; these syndromes include Angelmann, Dravet, Lennox-Gastaut, and Ohtahara syndromes, for which OMIM lists several causal genes identified prior to the genome era. The majority of cases of generalized epilepsy can be managed with one or more of 20 approved drugs, but these drugs may have long-term side effects, and surgery may be required in refractory cases.

Copy number variation is one contributor to epilepsy, with 1% of cases attributable to microdeletions in the vicinity of the *CHRNA7* acetylcholine receptor locus at 15q13.3 (Helbig et al. 2009). This deficiency, which is also associated with other types of neurological disorders to a lesser degree, shows incomplete penetrance, as it is often inherited from asymptomatic parents. A systematic survey by Cooper et al. (2011) of CNVs in the genomes of over 8000 healthy adults contrasted with almost 16,000 children with developmental delay and/or intellectual disability, and who were affected by a wide range of comorbid congenital diseases, found that deletions and duplications were more common in children with epilepsy than in those with ID or ASD, but less common than in children with craniofacial or cardio-vascular defects (**Figure 15.7**). They also noted that the enrichment in cases increased with the size of the CNVs, such that variants around 400 kb were observed in 35% of ID cases and 11% of healthy controls, whereas variants around 1.5 Mb were seen in 11% of cases and just 0.6% of healthy children.

Whole-exome sequencing has also revealed an excess of de novo mutations in children with severe forms of juvenile epilepsy. The Epi4K Consortium (2013) reported that around 10% of cases could be attributed to mutations in nine genes that were multiply mutated. They found that the top quartile of genes that are most intolerant of mutations, after various statistical adjustments, was significantly enriched for de novo epilepsy mutations, and they estimated that as many as 90 of those 4264 genes may be targets. Ion channel genes were also enriched, notably those involving GABA, as were genes previously implicated in ID and ASD, extending the shared genetic etiology of neurological disorders to epilepsy. At the same time, the EPICURE Consortium (2012) conducted GWAS of 3000 cases, noting a heritability of up to 30% for generalized epilepsy, and found four loci with genome-wide significance, two of which may be specific for juvenile absence and so-called myoclonic subclasses of

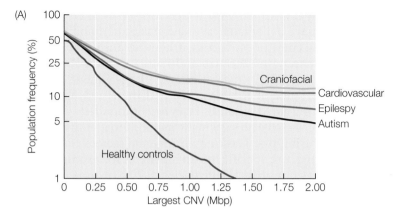

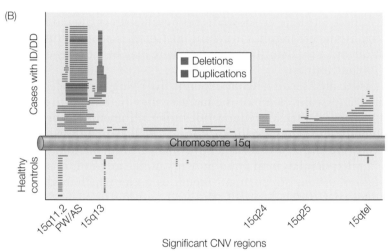

Figure 15.7 Copy Number Variation in Intellectual Disability/Developmental Delay (ID/DD). (A) A survey of over 15,000 children with intellectual disability and/or developmental delay found an increasing burden of longer copy number variants (CNVs) for a variety of diseases. Epilepsy was intermediate between autism and craniofacial or cardiac defects. (B) Portrait of deletions and duplications on chromosome 15q, showing six regions that are enriched in children with ID/DD, including the portion proximal to the telomere. (After Cooper et al. 2011.)

epilepsy. Thus, it seems that epilepsy, like schizophrenia, will be seen to fit the infinitesimal model, with relatively small individual contributions from common variants along with many large rare mutations that have variable penetrance and expressivity.

Evolving Mental Health

Despite the strong evidence for contributions by CNVs, rare variants, and de novo variants, as well as common polymorphisms, to the risk of neuropsychiatric illness, this field is perhaps the most contentious in human genetics. I have deliberately avoided stating heritability estimates for each of the five conditions discussed here because they vary by as much as 30% depending on the size and nature of the cohort and the analytic methods used. While there is clearly a large genetic component to these disorders, some observers argue that the environmental contribution is underestimated because it can be confounded with familial factors and because we have no systematic methods to identify individual environmental contributors. These environmental factors could include random stressors that are specific for each patient, for example, or they could involve toxins or nutrient deficiencies that have gone undetected. They could be related to behaviors we can change (perhaps chronic iPad use and electronic stimulation before kids can walk or catch a butterfly), or to ones that it is highly unlikely we would change even if they could be identified, let alone be shown to be causal (such as the modern norm of living as nuclear families in suburbia). Genetics is powerful in both a statistical and a mechanistic sense, but there may be a tendency to overstate its contribution.

An associated issue is what explains the rising incidence of neuropsychiatric disease. Some of the increase may be related to ascertainment and new diagnostic criteria, but strong arguments can be made for genotype-by-environment interactions, as well as mutation accumulation, contributing to a real rise in prevalence. Humans today grow up in cultural and behavioral environments unlike any in human history, and it is not difficult to postulate scenarios under which modern social hierarchies perturb the sense of self-worth and contribute to depression, for example (but it is extremely difficult to substantiate them!). Furthermore, it is almost certain that there is a slowly increasing burden of rare deleterious variants in the human gene pool as a consequence of relaxed selection and expanding population sizes. As we have seen, some of these variants are major risk factors for schizophrenia and autism, and they probably influence intelligence and mood as well. On the other hand, something called the Flynn effect describes a highly significant gain in IQ, as much as 20 points or more over the past two generations, so any general increase in risk is for the time being masked by other forces influencing human cognition.

It is thus difficult to foresee how the genetic contribution to mental health will change in the coming centuries. It would seem inevitable, though, that some parents will be inclined to make reproductive choices based on genetic data once they are apprised of tools that, for example, indicate that their future children may be at risk of disease, or that polygenic risk scores may classify embryos before implantation with respect to even slightly elevated odds of preferred mental functions. The emotional and financial burden of highly deleterious rare variants is such that rationalists may argue strongly for intervention, but in the absence of certainty, the morality of such decisions will be questioned, and the specter of so-called experts making decisions for families raises new ethical concerns. Pointedly, these are not issues that can be delayed until the next generation: the technology is, in primitive form, already with us, and it is likely to have the greatest impact in relation to choices surrounding mental health.

Summary

1. Autism, schizophrenia, and bipolar disorder are all neuropsychiatric conditions with very high heritability that collectively affect more than 5% of the population and share components of genetic risk.

2. Intellectual disability and epilepsy are neurological disorders that also have a large, and to some extent shared, genetic component.

3. Copy number variants make a substantial contribution in 10%–15% of the cases of all these conditions. Their effects are more severe when they are longer and affect more genes.

4. Autism is affected by disruption of genes affecting cell adhesion, axonal migration, and synaptic function. Only a handful of loci have been established, but up to a thousand genes may be involved.

5. Despite the evidence that de novo mutations are enriched in genetic interaction networks in children with autism, most cases are presumed to be at least oligogenic, requiring multiple genetic risk factors.

6. Gene expression profiling of postmortem brain tissue or of neuronal derivatives of induced pluripotent stem cells also implicate networks of genes involved in synapse formation and function in neuropsychiatric disorders.

7. Schizophrenia has more than 20 published common variant risk associations, and a substantial component of its heritability can be attributed to thousands of variants of small effect.

8. Novel pathways in the etiology of schizophrenia include calcium channels, the HLA complex, and targets of the miR137 microRNA.

9. Approximately 50% of intellectual disability, defined as an IQ of less than 70, can be attributed to very rare mutations, including aneuploidy, CNVs, de novo mutations, and inherited variants.

10. Bipolar disorder has a higher heritability than major depressive disorder, but the genetic basis for both diseases is as yet unclear.

11. Epilepsy is a collection of syndromic and generalized seizure conditions in which ion channels and various types of neurotransmission defects have been implicated.

12. Although neurological and psychiatric diseases are commonly thought to fit the rare alleles of major effect model, increasing evidence points to infinitesimal contributions, along with environmental triggers that will be very difficult to dissect.

References

Brennand, K. J., Simone, A., Jou, J., Gelboin-Burkhart, C., Tran, N., et al. 2011. Modeling schizophrenia using hiPSC neurons. *Nature* 473: 221–225.

Cannon, T. D. and Keller, M. C. 2006. Endophenotypes in the genetic analyses of mental disorders. *Annu. Rev. Clin. Psychol.* 2: 267–290.

Cole, S. W. 2013. Social regulation of human gene expression: Mechanisms and implications for public health. *Am. J. Public Health* 103(Suppl 1): S84–S92.

Cook, E. H. and Scherer, S. W. 2008. Copy number variations associated with neuropsychiatric conditions. *Nature* 455: 919–923.

Cooper, G. M., Coe, B. P., Girirajan, S., Rosenfeld, J. A., Vu, T. H., et al. 2011. A copy number variation morbidity map of developmental delay. *Nat. Genet.* 43: 838–846.

Cross-Disorder Group of the Psychatric Genomics Consortium (CDG-PGC). 2013a. Genetic relationship between five psychiatric disorders estimated from genome-wide SNPs. *Nat. Genet.* 45: 984–994.

Cross-Disorder Group of the Psychiatric Genomics Consortium (CDG-PGC). 2013b. Identification of risk loci with shared effects on five major psychiatric disorders: A genome-wide approach. *Lancet* 381: 1371–1379.

de Ligt, J., Willemsen, M. H., van Bon, B. W. M., Kleefstra, T., Yntema, H. G., et al. 2012. Diagnostic exome sequencing in persons with severe intellectual disability. *N. Engl. J. Med.* 367: 1921–1929.

Duncan, L. E. and Keller, M. C. 2012. A critical review of the first 10 years of candidate gene-by-environment interaction research in psychiatry. *Am. J. Psych.* 168: 1041–1049.

Epi4K Consortium; Epilepsy Phenome/Genome Project. 2013. *De novo* mutations in epileptic encephalopathies. *Nature* 501: 217–221.

EPICURE Consortium; EMINet Consortium. 2012. Genome-wide association analysis of genetic generalized epilepsies implicates susceptibility loci at 1q43, 2p16.1, 2q22.3 and 17q21.32. *Hum. Mol. Genet.* 21: 5359–5372.

Fromer, M., Pocklington, A. J., Kavanagh, D. H., Williams, H. J., Dwyer, S., et al. 2014. De novo mutations in schizophrenia implicate synaptic networks. *Nature* 506: 179–184.

Helbig, I., Mefford, H. C., Sharp, A. J., Guipponi, M., Fichera, M., et al. 2009. 15q13.3 microdeletions increase risk of idiopathic generalized epilepsy. *Nat. Genet.* 41: 160–162.

International Schizophrenia Consortium (ISC). 2009. Common polygenic variation contributes to risk of schizophrenia and bipolar disorder. *Nature* 460: 748–752.

Kendler, K. S., Thornton, L. M., and Gardner, C. O. 2000. Stressful life events and previous episodes in the etiology of major depression in women: An evaluation of the "kindling" hypothesis. *Am. J. Psychiatry* 157: 1243–1251.

McCarroll, S. A. and Hyman, S. E. 2013. Progress in the genetics of polygenic brain disorders: Significant new challenges for neurobiology. *Neuron* 80: 578–587.

McGrath, J., Saha, S., Chant, D., and Welham, J. 2008. Schizophrenia: A concise overview of incidence, prevalence, and mortality. *Epidemiol. Rev.* 30: 67–76.

Munafò, M. 2012. The serotonin transporter gene and depression. *Depress. Anxiety* 29: 915–917.

Najmabadi, H., Hu, H., Garshasbi, M., Zemojtel, T., Abedini, S. S., et al. 2011. Deep sequencing reveals 50 novel genes for recessive cognitive disorders. *Nature* 478: 57–63.

Neale, B. M., Kou, Y., Liu, L., Ma'ayan, A., Samocha, K. E., et al. 2012. Patterns and rates of exonic de novo mutations in autism spectrum disorders. *Nature* 485: 242–245.

Need, A. C., McEvoy, J. C., Gennarelli, M., Heinzen, E. L., Ge, D., et al. 2012. Exome sequencing followed by large-scale genotyping suggests a limited role for moderately rare risk factors of strong effect in schizophrenia. *Am. J. Hum. Genet.* 91: 303–312.

O'Roak, B. J., Vives, L., Girirajan, S., Karakoc, E., Krumm, N., et al. 2012. Sporadic autism exomes reveal a highly interconnected protein network of de novo mutations. *Nature* 485: 246–250.

Parikshak, N. N., Luo, R., Zhang, A., Won, H., Lowe, J. K., et al. 2013. Integrative functional genomic analyses implicate specific molecular pathways and circuits in autism. *Cell* 155: 1008–1021.

Pinto, D., Pagnamenta, A. T., Klei, L., Anney, R., Merico, D., et al. 2010. Functional impact of global rare copy number variation in autism spectrum disorders. *Nature* 466: 368–372

Piton, A., Redin, C., and Mandel, J. L. 2013. XLID-causing mutations and associated genes challenged in light of data from large-scale human exome sequencing. *Am. J. Hum. Genet.* 93: 368–383.

Purcell, S. M., Moran, J. L., Fromer, M., Ruderfer, D., Solovieff, N., et al. 2014. A polygenic burden of rare disruptive mutations in schizophrenia. *Nature* 506: 185–190.

Ripke, S., Neale, B. M., Corvin, A., Walters, J. T. R., Farh, K-H., et al. for the Schizophrenia Working Group of the Psychiatric Genomics Consortium. 2014. Biological insights from 108 schizophrenia-associated genetic loci. *Nature* 511: 421-427.

Ripke, S., Wray, N. R., Lewis, C. M., Hamilton, S. P., et al. 2013. A mega-analysis for the Major Depressive Disorder Working Group of the Psychiatric GWAS Consortium of genome-wide association studies for major depressive disorder. *Mol. Psychiatry* 18: 497–511.

Ropers, H. H. 2010. Genetics of early onset cognitive impairment. *Annu. Rev. Genomics Hum. Genet.* 11: 161–187.

Saha, S., Chant, D., Welham, J., and McGrath, J. 2005. A systematic review of the prevalence of schizophrenia. *PLoS Med.* 2: e141.

Sanders, S. J., Murtha, M. T., Gupta, A. R., Murdoch, J. D., Raubeson, M. J., et al. 2012. De novo mutations revealed by whole exome sequencing are strongly associated with autism. *Nature* 485: 237–241.

Sullivan, P. F., Daly, M. J., and O'Donovan, M. 2012. Genetic architectures of psychiatric disorders: The emerging picture and its implications. *Nat. Rev. Genet.* 13: 537–551.

Tarpey, P. S., Smith, R., Pleasance, E., Whibley, A., Edkins, S., et al. 2009. A systematic, large-scale resequencing screen of X-chromosome coding exons in mental retardation. *Nat. Genet.* 41: 535–543.

Topper, S., Ober, C., and Das, S. 2011. Exome sequencing and the genetics of intellectual disability. *Clin. Genet.* 80: 117–126.

Vissers, L. E., de Ligt, J., Gilissen, C., Janssen, I., Steehouwer, M., et al. 2010. A de novo paradigm for mental retardation. *Nat. Genet.* 42: 1109–1112.

Voineagu, I., Wang, X., Johnston, P., Lowe, J. K., Tian, Y., et al. 2011. Transcriptomic analysis of autistic brain reveals convergent molecular pathology. *Nature* 474: 380–384.

Wang, K., Zhang, H., Ma, D., Bucan, M., Glessner, J. T., et al. 2009. Common genetic variants on 5p14.1 associate with autism spectrum disorders. *Nature* 459: 528–531.

Weiss, L., Arkin, D. E., and Gene Discovery Project of Johns Hopkins, Autism Consortium. 2009. A genome-wide linkage and association scan reveals novel loci for autism. *Nature* 461: 802–807.

Willsey, A. J., Sanders, S. J., Li, M., Dong, S., Tebbenkamp, A. T., et al. 2013. Coexpression networks implicate human midfetal deep cortical projection neurons in the pathogenesis of autism. *Cell* 155: 997–1007.

Xu, B., Ionita-Laza, I., Roos, J. L., Boone, B., Woodrick, S., et al. 2012. De novo gene mutations highlight patterns of genetic and neural complexity in schizophrenia. *Nat. Genet.* 44: 1365–1369.

16

Genetics of Aging

It turns out that aging is one of the most genetically regulated of all biological processes (Finch and Tanzi 1997). This may not seem obvious, since most people die in their 70s or 80s—if they survive accidents or do not contract cancer or suffer a midlife heart attack. Yet we all know of nonagenarians who parachute out of planes for pleasure and of 65-year-olds who have the normal frailties of an octogenarian. It is difficult to place an estimate on the heritability of age of death, but such estimates nevertheless range from 25% for most of the population to 50% for centenarians. Life span has fluctuated throughout the ages, increasing in general by over a decade in the past century, but probably with a low mean close to 30 in the Middle Ages and Industrial Revolution. There is, though, good evidence that those who survived adolescence have generally lived well past 50, whether in Paleolithic or modern times. Similarly, desert nomads and remote rural villagers can expect to live almost as long as contemporary urban humans. The dominant factor influencing human life span has probably been infectious disease and accidents, though much discussion has also surrounded malnutrition.

One clear piece of evidence for a genetic influence on aging is the remarkable species specificity of life span. Dogs, for example, live to one-seventh the age of their human owners, even contracting the same complex diseases at the same relative ages. Even breeds differ: Irish wolfhounds live little more than 5 years, whereas terriers can reach well into their teens. Birds have an extraordinary range of life spans, no more than a few years for some passerine species but 80 years for parrots and swans. Our closest primate relatives, gorillas and chimpanzees, both live up to 60 years in captivity but not much past 40 in the wild, which suggests a significant extension of longevity in *Homo sapiens*. Alligators and elephants also have humanlike life spans, bowhead whales and koi fish live past two centuries, and some bristlecone pines in California have been around almost 5000 years.

Invertebrate geneticists have been very successful in identifying specific mutations that increase longevity (Finch and Ruvkun 2001). *Drosophila* fruit flies with the *methuselah* mutation, for example, live an extra 2 weeks, 35% longer than wild-type flies. The *methuselah* gene encodes a

G protein–coupled membrane receptor whose primary function is probably neuronal. Nematodes (*Caenorhabditis elegans*) have yielded dozens of aging-related mutations that define a pathway discussed in detail later in this chapter. Interestingly, long-lived mutant nematodes competing against normal relatives don't do well because they tend to produce progeny later in life and are outbred. In humans, Hutchinson-Gilford progeria syndrome, in which accelerated aging is evident early in life, reduces life expectancy to just 20 years. This condition, which is extremely rare (1 in 8 million births), has been traced to the gene encoding lamin A, which provides structural support for the nuclear envelope; hence, mutations lead to aberrant cell morphology. The longest-lived human was a French woman, Jeanne Calment, who died in 1997 at the age of 122. It is not known whether centenarians harbor life-extending mutations, or how these would be detected, but there is at least nominal evidence ($p = 0.005$) that mitochondrial haplotype associates with extended life span in Italians (De Benedictis et al. 1999).

Aging presents something of an evolutionary conundrum. Superficially, you might think that any genetic variants that extend life span would be selected for, and that it would generally be good to live longer. However, from a genetic perspective, what really matters is reproductive age (Gavrilov and Gavrilova 2002). Any genetic advantage to living past the age when their children are independent is minimal, and most humans do not have children after the age of 40. This reasoning has given rise to two evolutionary theories of aging, the antagonistic pleiotropy theory and the mutation accumulation theory (**Figure 16.1**). The first, most often attributed to George Williams, postulates that genetic variants that promote fecundity tend to reduce longevity. In other words, there is a trade-off between having children early and living to an old age. The second, promoted by Peter Medawar, states more simply that since there is a constant input of deleterious mutations into the gene pool, evolution tends to remove those that act early in life, before reproductive senescence, whereas those that act only in old age are subject to weak selection. These are not mutually exclusive ideas, and other viable hypotheses have also been proposed.

From the point of view of age-related decline in health, it is not clear how much selection has shaped risk profiles because the diseases that cause these declines may have reached high prevalence only in recent history. Cellular theories of senescence postulate that there is a finite life span for all cells and that eventually oxidative damage or other processes reduce cell viability (Martin et al. 1996). Decline is inevitable, but it arises in different tissues or organs in different people at different rates. Cognitive and neurological decline receives the most attention, but losses of eyesight and hearing afflict most elderly people, and failure of organs such as the kidneys, liver, and heart is a common cause of age-dependent morbidity and mortality. This chapter discusses some of the genetic features of these diseases of old age before concluding with a brief discussion of genomic surveys of aging in general and the extraordinary phenomenon of methyl age.

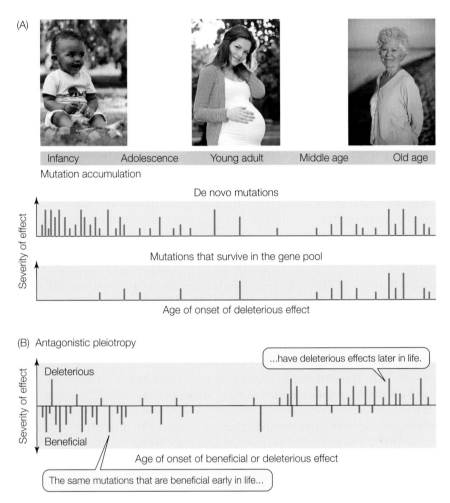

Figure 16.1 Two Theories of Aging. (A) The mutation accumulation theory postulates that new mutations can act at any part of the life span and are usually deleterious. Those that act prior to reproductive maturity are more strongly selected against, so they do not survive long enough to be observed in the gene pool. Deleterious alleles affecting aging can, however, accumulate. (B) The antagonistic pleiotropy theory instead postulates that genetic variation for aging is maintained by trade-offs between effects of variants that promote fecundity (more children and earlier in life) and their deleterious effects on the aging process.

Alzheimer's Disease

The pathology most often associated with aging is Alzheimer's disease (AD). This progressively worsening form of dementia generally starts with short-term memory loss and mild cognitive impairment, then moves through major changes in mood and cognitive function to impaired long-term memory and control of bodily functions. Death, which is most often caused by pneumonia or dehydration, comes an average of 7 years after diagnosis. That aging is the major risk factor for AD is clearly shown by the fact that prevalence approximately doubles every 5 years after the age of 65: whereas fewer than 0.3% of people have AD in their 50s, more than half have it in their 90s. Extraordinarily for a disease first recognized in 1906 (by the German psychiatrist Emil Alzheimer), there are now 25 million people diagnosed with AD worldwide, and over 1% of humanity will be afflicted by midcentury. According to the Centers for Disease Control and Prevention (CDC), in the United States in 2010, after adjustment for age, non-Hispanic Caucasians were 26% more likely to die of AD than African Americans, who in turn were 13% more likely to die of AD than Hispanic Caucasians. These differences were accentuated in women, who overall were 30% more likely to die of AD than men. Smoking and chronic inflammation are thought to elevate risk, while intellectual stimulation and social engagement are thought to at least delay onset, but despite thousands of clinical trials for different proposed therapies, there are no established cures. Furthermore, it has proved very difficult to predict dementia even in the elderly, though an eye-tracking task shows promise for positive diagnosis 3 years before clinical signs of cognitive decline (Zola et al. 2013).

Early-onset and late-onset Alzheimer's disease are known to have different genetic risk factors (**Figure 16.2**). Early-onset forms are familial and almost always trace to rare but highly penetrant mutations in one of three genes: those encoding amyloid precursor protein (APP), presenilin 1, or presenilin 2. The presenilins are enzymes, also known as γ-secretases, that can be thought of as molecular scissors. They perform one of the cleavage steps required to generate the extracellular amyloid β (Aβ) peptide that is the major component of amyloid plaques observed in the brains of people with AD. The longer $A\beta_{42}$ isoform of this peptide forms plaque-forming fibrils more readily, and many of the early-onset AD mutations result in an increased ratio of $A\beta_{42}$ to $A\beta_{40}$. These observations have given rise to the amyloid hypothesis, which posits that this peptide is a major contributor to dementia. There is conflicting evidence as to whether the plaques themselves are responsible for dementia, or perhaps just oligopeptides that are more diffusible and may inhibit normal synaptic processes when in excess. It is also possible that it is the loss of normal Aβ function that is most critical (Hardy and Selkoe 2002). Another biochemical hallmark of AD is the presence of neurofibrillary tangles, which are aggregates of the phosphorylated form of another neuronal protein, tau, within neurons. Adherents of the tau hypothesis argue that these accumulations disrupt neuronal microtubule structure (Farías et al. 2011).

Late-onset Alzheimer's disease (LOAD) has one major genetic risk factor, the apolipoprotein allele *APOE4*. Genome-wide association studies have identified 20 other common variant loci as risk factors, and a similar number of loci have been reported in the AlzGene database (www.AlzGene.org), assembled from 320 meta-analyses of 695 genes in 1395 studies (Bertram and Tanzi 2008). Apolipoprotein was initially recognized as a contributor to cardiovascular disease as a component of triglyceride metabolism, but *APOE* came to light in the early 1990s as the major AD gene through

(A) Early-onset Alzheimer's disease

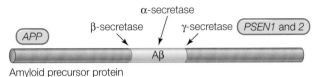

Amyloid precursor protein

(B) Late-onset Alzheimer's disease

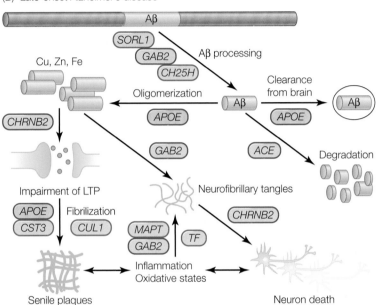

Figure 16.2 Genes Involved in the Pathology of Alzheimer's Disease.   (A) Most early-onset disease is attributed to rare but highly penetrant mutations in *APP*, which encodes the precursor for the Aβ peptide, as well as in *PSEN1* and *PSEN2*, which encode the γ-secretase enzymes. (B) Late-onset Alzheimer's disease, by contrast, is attributed to hundreds of variants of small effect, typified by the ten genes highlighted here that have roles in Aβ biochemistry, including its clearance from the brain, degradation, and oligo-merization (which may impair long-term potentiation and memory formation), as well as development of the plaques and neurofibrillary tangles that are associated with neuronal death. Red outlines imply that the minor allele is pathogenic, and blue that it is protective, indicating that a mix of effects is present. (After Bertram and Tanzi 2008.)

classical linkage studies (Pericak-Vance et al. 1991; Corder et al. 1993). The ApoE protein is involved in both the clearance of APP from the brain and oligomerization of the Aβ peptide; nine other proteins with roles in these pathways are highlighted in Figure 16.2.

There are three common alleles of *APOE*. The most common allele worldwide is *APOE3*, which has a frequency close to 0.8 in Europeans, so a majority of people are homozygous for this form. Nearly 25% of people, though, are at least heterozygous for *APOE4*, the ancestral allele, which increases their risk of AD as much as fivefold, while the 2% of homozygotes have double the risk again. By contrast, carriers of the less common *APOE2* allele are somewhat protected against AD, though they have elevated risk of coronary disease. Despite its large effect size, *APOE4* should not be seen as a determinant of disease, both because at least one-third of people with LOAD do not carry the allele and because not all carriers develop LOAD. Some studies have also suggested that there may be heterogeneity for the effects of *APOE4* in different ethnicities, but the epidemiology is confounded by differences among populations and cultures in age distribution and overall burden of Alzheimer's disease (Weiner 2008).

Only a few of the other 20 loci identified by GWAS have been reduced to known causal genes or variants (Lambert et al. 2013). Most of the loci can be plausibly related to the modulation of Aβ and/or tau levels in the brain, some through endosomal trafficking where the first APP cleavage event occurs, or to some aspect of neuronal function or neuroinflammation. The established loci are *ABCA7*, *BIN1*, *CLU*, *CR1*, *CD2AP*, *EPHA1*, *MS4A6A-MS4A4E*, and *PICALM*, each with small effect sizes typical of GWAS findings, while a rare coding variant in *TREM2* increases risk to approximately the same degree as *APOE4* (Guerreiro et al. 2012). Consistent with the notion that neurofibrillary tangles are causal in AD, GWAS for the level of phosphorylated tau protein in cerebrospinal fluid identified three novel loci and *TREM2*, and each SNP was also weakly associated with $A\beta_{42}$ levels as well as with AD in an independent study (Cruchaga et al. 2013). It is not yet clear whether a genetic risk score derived from these variants may classify individuals with respect to likely age of AD onset.

Much research in this field is aimed at understanding how genetic variation in apolipoproteins mediates cleavage of Aβ and, apparently independently, tau hyperphosphorylation, converging on a common pathology. An integrative genomics approach combining gene expression profiling and experimental manipulation of cell lines also provides evidence that there is a distinct biochemical network linking cholesterol metabolism and inflammation to neuropathology (Rhinn et al. 2013; **Figure 16.3A**). One finding was that many of the genes that are differentially expressed in the brains of *APOE4* carriers in the absence of Alzheimer's disease are differentially expressed in the same direction in LOAD relative to healthy brains of *APOE3* homozygotes. In other words, the genetic risk generates a molecular profile that is reminiscent of disease even in healthy adults, suggesting the existence of a "prodromal" state of pre-disease. Six of the twenty most connected nodes of the gene expression network connecting these genes can be plausibly related to APP processing, and knockdown of each of these genes indeed reduced the production of $A\beta_{42}$ in the presence of exogenous ApoE4 protein in culture (**Figure 16.3B**). Similarly, an anti-epileptic drug that inhibits one of the six proteins, SV2A (levetiracetam), reduced amyloid peptide production in neuronal cells derived from *APOE4* carriers (**Figure 16.3C**). An implication of these results is that some individuals have at-risk neuronal physiology that will begin to progress to disease given additional genetic or environmental perturbations. We have

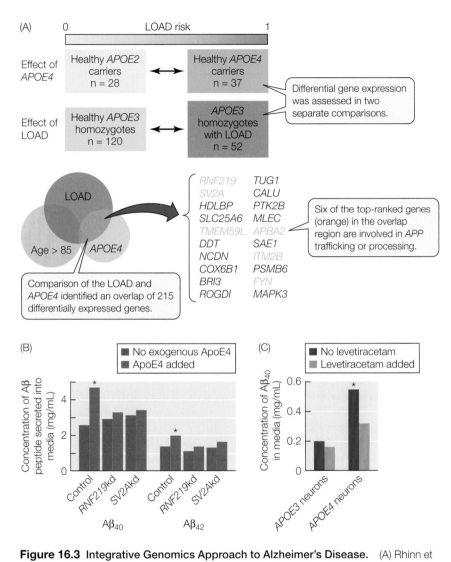

Figure 16.3 **Integrative Genomics Approach to Alzheimer's Disease.** (A) Rhinn et al. (2013) compared gene expression profiles of brain cortex from healthy adults carrying at least one *APOE2* allele and from healthy adults carrying at least one *APOE4* allele, then contrasted those results with differential expression between healthy adults and adults with LOAD, both carrying *APOE3*. There was significant overlap that defined a 215-gene LOAD "prodromal" signature; namely, an incipient Alzheimer's pattern. After ranking the genes according to their centrality to the network, they noticed that 6 of the genes have roles in APP trafficking or processing. (B) Incubation of a mouse neuroblastoma cell line that overexpresses human APP, N2a-APP, with ApoE4 results in elevated secretion of both $A\beta_{40}$ and $A\beta_{42}$, but this does not happen when two of the top genes, *RNF219* and *SV2A*, are knocked down by siRNA. (Asterisks indicate significant elevation of secretion only in control + ApoE4. (C) Similarly, a selective inhibitor of SV2A, levetiracetam, inhibits secretion of Aβ peptide specifically in fibroblast-derived hiN neurons from *APOE4*, but not *APOE3*, donors. (After Rhinn et al. 2013.)

no concrete conception of what the age-dependent triggers may be, but Franceschi et al. (2000) have argued that inflammation is a prime driver not only of cognitive, but also of general, physiological decline with age that has cumulative effects that promote disease in susceptible individuals.

Parkinson's Disease

A second neurological disorder associated with aging is Parkinson's disease (PD). Only 5%–10% of cases have an age of onset before late middle age (notable examples include Michael J. Fox and Mohammed Ali), and prevalence increases from approximately 1% at 60 to 4% at 80 years of age. The symptomatology is initially quite distinct from that of AD, beginning with tremors and slowness of movement, but progressing to rigidity and impaired gait and posture, often also including mood disorders (depression, anxiety) and loss of mental acuity. Parkinson's pathology includes the development of neuronal aggregates of the protein α-synuclein, called Lewy bodies, and the loss of function of dopaminergic neurons in the midbrain. Treatment with the dopamine precursor L-DOPA (levodopa medication) can control symptoms, extending the time from diagnosis to dependency on caregivers up to several decades in many cases. Gene and stem cell therapy are also being explored.

Most cases of PD are termed idiopathic, meaning that they have no known cause. However, it is now recognized that risk to first-degree relatives of people with PD is significantly elevated and that its heritability is around 25%. Pesticide exposure is also an established risk factor, and some forms of PD seem to be induced by closed-head injuries (de Lau and Breteler 2006). A dozen loci have been linked to familial forms of the disease. Causal mutations have been defined for six of these loci, which collectively explain 15% of cases, showing a mixture of autosomal dominant and recessive inheritance (Trinh and Farrer 2012; Coppedè 2012). The best known of these loci is the *SCNA* gene, which encodes α-synuclein and is disrupted both by very rare amino acid–altering mutations and by copy number duplications. More common, collectively, are over 100 mutations affecting dardarin, the product of the *LRRK2* gene, which is thought to participate in cytoskeletal organization through its kinase activity. Three of the recessive genes, *PARK2*, *PINK-1*, and *DJ-1*, are linked biochemically to mitochondrial function. Why the onset of disease is delayed until middle age is unclear. **Figure 16.4** shows a schematic of the suspected function of 11 genes, half discovered by GWAS, in the presynaptic dopaminergic neuron.

Recent GWAS have uncovered 11 well-replicated loci and a dozen more candidates (IPDGC 2011). After *SCNA* and *LRRK2*, the greatest genetic risk is attributed to *MAPT*, which encodes the microtubule-associated tau protein. The protective minor allele at *MAPT* is absent in Asians, as is the risk allele at *SYT11*. The effect sizes of the other variants appear to be fairly constant, although allele frequencies vary. Several studies have provided

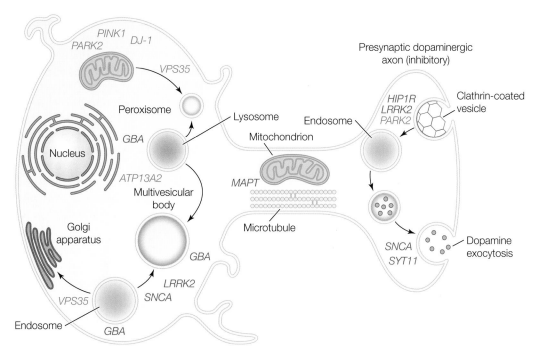

Figure 16.4 Schematic of the Role of PD-Associated Genes in the Presynaptic Dopaminergic Neuron.   Genes implicated by GWAS are shown in green, and those discovered in studies of familial cases of the disease are in red. Roles of the genes include synaptic vesicle function, endocytosis, cellular stress response, intracellular trafficking, and mitochondrial activity.

evidence for differential expression of many of the GWAS-identified loci as well as differential methylation associated with the risk genotype. Thus, gene dosage and epigenetic processes are likely to contribute to PD, and these factors may change with age. The observation that α-synuclein is implicated in alteration of DNA methyltransferase as well as histone deacetylase activity, and that drugs that inhibit the latter enzyme are neuroprotective in animal models, suggests novel therapeutic strategies.

The company 23andme has assembled a database of over 10,000 participants with PD, for whom the only clinical information is self-report through online surveys, yet most of the GWAS results generated by the International PD Consortium were replicated on the first 3426 cases (Do et al. 2011). The 23andme analysis also found a SNP-based heritability estimate of 27%, confirming a large polygenic component due to yet-to-be-discovered common variants. The known variants explain approximately 6% of liability, but genetic predictors achieve an area under the curve of just 0.6, which, given the rarity of the disease, is not yet clinically useful, even within families. No studies have yet reported associations with rate of progression or response to therapy.

Age-Related Organ Failure

This section considers age-related decline in organ function, using the eye and kidney as examples. Age-related macular degeneration (ARMD) afflicts almost a third of retirees, one-tenth seriously enough to threaten blindness. The macula is the central portion of the retina that is responsible for the visual acuity that supports reading, facial recognition, driving, and other basic daily activities. ARMD typically starts with the appearance of extracellular protein and lipid deposits, known as drusen, at the base of the macula and develops as either a "dry" form, with progressive retinal atrophy, or a "wet" form, in which neovascularization occurs as blood vessels invade the retina from the underlying choroid layer (**Figure 16.5A**). Histology, immunology, classical genetics, and most recently, GWAS all

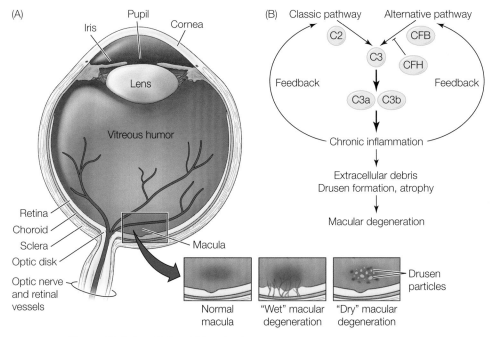

Figure 16.5 Age-Related Macular Degeneration. (A) The macula is the portion of the retina at the back of the eye that is responsible for the core of our vision. Age-related macular degeneration consists of the buildup of deposits, or drusen, and the atrophy of retinal tissue, which may be accompanied by neovascularization in the wet form of the disease. (B) The major genetic risk pathway consists of variation in four components that regulate complement via that classical (C2, immunogobulin) and alternative (CFB, bacterial cell wall) pathways. The proteins are regulators of, or co-factors for, generation of complement molecules that mediate such processes as inflammation and phagocytosis. Cleavage of C3 generates two smaller proteins that promote inflammation and feedback to the bacterial cell surface to promote lysis. CFH is also important for prevention of damage to "self." Common mutations affecting CFH lead to loss of the ability to inhibit complement in the retina, setting up a loop of chronic inflammation and damage that worsens with age.

support the hypothesis that ARMD is an inflammatory disease driven by hyperactivation of mostly innate immune responses to, and clearance of, accumulating cellular debris in the aging retina (Ambati et al. 2013; Ratnapriya and Chew 2013).

Age-related macular degeneration was the first disease to be successfully studied by genome-wide association, and alongside the MHC in type 1 diabetes, holds the record for the largest contribution of a single-locus common variant to a complex disease. The Tyr402His polymorphism of complement factor H (*CFH*) results in a form of the protein that is apparently less efficient at binding damaged cellular components and thus provides less inhibition of the alternative complement pathway. This single nucleotide change contributes to both the wet and dry forms of ARMD and accounts for as much as 50% of the heritability of the disease, as the risk to the two homozygote classes differs by more than seven times. Rare variants of even larger effect have also been described. Subsequently, other activating polymorphisms of proteins in the same pathway have been discovered, including variants in *HTRA1*, a duplicate copy of *CFH*, and *C3*, *C2-CFB*, and *CFI* (**Figure 16.5B**). In all, 19 loci have been discovered through meta-analysis of over 17,000 cases. There is evidence implicating variants affecting migration of macrophages and glial cells as well as oxidative stress response and lipid and collagen/extracellular matrix biosynthesis. One important component of the disease that was only weakly hinted by genetic analysis, but is central to therapeutic intervention, is the neovascularization observed in the wet form of ARMD, which is treated with inhibitors of the angiogenesis factor VEGF-A. Intriguingly, despite good knowledge of the genetics of ARMD, the sensitivity and specificity obtained with a combination of genetic and classical epidemiological risk factors are only around 70%, which is sufficient for risk classification but not yet of clinical utility (Spencer et al. 2011).

Chronic kidney disease (CKD) is another remarkably common disease, probably affecting over 10% of Western adults, a small fraction of whom will eventually develop end-stage renal disease. Chronic kidney disease is associated with diabetes and hypertension, but it can also have independent causes, such as glomerulonephropathy and fibrosis. It is defined with respect to glomerular filtration rate (GFR), which is negatively correlated with levels of blood urea nitrogen and serum creatinine. Genome-wide association studies in over 70,000 East Asians have identified 18 loci associated with at least one of these measures, seven of which are also associated with CKD, including one locus identified from elevated uric acid, which also reflects poor kidney function (Okada et al. 2012). There is good overlap among the associations, which thus help to define intrinsic risk for kidney failure. At least three of the identified genes, *MTX1*, *PAX8*, and *UNCX*, have critical regulatory roles in kidney development, implying lifelong risk of CKD that may build with time. Similarly, rare ciliopathies also often disrupt kidney development and are a cause of chronic kidney disease. A very rare condition known as karyomegalic interstitial nephritis has been traced to truncation mutations in the *FAN1* DNA damage repair gene, raising the possibility that genetic mediation of the response to genotoxic environmental agents may also contribute to CKD (Zhou et al. 2012).

Degenerative Bone Disease

Musculoskeletal disease really deserves its own chapter, but it is an understudied area of human genetics. Lower back pain may be the single largest cause of lost productivity and quality of life, yet it is barely a blip on the GWAS radar. Similarly, loss of muscle tone is one of the hallmarks of old age, but aside from research in the context of HIV, for example, suggesting accelerated muscle fibrosis in infected individuals (Kusko et al. 2012), it has received little attention. More progress has been made with respect to the two common degenerative bone diseases, osteoarthritis and osteoporosis.

Whereas nearly 100 genes have been associated with autoimmune rheumatoid arthritis, the tally for osteoarthritis is no more than 10, all of which have very small effect sizes. The two best-established of these genes are *GDF5*, a growth and differentiation factor of the bone morphogenetic protein (BMP) family, and *MCF2L*, an intracellular signaling guanine nucleotide exchange factor, but neither of these genes was replicated even in the largest GWAS to date, which included 7410 cases (arcOGEN Consortium 2012). That study identified seven new associations near genes plausibly related to bone biology as well as *FTO*, which probably appears solely because high BMI is an established risk factor for osteoarthritis. Part of the difficulty with this disease is that different joints, principally the hips, knees, and fingers, seem to have different genetic susceptibilities. Another is that definitions based on radiography or patient-reported symptoms (e.g., need for total joint replacement) result in different grades of disease that also vary by age and pain perception.

Osteoporosis has been similarly refractory to genetic dissection. Loss of bone mineral density, which is a major risk factor for bone fracture, is thought to trace to alteration of the balance of activity of the two cell types that promote bone deposition (osteoblasts) and bone resorption (osteoclasts). The largest known genetic polymorphism affects the expression of *RANKL* (*TNFSF11*), which encodes a member of the tumor necrosis factor family. This protein is a cytokine that promotes differentiation of osteoclasts. The gene that encodes its receptor, osteoprotegerin (OPG), is also associated with osteoporosis. Osteoblasts, on the other hand, are implicated through polymorphism in the *GREM2* antagonist of BMP signaling (Paternoster et al. 2013). Osteoporosis is more common in women than in men, typically appearing either postmenopause or after the age of 75, when it also contributes to general frailty. Exercise and diet are contributing factors, as are calcium and vitamin D levels, and it is likely that variation not only in bone density, but also in bone structure, contributes to fracture risk. Osteoporosis is one condition in which the analysis of human variability takes a back seat to classical molecular and cell biology, and the genetics of risk has not yet made a major impact.

Biochemical Pathways of Aging

Genetic analyses of nematodes, flies, and rodents have all converged on a conserved biochemical pathway that regulates longevity (Vijg and Suh 2005; **Figure 16.6**). Upstream inputs are insulin signaling and caloric restriction, which feed through a key transcription factor, FOXO, to regulate oxidative damage and metabolism. Embedded in this pathway are the sirtuins, a family of NAD-dependent deacetylases that regulate longevity in yeast, and which famously interact positively with the red wine metabolite resveratrol. While there has been much debate about the specific roles of individual components of this pathway due to the complex nature of the gene families involved and the ubiquitous impact of genetic background effects (Burnett et al. 2011), there is broad consensus that mechanisms linking

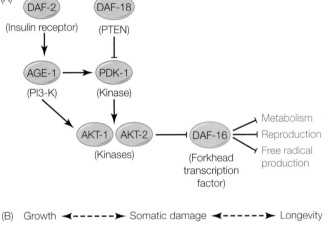

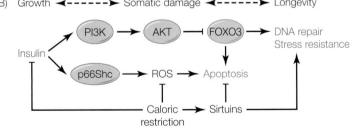

Figure 16.6 A Conserved Aging Pathway.   (A) The aging pathway in the nematode *Caenorhabditis elegans*. Loss of function of the genes highlighted in red leads to a reduction of life span, while loss of blue leads to an increase of life span. (B) The general architecture of this pathway is conserved in mammals, in which it is believed that caloric restriction in part acts through regulation of insulin signaling, free radical production in mitochondria, and the sirtuin family of proteins. There are multiple other inputs and loops that complicate analysis of a very complex system, but in a broad sense, the balance between reproduction and growth, on the one hand, and longevity on the other may reflect how this biochemical pathway modulates somatic damage and repair.

growth and reproductive development to preservation of cellular integrity and central metabolism can shed light on the aging process (Guarente 2013).

The aging pathway was first defined in *Caenorhabditis elegans* through the isolation of mutations that increase or decrease longevity. Losses of function of the genes *daf-2*, *age-1*, and *pdk-1* all extend life span from 10 to 20 or more days, whereas losses of *daf-18* or *daf-16* can halve life span. The *daf* genes are also known through their role in the formation of dauer larvae, a suspended developmental state that forms when the worms are starved. These genes have been ordered into a pathway (see Figure 16.6A), which starts with DAF-2, a homolog of the mammalian insulin receptor, then signals by way of the PDK-1, AKT-1, and AKT-2 kinases to a Forkhead family transcription factor (DAF-16, a homolog of the mammalian FOXO proteins). The AGE-1 and DAF-18 proteins are co-regulators homologous to phosphatidylinositol 3-kinase (PI3K) and PTEN, well-known mediators of insulin signaling. Phosphorylation of FOXO regulates its nuclear localization, facilitating transcriptional activation of target genes that on the one hand regulate metabolism and on the other prevent the accumulation of free radicals (H_2O_2) that cause oxidative damage within cells.

The delineation of this pathway was soon seen to help explain the phenomenon of caloric restriction (CR), wherein a period of reduced food consumption at some point in adulthood has been observed to extend life and reduce the incidence of metabolic disease and cancer in a wide variety of animals. It has not been shown conclusively to do so in humans, and experimental studies of macaques in captivity failed to demonstrate much of an effect (Mattison et al. 2012; Austad 2012). Caloric restriction improves insulin sensitivity, presumably increasing the activity of FOXO and enabling the metabolic changes and protection from oxidative damage that can extend life. The reduction in cancer incidence is thought to relate to promotion of cell death through p53-mediated apoptosis, but this process is further regulated by the sirtuins and by reactive oxygen species (ROS) through detailed mechanisms that now appear to be specific to each tissue type (Hursting et al. 2013). Thus, the regulation of aging in the adipose tissues, heart, liver, kidney, and brain is subject to different genetic and environmental influences, and these organs age at different rates in each individual.

The free radical theory of aging posits that ROS are toxic to cells because they cause oxidative damage to a wide variety of proteins and lipids and to DNA. Since most ROS are produced as a by-product of mitochondrial electron transport, mitochondrial activity is also implicated in the aging process, which is seen by some as accumulated oxidative stress (Hwang et al. 2012). Support for this idea comes from the life extension observed in animals with some mutations that affect electron transport and from genetic manipulation of the superoxide dismutase and catalase enzymes that help remove ROS. However, there is also some evidence that ROS are part of a retrograde signaling pathway from the mitochondria to the nucleus that induces protective gene expression, and in this sense ROS may also be partially pro-longevity. It is thus difficult to quantify the proportional contribution of oxidative damage to cell and eventually organismal aging.

Another process that is often linked to aging is the regulation of telomere length (Allsopp et al. 1992; Campisi et al. 2001). Although there is wide variation among humans in telomere length, telomeres generally shorten as people age. Lifestyle choices, including smoking, diet, and exercise, modulate the rate of telomere shortening. In a prospective study of 4500 Danish adults, 10-year change in telomere length was not associated with mortality or morbidity (Weischer et al. 2014). However, once telomeres are reduced below a threshold of between 4 and 7 kb per chromosome end (from up to 20 kb in germ cells), genome integrity is affected, and such reduction has been associated with multiple chronic diseases as well as cancer. Telomere integrity provides a checkpoint that is used to regulate cell senescence in a p53-dependent process that links this aspect of cell physiology to the FOXO-sirtuin pathway, but the relationship between cell and organismal senescence is again complex.

Genomics of Human Longevity

Twin studies suggest that the heritability of survival to old age increases with age. In other words, if one twin reaches 90, the other is more likely to do so than a random person born at the same time, and this ratio is greater than for 80-year-olds. A handful of GWAS on longevity, in which exceptionally long-lived European and American individuals (e.g., centenarians) were compared with others, have been published. With samples of around 1000 "cases," none of these studies was sufficiently powered to find any but the strongest effects, and concordantly, the only consistent association at $p < 10^{-8}$ was with the *APOE* locus (Brooks-Wilson 2013; Beekman et al. 2013). However, considering candidate genes, there is suggestive evidence for polymorphism in the progeria syndrome genes *LMNA* and *WRN*, and collectively for variation in the insulin, ROS, and telomere maintenance pathways. Sebastiani et al. (2012) also identified a 281 SNP signature of exceptional life span that replicated modestly in an independent cohort and was enriched for genes annotated to Alzheimer's disease, though these findings need to be independently validated.

A substantive question raised by these findings is whether survival to a very old age is simply a consequence of not carrying genetic risk for any of the major common diseases, or rather requires that an individual have protective alleles that buffer the effects of the known risk factors. Two arguments in favor of the latter alternative are (1) that in general, there does not seem to be a significant decrease in the burden of common risk alleles carried by long-lived individuals, and (2) that the children of very long-lived individuals, when matched for genetic risk of metabolic disease, have better metabolic profiles. The first published whole-genome sequences of centenarians have not revealed any obvious differences in their SNP or rare variant profiles, but it is not clear what the expectation is. If rare variants are responsible for protection, they will be extremely difficult to find, while finding protective common variants that are independent of known disease risk factors will require carefully designed and very large studies.

A related issue is whether the genetics of healthy aging is equivalent to that of general aging. There is some evidence that relatives of nonagenarians tend not to suffer from chronic diseases as much as might be expected (Westendorp et al. 2009). Individuals who reach retirement in good health are much more likely to live a comfortable and relatively disease-free retirement and to survive longer. Furthermore, since good mental health and a positive outlook on life are also positively correlated with longevity, some have suggested that the genetics of optimism and other temperaments may be enlightening, although to date, GWAS of mood have not been successful, probably because of very small allele effect sizes.

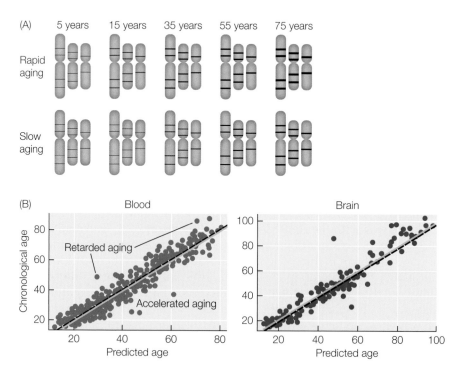

Figure 16.7 Age Prediction from Methylated DNA. (A) As a person ages, hundreds of sites throughout the genome become more heavily methylated: the average proportion of each of these sites in a person's cells that is methylated increases. At any given stage of life, people who age at a faster rate are expected to have a greater level of methylation than people who age less quickly. The thicknesses of the 10 bars on the schematic chromosomes correspond to average methylation level. (B) Age predicted from the average methylation level of 353 CpG sites plotted against actual chronological age in two different studies, one of blood and the other of cortical brain tissue. Both of these graphs are from training data sets; test data sets show lower correlations, but they are still highly significant. The red curves show the line of perfect fit, and the dashed blue curves, the observed regression. Most people are within 5 years of their predicted age, but some people show accelerated and others retarded aging relative to their methylation age. (B, after Horvath 2013.)

Interest has also turned to the possibility that aging is associated with a delay in the genome-wide rate of methylation, which tend to increase with age, but at a slower rate in the children of parents who lived to old age. By reanalyzing 8000 whole-genome methylation profiles in 84 publicly available data sets, Horvath (2013) identified 353 CpG sites that jointly formulate an "aging clock," since their weighted average level of methylation correlates very strongly with chronological age. It does so across dozens of different tissues with an average predictive error of less than 3.5 years in half the subjects (**Figure 16.7**). Even more remarkably, the difference between a person's actual age in years and his or her predicted DNA methyl age is highly heritable. First- and second-degree relatives are correlated in the extent to which the change in their epigenetic marker state with age is accelerated compared with the population average. Expression of adjacent transcripts is not so altered, so it is not clear what function, if any, specific methylation events have, but collectively, they provide a convenient marker with potential applications from forensics to clinical prediction.

Concluding Remarks

This brings to a conclusion our quick tour of the genetics of disease. Aging itself, of course, is not a disease. Rather, it is the gradual accumulation of no-longer-quite-right bodily functions, one or more of which will eventually become pathological. As with any other aspect of physiology, there are really four influences on the progression of decline: two genetic and two environmental. Certainly the impact of myriad common variants must contribute to how our cells and tissues age, but it is not clear whether this normal variation is what causes some people's bones or heart or brain or kidneys to give way earlier than other people's. It seems likely that private, or at least very rare, variants of large effect play a role as well, though they have yet to be identified. Similarly for the environment, how we choose to live our lives, what we eat, whether we exercise, and how much we stress our minds and bodies cumulatively contribute to our longevity. Then there are the unpredictable perturbations, traumatic accidents, deaths of friends and family, and sudden changes of circumstance that may act as triggers as potent as any genetic mutation. On top of this, we have evolved rapidly as a species, and we have changed our cultural environment dramatically over the past several generations. I have argued (Gibson 2009) that as a consequence of these twin perturbations, a major source of human morbidity is the mismatch between our evolved genome and the modern world. Whether humans are evolving greater resistance or susceptibility to disease is unclear, but it is a fascinating question that the next generation of researchers will address. Before long, we will have sequenced whole genomes of hundreds of thousands of people, and this will greatly accelerate the discovery phase of human genetics. Then the next really interesting scientific pursuit will continue as we model how all this variation works together to make us human, and to make us people.

Summary

1. Longevity differs by over two orders of magnitude among animal species, ranging from just weeks for small invertebrates to hundreds of years for some large mammals.

2. Heritability of life span is greatest for long-lived humans, and longevity tends to run in families.

3. Evolutionary theories of aging postulate either mutation accumulation of postreproductive traits or antagonistic pleiotropy between early-life benefits (such as early reproductive maturity) and deleterious effects in old age.

4. Early-onset Alzheimer's disease is due to high-penetrance mutations in the genes encoding amyloid precursor protein (APP) or either of the presenilin γ-secretase enzymes.

5. Prevalence of late-onset Alzheimer's disease (LOAD) doubles every 5 years after the age of 65. *APOE* genotype is a major genetic risk factor influencing age of onset.

6. Parkinson's disease arises from loss of function of dopaminergic neurons in the midbrain that is genetically regulated.

7. Age-related macular degeneration is an inflammatory disease with a large contribution from the alternative complement pathway that normally clears cellular debris from the retina.

8. Genetic analyses are just starting to dissect the mechanisms for age-dependent decline in function of organs such as the kidney, liver, skin, and muscle.

9. Osteoporosis risk is influenced by genetic factors that mediate the activity of osteoclasts and osteoblasts during growth and renewal of bone and cartilage.

10. A conserved signaling pathway from the insulin receptor to FOXO transcription factors links nutrient sensing and caloric restriction to generalized cellular aging.

11. It is not yet known whether long life requires inheritance of protective genetic factors that buffer individuals from disease-promoting variants or simply reflects a reduced burden of disease risk.

12. The rate at which methylation changes with age is genetically influenced, giving rise to an aging clock consisting of several hundred methylated CpGs.

References

Allsopp, R. C., Vaziri, H., Patterson, C., Goldstein, S., Younglai, E. V., et al. 1992. Telomere length predicts replicative capacity of human fibroblasts. *Proc. Natl. Acad. Sci. U.S.A.* 89: 10114–10118.

Ambati, J., Atkinson, J. P., and Gelfand, B. D. 2013. Immunology of age-related macular degeneration. *Nat. Rev. Immunol.* 13: 438–451.

arcOGEN Consortium and arcOGEN Collaborators, Zeggini, E., et al. 2012. Identification of new susceptibility loci for osteoarthritis (arcOGEN): A genome-wide association study. *Lancet* 380: 815–823.

Austad, S. N. 2012. Ageing: Mixed results for dieting monkeys. *Nature* 489: 210–211.

Beekman, M., Blanché, H., Perola, M., Hervonen, A., Bezrukov, V., et al. 2013. Genome-wide linkage analysis for human longevity: Genetics of Healthy Aging Study. *Aging Cell* 12: 184–193.

Bertram, L. and Tanzi, R. E. 2008. Thirty years of Alzheimer's disease genetics: The implications of systematic meta-analyses. *Nat. Rev. Neurosci.* 9: 768–778.

Brooks-Wilson, A. R. 2013. Genetics of healthy aging and longevity. *Hum. Genet.* 132: 1323–1338.

Burnett, C., Valentini, S., Cabriero, F., Goss, M., Somogyvári, M., et al. 2011. Absence of effects of Sir2 overexpression on lifespan in *C. elegans* and *Drosophila*. *Nature* 477: 482–485.

Campisi, J., Kim, S. H., Lim, C. S., and Rubio, M. 2001. Cellular senescence, cancer and aging: The telomere connection. *Exp. Gerontol.* 36: 1619–1637.

Coppedè, F. 2012. Genetics and epigenetics of Parkinson's disease. *ScientificWorldJournal* 2012: ID489830.

Corder, E. H., Saunders, A. M., Strittmatter, W. J., Schmechel, D. E., Gaskell, P. C., et al. 1993. Gene dose of apolipoprotein E type 4 allele and the risk of Alzheimer's disease in late onset families. *Science* 261: 921–923.

Cruchaga, C., Kauwe, J. S. K., Harari, A., Jin, S. C., Cai, Y., et al. 2013. GWAS of cerebrospinal fluid tau levels identifies risk variants for Alzheimer's disease. *Neuron* 78: 256–268.

De Benedictis, G., Rose, G., Carrieri, G., de Luca, M., Falcone, E., et al. 1999. Mitochondrial DNA inherited variants are associated with successful aging and longevity in humans. *FASEB J.* 13: 1532–1536.

de Lau, L. M. L. and Breteler, M. M. B. 2006. Epidemiology of Parkinson's disease. *Lancet* 5: 525–535.

Do, C. B., Tung, J. Y., Dorfman, E., Kiefer, A. K., Drabant, E. M., et al. 2011. Web-based genome-wide association study identifies two novel loci and a substantial genetic component for Parkinson's disease. *PLoS Genet.* 7: e1002141.

Farías, G., Cornejo, A., Jiménez, J., Guzmán, L., and Maccioni, R. B. 2011. Mechanisms of tau self-aggregation and neurotoxicity. *Curr. Alzheimer Res.* 8: 608–614.

Finch, C. and Ruvkun, G. 2001. The genetics of aging. *Annu. Rev. Genom. Hum. Genet.* 2: 435–462.

Finch, C. and Tanzi, R. E. 1997. Genetics of aging. *Science* 278: 407–411.

Franceschi, C., Bonafè, M., Valensin, S., Olivieri, F., de Luca, M., et al. 2000. Inflamm-aging. An evolutionary perspective on immunosenescence. *Ann. NY Acad. Sci.* 908: 244–254.

Gavrilov, L. A. and Gavrilova, N. S. 2002. Evolutionary theories of aging and longevity. *ScientificWorldJournal* 2: 339–356.

Gibson, G. 2009. *It Takes a Genome: How a Clash between Our Genes and Modern Life Is Making Us Sick*. FT Press Science, New York.

Guarente, L. 2013. Calorie restriction and sirtuins revisited. *Genes Dev.* 27: 2072–2085.

Guerreiro, R., Wojtas, A., Bras, J., Carrasquillo, M., Rogaeva, E., et al. 2012. *TREM2* variants in Alzheimer's disease. *N. Engl. J. Med.* 368: 117–127.

Hardy, J. and Selkoe, D. J. 2002. The amyloid hypothesis of Alzheimer's disease: Progress and problems on the road to therapeutics. *Science* 297: 353–356.

Horvath, S. 2013. DNA methylation age of human tissues and cell types. *Genome Biol.* 14: R115.

Hursting, S. D., Dunlap, S. M., Ford, N. A., Hursting, M. J., and Lashinger, L. M. 2013. Calorie restriction and cancer prevention: A mechanistic perspective. *Cancer Metab.* 1: 10.

Hwang, A. B., Jeong, D. E., and Lee, S-J. 2012. Mitochondria and organismal longevity. *Curr. Genomics* 13: 519–532.

International Parkinson Disease Genomics Consortium (IPDGC), Nalls, M. A., et al. 2011. Imputation of sequence variants for identification of genetic risks for Parkinson's disease: A meta-analysis of genome-wide association studies. *Lancet* 377: 641–649.

Kusko, R. L., Banerjee, C., Long, K. K., Darcy, A., Otis, J., et al. 2012. Premature expression of muscle fibrosis axis in chronic HIV infection. *Skelet. Muscle* 2: 10.

Lambert, J. C., Ibrahim-Verbaas, C. A., Harold, D., Naj, A. C., Sims, R., et al. 2013. Meta-analysis of 74,046 individuals identifies 11 new susceptibility loci for Alzheimer's disease. *Nat. Genet.* 45: 1452–1456.

Martin, G. M., Austad, S. N., and Johnson, T. E. 1996. Genetic analysis of ageing: Role of oxidative damage and environmental stresses. *Nat. Genet.* 13: 25–34.

Mattison, J. A., Roth, G. S., Beasley, T. M., Tilmont, E. M., Handy, A. M., et al. 2012. Impact of caloric restriction on health and survival in rhesus monkeys from the NIA study. *Nature* 489: 318–321.

Okada, Y., Sim, X., Go, M. J., Wu, J-Y, Gu, D., et al. 2012. Meta-analysis identifies multiple loci associated with kidney function–related traits in east Asian populations. *Nat. Genet.* 44: 904–909.

Paternoster, L., Lorentzon, M., Lehtimäki, T., Eriksson, J., Kähönen, M., et al. 2013. Genetic determinants of trabecular and cortical volumetric bone mineral densities and bone microstructure. *PLoS Genet.* 9: e1003247.

Pericak-Vance, M. A., Bebout, J. L., Gaskell, P. C., Jr., Yamaoka, L. H., Hung, W-Y., et al. 1991. Linkage studies in familial Alzheimer disease: Evidence for chromosome 19 linkage. *Am. J. Hum. Genet.* 48: 1034–1050.

Ratnapriya, R. and Chew, E. Y. 2013. Age-related macular degeneration-clinical review and genetics update. *Clin. Genet.* 84: 160–166.

Rhinn, H., Fujita, R., Qiang, L., Cheng, R., Lee, J. H., and Abeliovich, A. 2013. Integrative genomics identifies APOE e4 effectors in Alzheimer's disease. *Nature* 500: 45–50.

Sebastiani, P., Solovieff, N., Dewan, A. T., Walsh, K. M., Puca, A., et al. 2012. Genetic signatures of exceptional longevity in humans. *PLoS ONE* 7: e29848.

Spencer, K. L., Olson, L. M., Schnetz-Boutaud, N., Gallins, P., Agarwal, A., et al. 2011. Using genetic variation and environmental risk factor data to identify individuals at high risk for age-related macular degeneration. *PLoS ONE* 6: e17784.

Trinh, J. and Farrer, M. 2012. Advances in the genetics of Parkinson disease. *Nat. Rev. Neurol.* 9: 445–454

Vijg, J. and Suh, Y. 2005. Genetics of longevity and aging. *Annu. Rev. Med.* 56: 193–212.

Weiner, M. F. 2008. Perspective on race and ethnicity in Alzheimer's disease reseach. *Alzheimers Dement.* 4: 233–238.

Weischer, M., Bojesen, S. E., and Nordestgaard, B. G. 2014. Telomere shortening unrelated to smoking, body weight, physical activity, and alcohol intake: 4,576 general population individuals with repeat measurements 10 years apart. *PLoS Genet.* 10: e1004191.

Westendorp, R. G., van Heemst, D., Rozing, M. P., Frölich, M., Mooijart, S. P., et al. 2009. Nonagenarian siblings and their offspring display lower risk of mortality and morbidity than sporadic nonagenarians: The Leiden Longevity Study. *J. Am. Geriatr. Soc.* 57: 1634–1637.

Zhou, W., Otto, E. A., Cluckey, A., Airik, R., Hurd, T. W., et al. 2012. *FAN1* mutations cause karyomegalic interstitial nephritis, linking chronic kidney failure to defective DNA damage repair. *Nat. Genet.* 44: 910–915.

Zola, S. M., Manzanares, C. M., Clopton, P., Lah, J. J., and Levey, A. I. 2013. A behavioral task predicts conversion to mild cognitive impairment and Alzheimer's disease. *Am. J. Alzheimers Dis. Other Demen.* 28: 179–184.

Glossary

A

adaptation Evolutionary process whereby a species becomes more suited to the ecological and biogeographic niche it occupies, generally as a result of natural selection favoring certain genotypes that attain optimal frequencies in the species' gene pool.

additive trait Trait in which the genetic influences are a simple sum of the number of contributing alleles. For a monogenic trait, individuals have 0, 1, or 2 increasing alleles, and the heterozygotes are expected to have a phenotype intermediate between the two homozygote classes.

admixture The mixing of two or more previously somewhat isolated populations that had diverged in genotype frequencies, which tends to increase genetic variability in the next generations.

alleles In classical genetics, the different versions of a gene that contribute to a phenotype, which in molecular genetics are equivalent to different nucleotides (A, C, G, or T) at a particular location in a gene.

allelic heterogeneity Situation in which different alleles are responsible for a similar phenotype, for example because two or more mutations are present in the same gene that both have similar effects on the function of the gene.

allelic sum scores When more than one gene has been shown to influence a trait, the sum total of the number of "risk" alleles carried at all of the genes is referred to as the allelic sum. For example, if 10 genes are known to influence risk of a disease, in theory people could have allelic sums between 0 and 20, but in practice most people have an intermediate number, since the alleles segregate independently and have a range of frequencies.

allelic trend test Test of association between the number of each of the two types of allele at a locus and case-control status or a continuous phenotype. If the minor allele frequency is p, then the number of minor alleles is expected to be $[p^2 + 2p(1 - p)]$, and the number of major alleles should be 1 minus the number of minor alleles. An allelic trend exists if the proportions of these alleles differ between cases and controls or in relation to the phenotype.

analysis of variance (ANOVA) Standard statistical procedure for testing whether two groups differ with respect to some property. The idea is to evaluate whether there is more variation in the sample as a whole than in the sub-groups by comparing the ratio (F) of these two measures to that expected given the degrees of freedom in the analysis. ANOVA is a powerful and robust tool for evaluating whether the mean value of two distributions is large relative to the variation within the groups.

aneuploidy   Condition where one or more chromosomes in a cell have a different number of copies than expected. Somatic autosomes (non-sex chromosomes) are usually diploid, so cells with 1 or 3 (or more) copies of the chromosome are aneuploidy. Segmental aneuploidy is where a large portion of a chromosome has an abnormal copy number.

autoimmune disease   Pathological condition due to a person's immune system mistakenly attacking the person's own organs due to failure to distinguish their own cells (self) from non-self. Examples include type 1 diabetes, multiple sclerosis, and rheumatoid arthritis.

C

canalization   The evolution of developmental or physiological systems to a state of increased robustness and insensitivity to environmental or genetic perturbation. Homeostatic systems are well buffered; canalized systems have reached that state as a result of long-term stabilizing selection that ensures that most individuals are close to the optimum.

carriers   Heterozygous individuals who have one copy of an allele that, when homozygous, can cause disease. Carriers are usually asymptomatic but can pass a disease to their children if the other parent is also a carrier. Most Mendelian disease alleles are found in carriers, rather than affected individuals.

cases   Individuals with a disease or phenotype, who are compared to healthy (or "normal") controls.

ChIA-PET (chromatin interaction analysis by paired-end tag) sequencing method   Method for detecting physical interactions between short stretches of DNA in the nucleus genome-wide. Generally used to identify which genes are regulated by specific enhancers or regulatory polymorphisms, since these can operate over distances of hundreds of kilobases and jump over intervening genes. Some interactions also occur between genes on different chromosomes. *See also* chromosome conformation capture.

chromosome conformation capture   A series of methods for detecting physical interactions between short stretches of DNA in the nucleus. It is generally used to identify which genes are regulated by specific enhancers or regulatory polymorphisms, since these can operate over distances of hundreds of kilobases and jump over intervening genes. HiC and 5C are genome-wide methods, whereas 3C and 4C are targeted to individual loci.

***cis*-eSNP**   *See* local eSNP.

Cochran-Armitage trend test   Most commonly used test of association between genotypes and case-control status in a GWAS. The test requires that there is a linear trend due to each additional risk allele, so it can be thought of as a chi-square test constrained by the assumption that heterozygotes have risk intermediate to that of the two homozygote classes.

common disease–common variant (CD–CV)   Model of the genetic architecture of disease that postulates that a small number of relatively large effect common variants are responsible for cases of common diseases. Prior to GWAS, many expected this to be a typical model and assumed that dozens rather than thousands of genes would explain most of the genetic variance for chronic diseases.

complementary DNA (cDNA)   DNA copy of an RNA molecule. Single stranded cDNA is the opposite strand to the messenger RNA, since it is generated by reverse transcriptase, but the term generally refers to double-stranded cDNA.

complete linkage disequilibrium   Situation in which one allele is always found on the same haplotype (chromosome) as one of the alleles at another nearby site. The alleles do not have to be the same frequency, it is only necessary that no recombinants are found. Thus, if the two alleles at one site are A and G, and the two alleles at another site are C and T, then only two or three of the four possible combinations (A,T), (A,C), (G,T) and (G,C) are observed. Complete LD is represented by the statistic $D' = 1$.

compound heterozygous   Situation in which an individual is heterozygous for two different mutations, one on each chromosome inherited from either parent. Whereas homozygotes inherit both mutant copies from parents who carry the same allele, compound heterozygotes inherit two different copies from parents who carry different mutations. The term usually implies the individual has two different mutations at the same gene, but intergenic compound heterozygotes may have different mutations in two genes in the same pathway.

conditional association   Association between a genotype and a phenotype that remains after including the effect of another genotype. The process of conditioning a test on another genotype is used to establish that the two alleles independently influence the trait.

consanguinity   Close inbreeding, for example between first or second cousins, which increases the likelihood that children are homozygous for alleles transmitted from a common ancestor. Consanguineous individuals are also known as blood relatives, or kin.

controls   In a case-control study, individuals who do not have the disease. Super-controls are people who are at high risk, but still do not have the disease (e.g.,

elderly people who smoked a pack of cigarettes a day but do not have lung disease).

copy number variants (CNVs)   Regions of the genome shorter than a chromosome that have an abnormal number of copies either due to deletion on one chromosome or (usually tandem) duplication. Generally refers to inherited CNV or new germline mutations that are found in all cells of a person, but in cancer, somatic copy number alterations (SCNA) are the equivalent phenomenon that arises due to mutation only in the tumor.

CpG island   Short stretch of GC rich DNA, usually located within a few kilobases of the start site for transcription of a gene, that is often targets of DNA methylation.

cytogenetics   The study of the structure and visible content of the chromosomes in a cell using histology and microscopy, including fluorescent labeling.

D

de novo mutation   Spontaneously occurring genetic alteration that is present in offspring but is not possessed or transmitted by parents.

derived allele frequency (DAF) distribution   The distribution of the frequency in a population sample of the allele that appeared more recently and is thus derived. DAF is estimated using phylogenetic inference, typically where recent common ancestors all have the alternate (ancestral) allele.

distal eSNP   eSNP that is located a large distance (more than 250kb away) from the transcription start site. Also called a *trans*-eSNP.

DNA repair genes   Genes that encode proteins that are necessary for the maintenance of chromosome integrity and repair of damage to DNA sequences.

DNase hypersensitive sites (DHS)   Locations in the genome that are accessible to digestion by a DNA nuclease, since they are relatively exposed in the nucleus when they are in enhancers that are associated with active transcription. DHS are considered to be biomarkers of active regulatory regions.

dominance variance (V_D)   In quantitative genetic theory, genetic effects are partitioned into additive, dominance, and interaction terms. Dominance variance is the variance due to single genetic polymorphisms that are not accounted for by the additive component. That is, if the heterozygotes on average have a different phenotype or risk than the average of the two homozygotes, there is dominance variance attributable to the locus.

dominant haploinsufficiency   A dominant effect is observed where the mutant homozygote and heterozygote phenotypes are both similarly different from the wild-type homozygotes; this can either be due to gain of function or because the one normal copy is unable to provide sufficient gene activity. The latter situation is called dominant haploinsufficiency.

dominant trait   Trait in which the genetic influences include situations in which the heterozygotes have a similar phenotype as one of the homozygote classes, which is the dominant class (while the other class is recessive). Traits influenced by multiple genes can also be dominant if the dominant effects of some loci are strong enough to cause the phenotypes to appear non-additive.

driver mutations   Mutations that are thought to be causal in driving normal cells toward a cancerous state. Tumor cells have hundreds or thousands of mutations, only some of which cause cancer; these are often first inferred to be the ones that are observed in the same gene of different tumors more often than expected by chance.

E

enhancer   Short sequence of regulatory DNA that leads a nearby gene to be expressed in specific cells under specific circumstances. Usually orientation-independent and can be located at variable distances from the promoter in either direction.

epigenetic inheritance   Transmission of a phenotype from parent to offspring that is not a result of differences in the DNA sequence. The most likely mechanism for epigenetic inheritance is thought to be methylation of the DNA in the vicinity of relevant genes.

epigenetics   The study of epigenetic inheritance, as well as of mechanisms that change or maintain patterns of gene expression for long periods of time. Mechanisms such as DNA methylation, histone modification, and miRNA networks all combine to influence global gene expression states that can be transmitted through mitosis, maintained in long-lived cells, and in some cases transmitted through the germline.

epistasis   In quantitative genetics, any interaction between two or more genotypes that leads to a phenotype of susceptibility to disease that is different from that expected by simply adding the contributions of the contributing genes.

eQTL   Since the identity of the SNP that actually causes the eSNP effect is not usually known, the term eQTL is used to imply that the SNP genotype is within a locus that regulates transcript abundance. Some authors prefer to use this term in the context of linkage studies (for example, in pedigrees or artificial crosses of model organisms), and to use eSNP for sites identified by GWAS. *See also* eSNP.

eSNP   Single nucleotide polymorphism that is associated with the expression of a gene, namely the abundance of a transcript. *See also* local eSNP and distal eSNP.

exons   Segments of a primary transcript that are spliced together and retained in the mature messenger RNA (mRNA). Most exons encode part of the protein, but some 5′ or 3′ exons may not. All of the exons in a genome are collectively called the exome.

expressivity   The severity of the phenotype that is observed in individuals that have a disease or condition attributed to a large-effect allele.

F

familial cancer susceptibility syndromes   Types of cancer that tend to be more prevalent in families, and are typically associated with a high risk of developing multiple independent tumors. They account for between 5% and 10% of all cancers, and are often attributed to tumor suppressor mutations that are inherited in a Mendelian manner.

familial cancers   Cancers that tend to be more prevalent in certain families than in the general population, though not necessarily associated with a syndrome of cancer susceptibility.

founder effects   When individuals leave an established population, and found a new group of people, they typically carry only a subset of the variation in the source population and some alleles have altered frequencies. These are called founder effects, as they shape the genetics of the new population into the future.

G

gene expression profiling   Process of measuring the abundance of all of the transcripts that are found in a tissue in an individual. Microarray analysis and RNA-Seq are the two most common methodologies.

gene ontology (GO)   Controlled vocabulary that aims to provide some level of biological, molecular, or cellular annotation to each gene product. Gene ontologies are hierarchical, including hundreds or thousands of genes in the broadest categories, and successively refining to just a few dozen or a handful of genes in the finest categories.

gene set enrichment (GSE) analysis   Bioinformatic strategy for evaluating whether specific genetic or biochemical pathways are enriched in a set of genes that share some feature, such as being co-expressed or having binding sites for a transcription factor.

genetic counseling   Profession involving experts giving advice to individuals or families on the likelihood of their genes causing diseases or abnormalities, usually in the context of family planning.

genetic drift   Process whereby isolated populations tend to diverge in allele frequencies. It is generally a result of gradual loss of heterozygosity by random sampling in finite populations.

genetic heterogeneity   Situation in which different genes are responsible for a similar phenotype, for example because two or more mutations are present in different genes that both have similar effects on the trait. Allelic heterogeneity is the analogous situation where different mutations in the same gene produce a similar phenotype.

genetic hitchhiking   Process whereby positive selection that increases the frequency of one adaptive allele, also increases the frequencies of other alleles in linkage disequilibrium that are carried along for the ride, as it were.

genetic risk score (GRS)   Count of the number of alleles that a person has that tend to increase their risk for a disease. Most simply a sum of the number of increasing alleles, but can also be a weighted sum relative to the effect sizes, or modified with respect to the allele frequencies.

genetical genomics   Term used to describe the study of the genetic basis of variation in functional genomic profiles, for example genetic regulation of transcript abundance.

genome-wide association studies (GWAS)   Studies designed to take an unbiased approach to discovery genes that contribute to disease risk or phenotypic variation, by measuring the genotypes of hundreds of thousands of genotypes in thousands of people. First introduced in the mid-2000s, GWAS have led to the discovery of several thousands of associations of single genes with diverse traits.

genomic control (GC)   Statistical genetic procedure used to control for artificial inflation of GWAS test statistics genome-wide, for example due to population structure.

genotype   Genetic state of an individual at a single locus, reflecting whether the alleles tend to increase or decrease a trait. In molecular terms, the status of an individual as heterozygous or homozygous for either the minor or major allele

genotype relative risk (GRR)   Risk that a person with a specific genotype has of contracting a disease, relative to the risk in the rest of the population. It is often presented as an odds ratio, namely the ratio of the prevalence of disease in people with and without the genotype, divided by the ratio of the prevalence of healthy people with and without the genotype.

genotypic trend test   Test of association between the number of each of the three types of genotype at a locus, and case-control status or a continuous phenotype. Unlike the Cochran-Armitage trend test, a genotypic trend test does not require that the heterozygotes have intermediate risk, so is essentially a chi-square test of differences in frequency of the three genotypes in the cases and controls. In theory, genotypic trend tests are more powerful to detect nonadditive associations, but in practice they are not widely used.

glycomics   Study of the distribution of sugars (glycan structures) on or in cells.

grade   Description of a tumor based on how abnormal the morphology is under a microscope, which is regarded as a good indicator of how quickly the tumor is likely to grow or spread.

H

haplotype   Multilocus genotype on a single chromosome. For example, if three successive genotypes at polymorphic sites are AT, CG, and AC, the corresponding haplotypes may be ACA and TGC (or ACC and TGA, AGC and TCA, or AGA and TCC).

haplotype blocks   Long stretches of DNA where linkage disequilibrium results in a small number of haplotypes typically up to 100 kb long in humans. Consequently the genome can be regarded as an assembly of several hundred thousand haplotype blocks.

hard selection   Natural selection acting on a new advantageous mutation that gradually increases in frequency until it becomes fixed in the population or species. Contrast with soft selection.

heat map   Way of representing thousands of integers as a grid of colors where high values are represented as a warm color (typically red) and low values as a cool color (typically blue).

heritability   Proportion of phenotypic variation in a population that can be attributed to genetic variation. Formally, the ratio of the genetic to the genetic plus environmental variance.

heterochromatin   Physically tightly packed and inactive DNA that stains darkly in chromosome spreads and generally consists of highly repetitive noncoding DNA.

Hi-C chromatin conformation capture
One of the chromatin conformation capture profiling method for detecting long-distance physical contacts between stretches of DNA in nuclei. Thought to represent active associations between enhancers and promoters.

I

identical by descent (IBD)   If a haplotype is the same in two individuals as a result of transmission from a common ancestor, it is said to be identical by descent.

idiopathic   Having unknown origin, either arising spontaneously due to a de novo mutation or otherwise not known to trace to a known cause.

imaging mass spectroscopy   Method for measuring metabolites or proteins by mass spectrometry by ionizing molecules directly from a tissue section so that the cellular location of the profile can be imaged.

imputation   Process of inferring the most likely genotypes at polymorphisms that were not directly measured on a genotyping array, using the known pattern of linkage disequilibrium in the population in the vicinity of the region being imputed.

indels   Short insertion-deletion polymorphisms, typically just one or a few bases in length. Indels in the coding region that are multiples of three nucleotides may preserve the reading frame.

infinitesimal model   One of the major theoretical models of the genetic architecture of complex traits that posits that average genetic effects are on average very small, and therefore it takes several thousand polymorphisms to account for all of the genetic variance for a trait. Most polymorphisms are assumed to be common, but rare variants may also contribute.

inheritance   Transmission of a trait from parent to offspring, which may have a genetic or an environmental or cultural basis. Not to be confused with heritability, which refers to the relative magnitude of the genetic influence.

insulators   Sequences of DNA that ensure that regulatory sequences only act within the limits bound by the insulators, thereby ensuring that enhancers act locally on a chromosome.

integrative genomics   A scientific research strategy based on combining multiple types of omic data such as DNA sequence, transcriptome, proteome, and metabolome, using statistical and computational biology.

interaction variance (V_I)   Genetic variation due to epistasis, namely interactions between genes that cause departures from additive effects. An important component of what is called broad sense heritability (H^2), compared with narrow sense heritability (h^2), which excludes interactions.

introns   Stretches of DNA that are spliced out of the primary transcript during the production of the mature messenger RNA. Introns do not encode amino acids. They can be tens of kilobases long.

K

karyotype analysis   Study of the number and content of chromosomes by visualizing stained metaphase spreads, possibly including fluorescent in situ hybridization (FISH).

L

linkage disequilibrium (LD)   Nonrandom association of linked genotypes on a chromosome due to the history of new mutations and the time taken for recombination to generate new haplotypes. Measured as the correlation of genotype frequencies, which may be scaled relative to the maximum possible association given the frequencies.

lipidomics   High throughput profiling of the abundance of dozens to thousands of different lipids, including ceramides, shingomyelins, and glycophospholipids.

local eSNP   Formally, local refers simply to the SNP being within a certain distance of the transcript (often taken arbitrarily as 250 kb), whereas *cis*-eSNP implies that the eSNP is actually on the same chromosome as the transcribed region. Contrast with distal eSNP.

locus   (Plural: loci) Region of the genome containing one or more genes that may influence one or more traits. It is used to refer to a mapped or physical region of a chromosome without actually specifying what the gene responsible for the effect is.

logarithm of the odds (LOD)   Term used in pedigree mapping to quantify the significance of the linkage between a polymorphism and a trait. As a rule of thumb, LOD scores greater than 3 are suggestive, while those greater than 5 provide strong evidence that a gene in the region is involved in regulation of the trait.

long noncoding RNAs (lncRNAs)   Long transcribed RNAs that do not encode a protein but are likely to have biological function by virtue of the secondary structure of the lncRNAs. Some authors include "intergenic" in the acronym lincRNA, implying that the transcript is derived from intergenic regions.

loss of heterozygosity (LOH)   In cancer, the observation that sequences that are heterozygous in healthy somatic tissue of an individual become homozygous in the tumor, most often due to deletion of the sequence on one of the homologous chromosomes. Copy number neutral LOH can also occur by the process of gene conversion.

M

major allele   Allele that has the highest frequency in the population at a polymorphic site. Most sites have just two alleles, for example A and G, and the major allele is the one whose frequency is greater than 0.5.

mean   Average value of a parameter, such as a measured trait.

median   Value of the measure that is in the center of a distribution, the middle number. For skewed distributions it is a better estimate of the parameter than the mean for some purposes.

mega-analysis   Reanalysis of multiple studies by combining all of the individual data points into a single analysis.

Mendelian randomization   Strategy used to attempt to distinguish causation from correlation. If a biochemical causally contributes to a phenotype and is also associated with a genotype, that genotype should also associate with the phenotype. In Mendelian randomization, this proposition is tested using a technique known as instrumental variable analysis.

Mendelian trait   Any trait that is largely determined by the genotype at a single gene, and typically has just two or three common phenotypes. If both parents are heterozygotes, the genotypes segregate in their progeny with ratio of 1:2:1, and the phenotypes may either follow the same pattern if the trait is additive, or have the ratio of 3:1 if it is dominant. Sex-linked Mendelian traits show different transmission patterns.

meta-analysis   Reanalysis of multiple studies by combining the individual *p*-values from each contributing study, without needing to combine all of the individual data points. Meta-analysis is expected to converge on the true strength of association between a genotype and phenotype according to the central limit theorem.

metabolomics   High throughput profiling of the abundance of hundreds to thousands of different organic compounds, including lipids, sugars, amino acids, and peptides. Metabolomics may include metabolites generated by the organism itself or those absorbed from the environment as xenobiotics.

metagenomics   High throughput surveys of the microbial content of a sample, either using representative ribosomal DNA sequences or whole genome sequencing of all of the microbial DNA in a sample. Metagenomics is also known as microbiome analysis.

methylation analysis   Quantification of the proportion of sites in the genome that are methylated, most often CpG dinucleotides in the vicinity of genes. A common technique is bisulfite conversion of cytosine to uracil, which does not happen if the cytosine is methylated, followed either by sequencing or genotyping of the site.

microarray analysis   Method for simultaneous measurement of the abundance of thousands of transcripts by hybridization of fluorescently labeled cDNA to an array of complementary probes.

microRNAs (miRNAs)   Small noncoding RNAs 21–24 nucleotides long that have important regulatory functions mediated by binding to 3′ untranslated regions of transcripts and either repressing translation or promoting degradation of the mRNA.

minor allele   Allele that has a frequency in the population that is lower than at least one other allele at the site. Most sites have just two alleles, for example A and G, and the minor allele is the one whose frequency is less than 0.5.

missense mutation   Mutation that changes the amino acid that is encoded by a codon. A missense mutation may be conservative or non-conservative, benign or deleterious.

monogenic trait   Trait influenced by a single gene, essentially equivalent to a Mendelian trait.

morpholino   Chemically stabilized antisense oligonucleotide that is injected into cells early in development so that it can knock down the activity of the complementary gene.

multiplexing   Combination of two or more samples in a single sequencing reaction, using molecular barcodes to deconvolute which sample each sequence belongs to. It is used to reduce the costs of sequencing, for example including six multiplexed RNASeq samples on one lane of a HiSeq2000 generates 50 million reads each.

mutational burden   Cumulative deleterious effects of all of the mutations that appear in an individual (or population, depending on the context).

N

negative predictive value (NPV)   Ratio of true negatives to the number of individuals called negative in a sample. It is a measure of how good a test, such as a genetic risk score, is at identifying who will not contract a disease: in general, it is desirable that the NPV be over 99%, so that only a small proportion of individuals called negative actually develop the disease.

nonsense mutation   Mutation that introduces a premature stop codon and hence leads to truncation of the protein.

nonsynonymous mutation   Mutation that changes the amino acid that is encoded by a codon, or introduces a premature stop codon. Also called a **nonsynonymous substitution**.

normalization   In gene expression profiling, the process of systematically removing technical artifacts and biological covariates from a dataset such that the true biological signal of interest can be recognized while minimizing the likelihood of false findings driven by technical biases. Methods range from simple mean centering to variance transforms and supervised approaches.

nucleotide diversity   Measure of the amount of polymorphism in a population from a sample of nucleotide sequences. A common measure, π, is the average number of differences per nucleotide site between any two randomly chosen sequences in a sample.

O

oligogenic trait   Trait influenced by a small number of genes, more than one but less than a dozen, each of which has a moderate effect.

oncogene   Mutated copy of a gene that promotes cancer development, usually in a dominant fashion. The normal gene, often referred to as a proto-oncogene, has a role in regulation of cell division or survival, but the mutation makes it over-active so that the cell either divides without the normal controls or fails to undergo apoptosis.

Online Mendelian Inheritance in Man (OMIM)   Website that provides a compendium of knowledge concerning the roles that thousands of genes have in promoting disease.

open reading frames (ORFs)   Stretch of nucleotides that are capable of encoding a peptide. Since 3 of the 64 codons are stop codons, random sequence should have a potential stop codon (TAA, TAG, or TGA) approximately once every 20 codons, so stretches much longer than 60 nucleotides without one of these are potential ORFs.

P

paired-end reads   In high throughput sequencing, the two sequence reads, typically 50–150 nucleotides long, that are read from opposite strands starting at either end of a fragment. The reads are thus paired and from the ends of a fragment.

parent-offspring regression   One of the methods for estimating the heritability of a trait, based on linear regression of the phenotypes of a set of offspring on the phenotypes of their parents, or the midparent phenotype (i.e., the average of the two parents).

parental imprinting   Epigenetic modification of a gene such that only the copy transmitted from either the mother or the father is expressed in the offspring. The modified locus is said to be imprinted.

partial LD   Incomplete linkage disequilibrium, namely a correlation between two genotypes that is nonrandom but not as strong as it could be.

passenger mutations   Mutations that do not contribute to tumor development or progression but are nevertheless observed at high frequency in a tumor because they have hitch-hiked along with a linked causal mutation (driver) that has been selected for as the cancer grows.

pedigree mapping   Process of identifying regions of the genome that contain genetic risk factors for a disease by tracing the frequency of genetic markers in large pedigrees. Alleles that segregate with the disease are said to be linked to a causal gene.

penetrance   Proportion of individuals with a genotype that promotes a disease who actually develop the disease. Mendelian mutations are 100% penetrant, but many rare alleles of large effect have reduced penetrance to the influence of the environment or genetic background.

perfect linkage disequilibrium   *See* complete linkage disequilibrium.

personal genome projects (PGPs)   Studies or services designed to provide individuals with information about the content of their own genome and how genetic variants they possess may influence their traits and disease risk.

pharmacodynamics   Branch of pharmacogenetics concerned with the impact of a drug or other agent on an organ of the body, which can be modified by genetic variation affecting recognition of the drug or modulation of biochemical activities once it is bound.

pharmacogenetics   Study of the effectiveness of pharmacological agents as a function of genetic variation. Polymorphisms that affect the kinetics or dynamics of a drug influence the dosage that is required for it to be effective, as well as the likelihood of adverse side effects. Pharmacogenetic variation is a major cause of drug failure.

pharmacokinetics   Branch of pharmacogenetics concerned with the fate of drugs or other agents once they are ingested, specifically focusing on the effects of absorption, transport, and metabolism, all of which affect how much of the drug is available to the target tissue.

phasing   Process of determining the haplotypes that correspond to two or more genotypes. For example, two genotypes AA and GC are phased as AG and AC haplotypes. Double heterozygotes are ambiguous since AT and GC could be phased either as AG and TC or AC and TG.

phenotype   Observed state of an individual. Phenotypes can be continuous (height), categorical (has a disease), or numeric (count of digits) and can be visible or measured by some type of biochemical assay.

Piwi-interacting RNAs (piRNAs)   Large class of small RNAs 26–31 nucleotides in length that repress retrotransmission of transposable elements through the germline, especially by complexing with Piwi proteins during spermatogenesis.

pleiotropic   Having two or more distinct biological consequences as a result of gene activity in different tissues or at different times of development. Most genes are now thought to be pleiotropic to some degree.

point estimate   In statistics, an estimate of a parameter from an observed distribution of a sample, such as the effect of a genetic polymorphism on a trait. Point estimates are a "best guess" and can be contrasted with interval estimates that specify the likely range of the parameter.

polygenic trait   Trait influenced by a large number of genes. If the number is greater than 10, it is likely to more than 100, so each of the contributing genes has a very small effect.

polymorphisms   Nucleotide sites that have two or more alleles at a detectable frequency in a population. Sometimes a lower "minor allele frequency" threshold, such as 1% or 5%, is used to distinguish rare and common polymorphisms.

population stratification   The influence of population structure on genetic association, which can lead to false positives where there is a difference both in the allele frequency and the trait or disease prevalence.

population structure   The existence of allele frequency differences that define sub-populations as founder effects, genetic drift, migration, and natural selection have caused groups of people to diverge genetically.

positive predictive value (PPV)   Ratio of true positive to the number of individuals called positive in a sample. It is a measure of how good a test, such as a genetic risk score, is at identifying who will contract a disease: even a test that identifies most cases may nevertheless have a low PPV if the disease is rare, since there will be many more controls with the risk genotype who do not get the disease.

predictive health   New branch of medicine concerned with predicting the likely onset and progression of disease based on genetic and clinical profiles as well as epidemiological and environmental data.

preimplantation diagnosis   Process of diagnosing a high likelihood of congenital abnormality in an in vitro–fertilized fetus prior to implantation in the womb. Single cell genotyping and DNA sequencing can be used to determine the genotypes of a series of embryos, allowing one to be selected for implementation.

primary immune deficiency (PID)   Series of rare genetic disorders characterized by deficient immune function as a result of abnormal development of one or more immune cell types, or abnormal activity of key immune regulators and recognition molecules.

private variants   Genetic variants that are unique either to an individual or, depending on the context, to one of several populations that have been sampled.

proband   Person who is being reported on in a medical genetic study, often the first member of a family to be identified with a disease or condition.

promoters   Stretches of DNA within a few hundred nucleotides of the transcription start site that regulate the assembly and release of the RNA polymerization complex. A minority of human genes have a classical TATA box motif in their promoter, while many more are characterized by a series of CpG dinucleotides upstream of the transcription start site.

proteomics   High throughput profiling of the abundance of hundreds to thousands of different proteins or peptides in a sample. The two most common methods are mass spectrometry and two-dimensional gel electrophoresis.

proto-oncogenes   Genes that normally play a role in the regulation of cell division or survival, which can become cancer-promoting genes if specific activating mutations occur.

pseudogenes   No longer normally functional genes that are related in sequence to a normal gene. They may be expressed, and may have reduced or abnormal activity, as result of the accumulation of mutations.

purifying selection   Natural selection that tends to remove deleterious alleles from the gene pool because individuals who carry the allele are less likely to produce viable offspring. The process ensures that deleterious mutations are generally at low frequencies.

Q

Q-Q plots   Plots of the observed against expected test statistic. Used to demonstrate that even though no polymorphisms may be significant genome-wide, there may be an excess of nominally significant associations, or that the spectrum of p-values is reduced by a technical bias such as population stratification.

quantile normalization   In gene expression analysis, an approach to normalization that forces the profile of gene expression values to have the same frequency distribution in each individual. Basically, the rank of each gene in each individual is assigned the average value of that rank. It is best used when only a small fraction of genes are differentially expressed.

quantitative trait locus (QTL)   Location in the genome where genotype differences are associated with variation for a quantitative trait.

R

rare alleles of major effect (RAME)
Genetic variants that tend to be associated with a larger genetic risk than common variants but are found in fewer individuals. They are thought to be particularly important for neuropsychiatric disease, for example.

recurrence risk   Risk that a sibling or twin has of developing a disease given that their sibling has it. For example, for autism spectrum disorders, the recurrence risk is estimated to be between 3% and 10%, about an order of magnitude higher than the general population risk.

repetitive elements   Sequences of DNA that are made up of repeats of from two to several hundred bases. If they are located in tandem copies of DNA, they are microsatellites or variable number tandem repeats (VNTRs), but most repetitive DNA is derived from transposable elements.

ribosomal RNA (rRNA) genes   Genes that encode RNA molecules that are components of ribosomes.

S

sensitivity   Refers to the true positive rate of a genetic test. It is the ratio of the number of people who are called positive (likely to get a disease) and who get it (that is, the true positives) to all of those who are positive for the condition (including false negatives). Thus, a test that predicts 80 out of the 100 people who get a disease has a sensitivity of 80%.

simple nucleotide polymorphism   Sometimes used in place of single nucleotide polymorphism (SNP) to include single base insertions or deletions as well as substitutions.

small nuclear RNAs (snRNAs)   Small RNAs approximately 150 nucleotides in length that complex with ribonuclear proteins to regulate splicing and messenger RNA processing.

small nucleolar RNA (snoRNA)   Family of small RNAs found in the nucleolus that are involved in processing of ribosomal and transfer RNAs, for example addition of methyl or pseudouridyl groups.

SNP-based heritability estimation
Approach to estimating the amount of genetic variance explained by common polymorphisms collectively. Equivalent to a regression of phenotypic on genetic relatedness. Used to estimate heritability explained by SNPs that do not individually exceed the genome-wide significance threshold.

social genomics   Genomic analysis of social traits such as loneliness or education level.

soft selection   Natural selection acting in a more complicated manner than the simple hard selection model of directional selection on a single mutation. It can refer either to selection acting simultaneously on more than one polymorphism in the same gene, or to selection on previously cryptic variation that is only functional or advantageous in a new environment. Contrast with hard selection.

somatic copy number alteration (SCNA)
Regions of the genome that are deleted or duplicated on one or both chromosomes as a result of mutations in somatic cells, consequently having an abnormal number of copies. Thought to be an important contributor to cancer progression as tumor cells accumulate mutations.

specificity   Refers to the true negative rate of a genetic test. It is the ratio of the number of people who are called negative (not likely to get a disease) and who do not get it (that is, the true negatives) to all of those who are negative for the condition (including false positives).

Thus, if a 1000 people in a study do not get a disease, and a test predicted that 900 of them would not get it, the test has a specificity of 90%.

spliceosome   Molecular complex in the nucleus where splicing of primary transcripts into mature messenger RNAs occurs. Visible under the microscope as nuclear speckles.

stage   Classification of cancers according to the size and distribution of the tumors in a person, reflecting how advanced the disease is and hence an indicator of prognosis.

standard deviation   Measure of the spread of values around the sample average, technically the square root of the variance. It is obtained by taking the square root of the average squared deviation of each measure from the mean. Thus, if three measures are 5, 7, and 12, then the mean (average) is $24/3 = 8$, the variance is $[(5 - 8)^2 + (7 - 8)^2 + (12 - 8)^2]/3 = 26/3 = 8.67$, and the standard deviation is the square root of $8.67 = 2.94$.

statistical burden tests   Test of association between all of the rare variants in a gene observed in a sample of cases and controls, with disease status. Since there is low power to test individual rare variants, they are pooled as a group and evaluated jointly.

substitution effect   Expected average change in phenotype due to substitution of one allele for another. Useful in animal and plant breeding.

supervised normalization   In gene expression analysis, an approach to normalization that involves the researcher supervising the choice of variables that are of biological interest or may be confounding adjustment factors and normalizing each gene relative to these factors.

suppressors   Short sequences of regulatory DNA that prevent a nearby gene from being expressed in specific cells under specific circumstances. They are usually orientation independent and can be located at variable distances from the promoter in either direction. Similar to enhancers but with the opposite effect.

syndrome   Series of symptoms that tend to be observed together in individuals with a disease.

synonymous mutation   Mutation in a coding region that does not change the amino acid. Due to the redundancy of the genetic code, many nucleotides substitutions do not affect the protein sequence. Also called a **synonymous substitution**.

synthetic association   Association between a common polymorphism and a disease that is actually driven by a series of relatively rare variants that are in linkage disequilibrium with the risk-associated allele. The rare variants may or may not be actually observed.

systems biology   Iterative approach to scientific discovery involving the use of genomics to generate hypotheses, which are tested by perturbation and then reanalyzed using genomic assays.

T

tandem affinity purification (TAP)   Method used to study protein-protein interactions using an "affinity tag" engineered onto one protein to help purify that protein and other proteins to which it binds, from cell extracts.

tandem MS (MS/MS)   Two-step mass spectrometry method for high resolution measurement of the protein or metabolite content of a sample. Analytes that are ionized in the first MS step are randomly fragmented in a second step, allowing disambiguation of a peak that could be one of several molecules as the most probable sum of its parts.

targeted genome sequencing (TGS)   High throughput resequencing of desired genes or genome intervals, which are isolated either by massively parallel PCR or by hybridization to magnetic beads coated with oligonucleotides complementary to the targeted DNA.

text mining   Systematic computational scanning of published abstracts in search of reports that mention two genes together, allowing them to be clustered in networks.

therapeutic drug monitoring   Measure of the concentration of drug in a bodily fluid (such as serum or urine) following administration in order to assess its possible bioavailability.

threshold liability model   Model of disease risk that postulates that individuals who have more than a threshold value of a genetic risk score (liability) are at higher risk of disease.

toxicogenomics   Use of genomic approaches to investigate the causes of toxicity or of variation in how the body responds to toxins.

trans-eSNP   *See* distal eSNP.

transfer RNA (tRNA) genes   Genes that encode the small RNA molecules that act as adaptors during translation.

transitions   Nucleotide substitutions between two purines (A or G) or between two pyrimidines (T or C). There are four possible transitions.

transversions   Nucleotide substitutions between a purine and a pyrimidine, namely A or T for G or C. There are eight possible transversions.

trinucleotide-repeat disorders   Set of neurological diseases that are attributed to expansion of microsatellite repeats, for example CAG in the spinocerebellar ataxia genes.

tumor suppressor genes   One of the major classes of cancer-related genes, tumor suppressors require inactivating mutations of both copies so that there

is insufficient activity of the encoded proteins which are required to suppress cell division.

two-hit hypothesis   Hypothesis first formally formulated by Alfred Knudson in 1971 to account for the rate of occurrence of tumors in certain familial cancers such as retinoblastoma. It proposes that at least two mutations are required for cancer to appear, one of which might be inherited.

U

upstream open reading frame (uORF)   Short reading frame beginning with an ATG that precedes the true start site of translation, in the 5′ untranslated region of a transcript.

V

variance   Measure of the spread of values around the sample average. In quantitative genetics the term is generally used in the context of an estimate variation in the phenotypes or of the magnitude of environmental or genetic contributions to the phenotype.

viral metagenomics   High throughput surveys of the viruses present in a sample by next generation sequencing.

volcano plot   Representation of the results of a gene expression profiling experiment plotting the significance of differential expression between two conditions on the y-axis, against the magnitude of the difference on the x-axis.

W

whole-exome resequencing (WER)   Determination of the sequences of all of the genes in an individual after first enriching for just the exome (all of the exons). It is called resequencing because the sequences are determined by alignment to the existing human reference genome, currently HuRef19.

whole-exome sequencing (WES) *See* whole-exome resequencing.

whole-genome sequencing (WGS) Determination of the complete sequence of the non-repetitive portion of an individual's genome. Currently performed by alignment of hundreds of millions of paired end sequence reads to a reference genome, currently HuRef19.

Y

yeast two-hybrid analysis Method for measuring potential protein-protein interactions. The idea is that when a bait protein makes contact with a prey protein, the interaction generates a biological activity such as transcription activation that reports that the two proteins can bind to one another. Libraries of bait and prey are constructed in strains of yeast, and crossed together in massively parallel fashion to assay millions of possible interactions.

Index